THE EQUATIONS OF OCEANIC MOTIONS

Modeling and prediction of oceanographic phenomena and climate are based on the integration of dynamic equations. *The Equations of Oceanic Motions* derives and systematically classifies the most common dynamic equations used in physical oceanography, from those describing large-scale circulations to those describing small-scale turbulence.

After establishing the basic dynamic equations that describe all oceanic motions, Müller then derives approximate equations, emphasizing the assumptions made and physical processes eliminated. He distinguishes between geometric, thermodynamic, and dynamic approximations and between the acoustic, gravity, vortical, and temperature–salinity modes of motion. Basic concepts and formulae of equilibrium thermodynamics, vector and tensor calculus, curvilinear coordinate systems, and the kinematics of fluid motion and wave propagation are covered in appendices.

Providing the basic theoretical background for graduate students and researchers of physical oceanography and climate science, this book will serve as both a comprehensive text and an essential reference.

PETER MÜLLER studied physics at the University of Hamburg. He received his Ph.D. in 1974 and his Habilitation in 1981. He worked at Harvard University before moving to the University of Hawaii in 1982, where he is now Professor of Oceanography in the School of Ocean and Earth Science and Technology. His research interests cover a broad range of topics in oceanography, climate dynamics, and philosophy, including wave dynamics, stochastic (climate) models, and foundations of complex system theories. He has published widely on these topics. He is co-author (with Hans von Storch) of the book *Computer Modelling in Atmospheric and Oceanic Sciences: Building Knowledge.* Peter Müller is the organizer of the ‘Aha Huliko’a Hawaiian Winter Workshop series and the chief editor of the *Journal of Physical Oceanography.*

THE EQUATIONS OF OCEANIC MOTIONS

PETER MÜLLER
University of Hawaii

CAMBRIDGE
UNIVERSITY PRESS

CAMBRIDGE
UNIVERSITY PRESS

University Printing House, Cambridge CB2 8BS, United Kingdom

Cambridge University Press is part of the University of Cambridge.

It furthers the University's mission by disseminating knowledge in the pursuit of education, learning and research at the highest international levels of excellence.

www.cambridge.org
Information on this title: www.cambridge.org/9780521855136

First published 2006
First paperback edition 2012

A catalogue record for this publication is available from the British Library

ISBN 978-0-521-85513-6 Hardback
ISBN 978-1-107-41060-2 Paperback

Contents

Preface

This book about the equations of oceanic motions grew out of the course "Advanced Geophysical Fluid Dynamics" that I have been teaching for many years to graduate students at the University of Hawaii. In their pursuit of rigorous understanding, students consistently asked for a solid basis and systematic derivation of the dynamic equations used to describe and analyze oceanographic phenomena. I, on the other hand, often felt bogged down by mere "technical" aspects when trying to get fundamental theoretical concepts across. This book is the answer to both. It establishes the basic equations of oceanic motions in a rigorous way, derives the most common approximations in a systematic manner and uniform framework and notation, and lists the basic concepts and formulae of equilibrium thermodynamics, vector and tensor analysis, curvilinear coordinate systems, and the kinematics of fluid flows and waves. All this is presented in a spirit somewhere between a textbook and a reference book. This book is thus not a substitute but a complement to the many excellent textbooks on geophysical fluid dynamics, thermodynamics, and vector and tensor calculus. It provides the basic theoretical background for graduate classes and research in physical oceanography in a comprehensive form.

The book is about equations and theorems, not about solutions. Free wave solutions on a sphere are only included since the emission of waves is a mechanism by which fluids adjust to disturbances, and the assumption of instantaneous adjustment and the elimination of certain wave types forms the basis of many approximations.

Neither does the book justify any of the approximations for specific circumstances. It sometimes motivates but mostly merely states the assumptions that go into a specific approximation. The reason is that I believe (strongly) that one cannot justify any approximation for a specific oceanographic phenomenon objectively. The adequacy of an approximation depends not only on the object, the phenomenon, but also on the subject, the investigator. The purpose of the investigation, whether aimed at realistic forecasting or basic understanding, determines the choice of approximation as much as the phenomenon. The question is not whether a particular approximation is *correct* but whether it is *adequate* for a specific purpose. This book is intended to help a researcher to understand which assumptions go into a particular approximation. The researcher must then judge whether this approximation is adequate for their particular phenomenon and purpose.

All of the equations, theorems, and approximations covered in this book are well established and no attempt has been made to identify the original papers and contributors. Among the people that contributed to the book I would like to acknowledge foremost Jürgen Willebrand. We taught the very first "Advanced Geophysical Fluid Dynamics" course together and our joint encyclopedia article "Equations for Oceanic Motions" (Müller and Willebrand, 1989) may be regarded as the first summary of this book. Vladimir Kamenkovich's book *Fundamentals in Ocean Dynamics* helped me to sort out many of the theoretical concepts covered in this book. I would also like to thank Frank Henyey, Rupert Klein, Jim McWilliams, and Niklas Schneider for constructive comments on an earlier draft; Andrei Natarov and Sönke Rau for help with LaTeX; Martin Guiles, Laurie Menviel-Hessler, and Andreas Retter for assistance with the figures; and generations of students whose quest for rigor inspired me to write this book.

1
Introduction

This book derives and classifies the most common dynamic equations used in physical oceanography, from the planetary geostrophic equations that describe the wind and thermohaline driven circulations to the equations of small-scale motions that describe three-dimensional turbulence and double diffusive phenomena. It does so in a systematic manner and within a common framework. It first establishes the basic dynamic equations that describe all oceanic motions and then derives reduced equations, emphasizing the assumptions made and physical processes eliminated.

The basic equations of oceanic motions consist of:

- the thermodynamic specification of sea water;
- the balance equations for mass, momentum, and energy;
- the molecular flux laws; and
- the gravitational field equation.

These equations are well established and experimentally proven. However, they are so general and so all-encompassing that they become useless for specific practical applications. One needs to consider approximations to these equations and derive equations that isolate specific types or scales of motion. The basic equations of oceanic motion form the solid starting point for such derivations.

In order to derive and present the various approximations in a systematic manner we use the following concepts and organizing principles:

- distinction between properties of fluids and flows;
- distinction between prognostic and diagnostic variables;
- adjustment by wave propagation;
- modes of motion;
- Reynolds decomposition and averaging;
- asymptotic expansion;
- geometric, thermodynamic, and dynamic approximations; and
- different but equivalent representations,

which are discussed in the remainder of this introduction.

First, we distinguish between properties of the fluid and properties of the flow. The equation of state, which determines the density of sea water, is a fluid property. To understand the impact that the choice of the equation of state has on the fluid flow we consider, for ideal fluid conditions, five different equations of state:

- a two-component fluid (the density depends on pressure, specific entropy and salinity);
- a one-component fluid (the density depends on pressure and specific entropy);
- a homentropic fluid (the density depends on pressure only);
- an incompressible fluid (the density depends on specific entropy only); and
- a homogeneous fluid (the density is constant).

Sea water is of course a two-component fluid, but many flows evolve as if sea water were a one-component, homentropic, incompressible, or homogeneous fluid, under appropriate conditions.

We further distinguish between:

- prognostic variables; and
- diagnostic variables.

A prognostic variable is governed by an equation that determines its time evolution. A diagnostic variable is governed by an equation that determines its value (at each time instant).

When a fluid flow is disturbed at some point it responds or adjusts by emitting waves. These waves communicate the disturbance to other parts of the fluid. The waves have different restoring mechanisms: compressibility, gravitation, stratification, and (differential) rotation. We derive the complete set of linear waves in a stratified fluid on a rotating sphere, which are:

- sound (or acoustic) waves;
- surface and internal gravity waves; and
- barotropic and baroclinic Rossby waves.

A temperature–salinity wave needs to be added when both temperature and salinity become dynamically active, as in double diffusion. The assumption of instantaneous adjustment eliminates certain wave types from the equations and forms the basis of many approximations. We regard these waves as linear manifestations of acoustic, gravity, Rossby, and temperature–salinity *modes of motion* though this concept is not well defined. A general nonlinear flow cannot uniquely be decomposed into such modes of motion. A well-defined property of a general nonlinear flow is, however, whether or not it carries potential vorticity. We thus distinguish between:

- the zero potential vorticity; and
- the potential vorticity carrying or vortical mode of motion.

This is a generalization of the fluid dynamical distinction between irrotational flows and flows with vorticity. Sound and gravity waves are linear manifestations of the zero potential vorticity mode and Rossby waves are manifestations of the vortical mode. We follow carefully the structure of the vorticity and potential vorticity equations and the structure of these two modes through the various approximations of this book.

The basic equations of oceanic motions are nonlinear. As a consequence, all modes and scales of motion interact, not only among themselves but also with the surrounding atmosphere and the solid earth. Neither the ocean as a whole nor any mode or scale of motion within it can be isolated rigorously. Any such isolation is approximate at best. There are two basic techniques to derive approximate dynamical equations:

- Reynolds averaging; and
- asymptotic expansions.

Reynolds averaging decomposes all field variables into a mean and a fluctuating component, by means of a space-time or ensemble average. Applying the same average to the equations of motion, one arrives at a set of equations for the mean component and at a set of equations for the fluctuating component. These two sets of equations are not closed. They are coupled. The equations for the mean component contain eddy (or subgridscale) fluxes that represent the effect of the fluctuating component on the mean component. The equations for the fluctuating component contain background fields that represent the effect of the mean component on the fluctuating component. Any attempt to decouple these two sets of equations must overcome the closure problem. To derive a closed set of equations for the mean component one must parametrize the eddy fluxes in terms of mean quantities. To derive a closed set for the fluctuating component one must prescribe the background fields. Prescribing background fields sounds less restrictive than parametrizing eddy fluxes, but both operations represent closures, and closures are approximations.

Asymptotic expansions are at the core of approximations. Formally, they require the scaling of the independent and dependent variables, the non-dimensionalization of the dynamical equations, and the identification of the non-dimensional parameters that characterize these equations. The non-dimensional equations can then be studied in the limit that any of these dimensionless parameters approaches zero by an asymptotic expansion with respect to this parameter. Often, these asymptotic expansions are short-circuited by simply neglecting certain terms in the equations. High-frequency waves are not affected much by the Earth's rotation and can be studied by simply neglecting the Earth's rotation in the equations, rather than by going through a pedantic asymptotic expansion with respect to a parameter that reflects the ratio of the Earth's rotation rate to the wave frequency. Similarly, when encountering a situation where part of the flow evolves slowly while another part evolves rapidly one often eliminates the fast variables by setting their time derivative to zero. The fast variables adjust so rapidly that it appears to be instantaneous when viewed from the slowly evolving part of the flow. Nevertheless, any of these heuristic approximations is justified by an underlying asymptotic expansion.

The basic equations of oceanic motions are characterized by a large number of dimensionless parameters. It is neither useful nor practical to explore the complete multi-dimensional parameter space spanned by the parameters. Only a limited area of this space is occupied by actual oceanic motions. To explore this limited but still fairly convoluted subspace in a somewhat systematic manner we distinguish between:

- geometric approximations;
- thermodynamic approximations; and
- dynamic approximations.

Geometric approximations change the underlying geometric space in which the oceanic motions occur. This geometric space is a Euclidean space, in the non-relativistic limit. It can be represented by whatever coordinates one chooses, Cartesian coordinates being the simplest ones. Moreover, the dynamic equations can be formulated in coordinate-invariant form using vector and tensor calculus, and we use such invariant notation wherever appropriate. It is, however, often useful (and for actual calculations necessary) to express the dynamic equations in a specific coordinate system. Then metric coefficients (or scale factors) appear in the equations of motions. These coefficients are simply a consequence of introducing a specific coordinate system. A geometric approximation is implemented when these metric coefficients and hence the underlying Euclidean space are altered. Such geometric approximations include the spherical, *beta*-plane, and f-plane approximations. They rely on the smallness of parameters such as the eccentricity of the geoid, the ratio of the ocean depth to the radius of the Earth, and the ratio of the horizontal length scale to the radius of the Earth. If the smallness of these parameters is only exploited in the metric coefficients, but not elsewhere in the equations, then one does not distort the general properties of the equations. The pressure force remains a gradient; its integral along a closed circuit and its curl vanish exactly.

Thermodynamic approximations are approximations to the thermodynamic properties of sea water. Most important are approximations to the equation of state. They determine to what extent sea water can be treated as a one-component, incompressible, homentropic, or homogeneous fluid, with profound effects on the dynamic evolution and permissible velocity and vorticity fields. Auxiliary thermodynamic approximations assume that other thermodynamic (and phenomenological) coefficients in the basic equations of oceanic motions are not affected by the flow and can be regarded as constant. This eliminates certain nonlinearities from the equations.

Dynamical approximations affect the basic equations of oceanic motions more directly. They eliminate certain terms such as the tendency, advection, or friction term and the associated processes. A common thread of these approximations is that flows at large horizontal length scales and long time scales are approximately in hydrostatic and geostrophic balance and two-dimensional in character, owing to the influence of rotation and stratification. These balanced flows are characterized by small aspect ratios and small Rossby and Ekman numbers (which are ratios of the advective and frictional time scales to the time scale of rotation). As the space and time scales become smaller and the Rossby and Ekman numbers larger, the flow becomes less balanced and constrained and more three-dimensional in character. At the smallest scales the influence of rotation and stratification becomes negligible and the flow becomes fully three-dimensional.

The various approximations are depicted in more detail in Figure 1.1. We first separate acoustic and non-acoustic motions. The acoustic mode could be obtained by considering irrotational motion in a homentropic ocean. Similarly, the non-acoustic modes could be isolated by assuming sea water to be an incompressible fluid. Both these assumptions are much too strong for oceanographic purposes. Instead, one separates acoustic and non-acoustic modes by assuming that the (Lagrangian) time scale of the acoustic mode is much shorter than the (Lagrangian) time scale of the non-acoustic mode. Acoustic motions are fast motions and non-acoustic motions are slow motions in a corresponding two-time scale expansion. Such an expansion implies that when one considers the evolution of the acoustic pressure field one may neglect the slow temporal changes of the background non-acoustic pressure field. This, together with some ancillary assumptions, leads to the acoustic wave equation that forms the basis for acoustic studies of the ocean.

When one considers non-acoustic motions, the two-time scale expansion implies that one can neglect the fast temporal changes of the acoustic pressure in the pressure equation. The pressure field adjusts so rapidly that it appears to be instantaneous when viewed from the slow non-acoustic motions. This elimination of sound waves from the equations is called the anelastic approximation. The resulting equations do not contain sound waves but sea water remains compressible. A consequence of the anelastic approximation is that the pressure is no longer determined prognostically but diagnostically by the solution of a three-dimensional Poisson equation. The anelastic approximation is augmented by approximations that utilize the facts that the density of the ocean does not vary much at a point and from the surface to the bottom. Together, these approximations comprise the Boussinesq approximation, which is at the heart of all that follows. Its major result is that the velocity field is non-divergent or solenoidal. The flow (as opposed to the fluid) is now incompressible.

Next we introduce the shallow water approximation. It assumes that the aspect ratio, i.e., the ratio between the vertical and horizontal length scale of the flow,

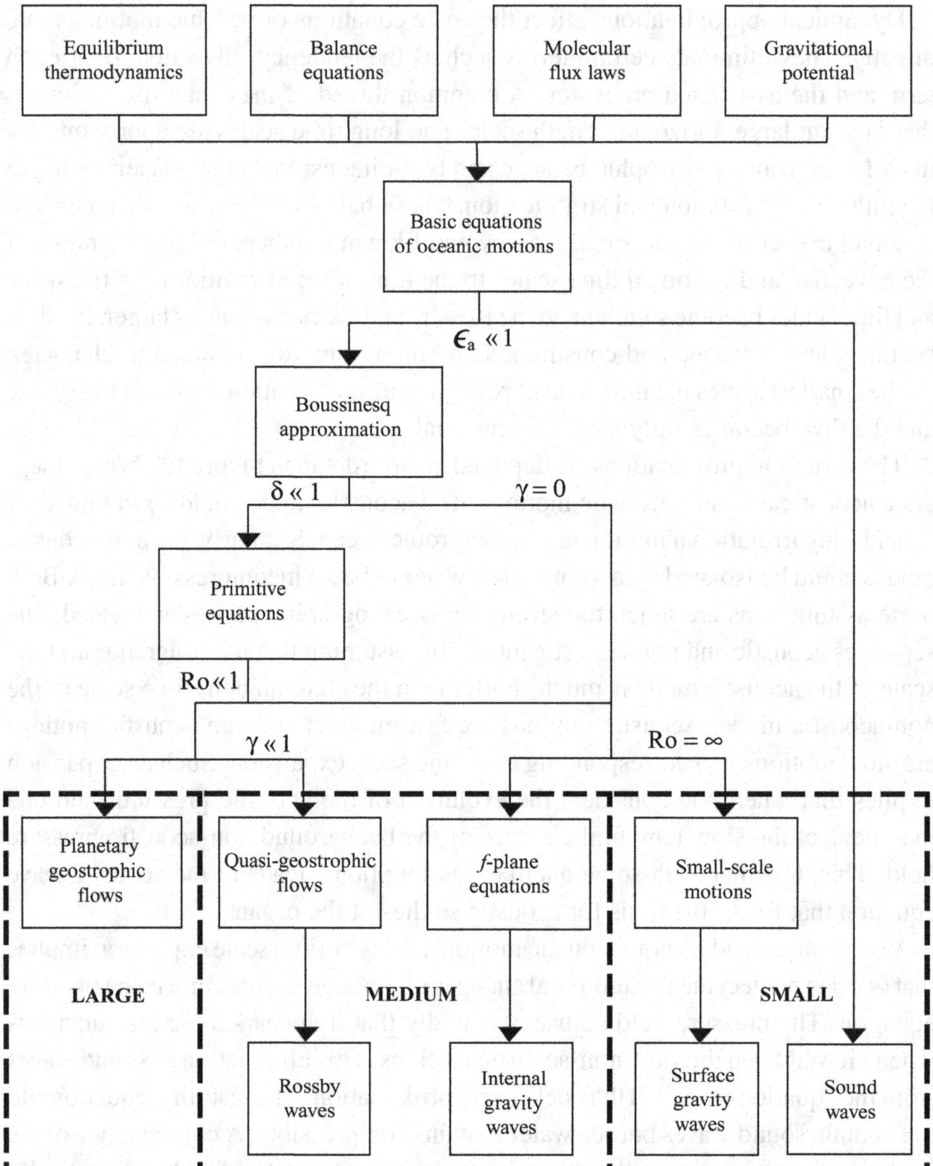

Figure 1.1. Diagram of the overall organization of this book. The parameter ϵ_a represents the ratio of the acoustic to the non-acoustic time scale, δ the aspect ratio, γ the ratio of the horizontal length scale to the radius of the Earth, and Ro the Rossby number.

is small. It leads to the primitive equations. The major simplification is that the vertical momentum balance reduces to the hydrostatic balance where the vertical pressure gradient is balanced by the gravitational force. The pressure is thus no longer determined by the solution of a three-dimensional Poisson equation but by an ordinary differential equation. The shallow water approximation also eliminates the local meridional component of the planetary vorticity, a fact referred to as the traditional approximation.

For small Rossby and Ekman numbers, the horizontal momentum balance can be approximated by the geostrophic balance, the balance between Coriolis and pressure force. This geostrophic approximation has far-reaching consequences. It eliminates gravity waves or the gravity mode of motion. The velocity field adjusts instantaneously. Geostrophic motions carry potential vorticity. They represent the vortical mode of motion. Their evolution is governed by the potential vorticity equation. One has to distinguish between planetary or large-scale geostrophic flows and quasi-geostrophic or small-scale geostrophic flows. For planetary geostrophic flows the potential vorticity is given by f/H, where f is the Coriolis frequency and H the ocean depth. Quasi-geostrophic flows employ two major additional assumptions. One is that the horizontal length scale of the flow is much smaller than the radius of the Earth. This assumption is exploited in the beta-plane approximation. The second assumption is that the vertical displacement is much smaller than the vertical length scale (or that the vertical strain is much smaller than 1). This assumption implies a linearization of the density equation and the flow becomes nearly two-dimensional. The quasi-geostrophic potential vorticity consists of contributions from the relative vorticity, the planetary vorticity, and the vertical strain.

As the length and time scales of the flow decrease further (or as the Rossby number, aspect ratio, and vertical strain all increase) one arrives at a regime where rotation and stratification are still important but not strong enough to constrain the motion to be in hydrostatic and geostrophic balance and nearly two-dimensional. The horizontal length scales become so small that the f-plane approximation can be applied. These f-plane equations contain all modes of motion except the acoustic one. The f-plane motions offer the richest variety of dynamical processes and phenomena. The zero potential vorticity and vortical mode can be isolated in limiting cases.

At the smallest scales one arrives at the equations of regular fluid dynamics. The fact that the motion actually occurs on a rotating sphere becomes inconsequential. The motions may be directly affected by molecular friction, diffusion, and conduction. The temperature–salinity mode (which is at the heart of double diffusive phenomena) again emerges since the molecular diffusivities for heat and salt differ. A major distinction is made between laminar flows for low Reynolds numbers and turbulent flows for high Reynolds numbers. If all buoyancy effects are neglected, one obtains the Navier–Stokes equations that are used to study motions such as three-dimensional isotropic turbulence and nonlinear water waves.

Tidal motions fall somewhat outside this classification since they are defined not by their scales but by their forcing. They are caused by the gravitational potential of the Moon and Sun. The tidal force is a volume force and approximately constant throughout the water column. It only affects the barotropic component of the flow. Tidal motions are thus described by the one-layer shallow water equations, generally called Laplace tidal equations. Since the tidal force is the gradient of the tidal potential it does not induce any vorticity into tidal flows. Tidal flows represent the zero potential vorticity mode though this fact has not been exploited in any systematic way. The tidal potential also causes tides of the solid but elastic earth, called Earth tides. The moving tidal water bulge also causes an elastic deformation of the ocean bottom, called the load tides. The moving bulge modifies the gravitational field of the ocean, an effect called gravitational self-attraction. All these effects need to be incorporated into Laplace tidal equations.

These approximations are overlaid with a triple decomposition into large-, medium-, and small-scale motions that arises from two Reynolds decompositions: one that separates large- from medium- and small-scale motions and one that separates large and medium motions from small-scale motions. For large-scale motions one must parametrize the eddy fluxes arising from medium- and small-scale motions; for medium-scale motions one must parametrize the eddy fluxes from small-scale motions and specify the large-scale background fields; for small-scale motions the subgridscale fluxes are given by the molecular flux laws and one must only specify the large- and medium-scale background fields. The medium-scale motions are the most challenging ones from this point of view. Eddy fluxes have to be parametrized and background fields have to be specified. In this book we only introduce the standard parametrizations of the eddy fluxes in terms of eddy diffusion and viscosity coefficients. Efforts are underway to improve these parametrizations but have not arrived yet at a set of canonical parametrizations.

Of course, the geometric, thermodynamic, and dynamic approximations do not match exactly the triple Reynolds decomposition since the Reynolds averaging scales are not fixed but can be adapted to circumstances. The same dynamical kind of motion might well fall on different sides of a Reynolds decomposition. Surface gravity waves might neglect rotation and the sphericity of the Earth and may be regarded as small-scale motions dynamically. Their subgridscale mechanisms may, however, be either molecular friction or turbulent friction in bottom boundary layers and wave breaking, depending on whether they are short or long waves. In Figure 1.1, we overlaid the Reynolds categories for general orientation only (and put surface gravity waves in the small-scale category).

The systematic representation of the equations of oceanic motions is further complicated by the fact that the equations can be represented in many different but equivalent forms. One can express them in different coordinate systems or in different but equivalent sets of state variables. Of particular relevance is the representation of the vertical structure of the flow. Here, we discuss:

- the decomposition into barotropic and baroclinic components;
- the representation in isopycnal coordinates;
- the representation in sigma coordinates;
- layer models; and
- the projection onto vertical normal modes.

All these representations fully recover the continuous vertical structure of the flow field, except for the layer models that can be obtained by discretizing the equations in isopycnal coordinates. These representations do not involve any additional approximations but their specific form often invites ancillary approximations.

These are the basic concepts and organizing principles to present the various equations of oceanic motions covered in this book. The basic concepts and formulae of equilibrium thermodynamics, of vector and tensor analysis, of orthogonal curvilinear coordinate systems, and of the kinematics of fluid motion and waves, and conventions and notation are covered in appendices.

2

Equilibrium thermodynamics of sea water

The basic equations of oceanic motion assume local thermodynamic equilibrium. The ocean is viewed as consisting of many fluid parcels. Each of these fluid parcels is assumed to be in thermodynamic equilibrium though the ocean as a whole is far from thermodynamic equilibrium. Later we make the continuum hypothesis and assume that these parcels are sufficiently small from a macroscopic point of view to be treated as points but sufficiently large from a microscopic point of view to contain enough molecules for equilibrium thermodynamics to apply. This chapter considers the equilibrium thermodynamics that holds for each of these fluid parcels or points. The thermodynamic state is described by thermodynamic variables. Most of this chapter defines these thermodynamic variables and the relations that hold among them. An important point is that sea water is a two-component system, consisting of water and sea salt. Gibbs' phase rule then implies that the thermodynamic state of sea water is completely determined by the specification of three independent thermodynamic variables. Different choices can be made for these independent variables. Pressure, temperature, and salinity are one common choice. All other variables are functions of these independent variables. In principle, these functions can be derived from the microscopic properties of sea water, by means of statistical mechanics. This has not been accomplished yet. Rather, these functions must be determined empirically from measurements and are documented in figures, tables, and numerical formulae. We do not present these figures, tables, and formulae in any detail. They can be found in books, articles, and reports such as Montgomery (1957), Fofonoff (1962, 1985), UNESCO (1981), and Siedler and Peters (1986). We list the quantities that have been measured and the algorithms that can, in principle, be used to construct all other thermodynamic variables. In the final section we discuss the mixing of water parcels at constant pressure. An introduction into the concepts of equilibrium thermodynamics is given in Appendix A.

2.1 Salinity

Sea water consists of water and sea salt. The sea salt is dissociated into ions. Consider a homogeneous amount of sea water. Let M_1 be the mass of water, $M_2, \ldots, M_N$ the masses of the salt ions, and $M = \sum_{i=1}^{N} M_i$ the total mass. The composition of sea water is then characterized by the N *concentrations*

$$c_i := \frac{M_i}{M} \qquad i = 1, \ldots, N \tag{2.1}$$

Only $N - 1$ of these concentrations are independent since $\sum_{i=1}^{N} c_i = 1$.

The composition of sea water changes mainly by the addition and subtraction of fresh water. The composition of sea salt thus remains unchanged. Only the concentration $c_w := c_1$ of water and the concentration of sea salt

$$c_s := \sum_{i=2}^{N} c_i \tag{2.2}$$

change. Sea water can thus be treated as a two-component system consisting of sea salt and water with concentrations c_s and $c_w := 1 - c_s$.

The concentration of sea salt c_s is called *salinity* and customarily denoted by the symbol S, a convention that we follow. The salinity is a fraction. Often it is expressed in parts per thousand or in *practical salinity units* (psu).

2.2 Equilibrium thermodynamics of a two-component system

Here we list the basic elements and relations of the equilibrium thermodynamics of a two-component system.

Thermodynamic variables The thermodynamic state of a system is described by *thermodynamic variables*. The most basic variables (with their units in brackets) are:

- pressure p [$\mathrm{Pa} = \mathrm{N\,m^{-2}}$]
- specific volume v [$\mathrm{m^3\,kg^{-1}}$]
- temperature T [K]
- specific entropy η [$\mathrm{m^2\,s^{-2}\,K^{-1}}$]
- salinity S
- chemical potential of water μ_w [$\mathrm{m^2\,s^{-2}}$]
- chemical potential of sea salt μ_s [$\mathrm{m^2\,s^{-2}}$]
- specific internal energy e [$\mathrm{m^2\,s^{-2}}$]

These variables are all intensive variables. (Intensive variables remain the same for each subsystem of a homogeneous system; extensive variables are additive.) The adjective "specific" denotes amounts per unit mass. This is a convention throughout the book. Amounts per unit volume are referred to as densities. The most important density is the

- mass density $\rho := v^{-1}$ [kg m^{-3}]

simply called the density in the following.

Thermodynamic representations Gibbs' phase rule states that the intensive state of a two-component, one-phase system is completely determined by the specification of three independent thermodynamic variables. Different choices can be made for these three independent variables and lead to different *thermodynamic representations*. The four most common choices are (v, η, S), (p, η, S), (v, T, S), and (p, T, S).

Thermodynamic potentials For each thermodynamic representation there exists a function of the independent variables that completely determines the thermodynamic properties of the system. This function is called the *thermodynamic potential* of the representation. If (v, η, S) are chosen as the independent variables, then the internal energy function $e(v, \eta, S)$ is the thermodynamic potential. The other thermodynamic variables are then given by

$$p = -\left(\frac{\partial e}{\partial v}\right)_{\eta,S} \tag{2.3}$$

$$T = \left(\frac{\partial e}{\partial \eta}\right)_{v,S} \tag{2.4}$$

$$\Delta\mu = \left(\frac{\partial e}{\partial S}\right)_{v,\eta} \tag{2.5}$$

where

$$\Delta\mu := \mu_s - \mu_w \tag{2.6}$$

is the chemical potential difference. The subscripts on the parentheses indicate the variables that are held constant during differentiation. This is the standard convention of most books on thermodynamics. The relations (2.3) to (2.5) imply the differential form

$$\mathrm{d}e = -p\,\mathrm{d}v + T\,\mathrm{d}\eta + \Delta\mu\,\mathrm{d}S \tag{2.7}$$

which is the first law of thermodynamics for reversible processes.

Euler's identity Extensive dependent variables must be homogeneous functions of first order of extensive independent variables. This fact implies *Euler's identity*

$$e = -pv + T\eta + \mu_w(1 - S) + \mu_s S \tag{2.8}$$

for the internal energy function. The individual chemical potentials for water and sea salt can be inferred from this identity.

Gibbs–Durham relation The complete differential of Euler's identity minus the first law in the form (2.7) results in the *Gibbs–Durham relation*

$$v\,\mathrm{d}p - \eta\,\mathrm{d}T = S\,\mathrm{d}\mu_{\mathrm{s}} + (1-S)\,\mathrm{d}\mu_{\mathrm{w}} \tag{2.9}$$

Alternative thermodynamic potentials If one introduces the

- specific enthalpy $h := e + pv$
- specific free energy $f := e - T\eta$
- specific free enthalpy $g := e + pv - T\eta$

then the functions $h(p, \eta, S)$, $f(v, T, S)$, and $g(p, T, S)$ become the thermodynamic potentials for the respective independent variables.

Further definitions and relations for the potentials $e(v, \eta, S)$, $h(p, \eta, S)$, $f(v, T, S)$, and $g(p, T, S)$ are listed in Table 2.1.

2.3 Potential temperature and density

For a variety of practical reasons oceanographers introduced additional non-standard thermodynamic variables. The two most important ones are the *potential temperature* and the *potential density*. These variables share with the specific entropy the property that they can only be changed by irreversible processes.

Potential temperature If a fluid particle moves at constant specific entropy and salinity across a pressure surface its temperature changes at a rate given by the *adiabatic temperature gradient* (or lapse rate)

$$\Gamma := \left(\frac{\partial T}{\partial p}\right)_{\eta,S} \tag{2.10}$$

To remove this effect of pressure on temperature oceanographers introduce the potential temperature

$$\theta(p, \eta, S; p_0) := T(p_0, \eta, S) \tag{2.11}$$

It is the temperature a fluid particle would attain if moved adiabatically, i.e., at constant η and S, to a reference pressure p_0.[1] The potential temperature depends on the choice of the reference pressure. Usually, $p_0 = p_a = 1.013 \cdot 10^5$ Pa ($= 1$ atm). Processes that do not change η and S also do not change θ. For

[1] Here we define as adiabatic those processes that do not change η and S and as diabatic those that do. This is a generalization of the definition given in Appendix A for a closed system.

Table 2.1. *Basic definitions and relations of the equilibrium thermodynamics of sea water*

The six columns represent different thermodynamic representations characterized by different independent variables. The partial derivatives are always taken at constant values of the other two independent variables of the respective column. The sub- or superscript "pot" denotes the value of a variable at the same specific entropy η and salinity S but at a reference pressure p_0.

Independent variables	v, η, S	p, η, S	v, T, S	p, T, S	p, θ, S	p, ρ_{pot}, S
Thermo-dynamic potential	e	$h = e + pv$	$f = e - T\eta$	$g = e + pv$ $-T\eta$		
Dependent variables	$p = -\frac{\partial e}{\partial v}$ $T = \frac{\partial e}{\partial \eta}$ $\Delta\mu = \frac{\partial e}{\partial S}$	$v = \frac{\partial h}{\partial p}$ $T = \frac{\partial h}{\partial \eta}$ $\Delta\mu = \frac{\partial h}{\partial S}$	$p = -\frac{\partial f}{\partial v}$ $\eta = -\frac{\partial f}{\partial T}$ $\Delta\mu = \frac{\partial f}{\partial S}$	$v = \frac{\partial g}{\partial p}$ $\eta = -\frac{\partial g}{\partial T}$ $\Delta\mu = \frac{\partial g}{\partial S}$		
Euler's identity	$e = T\eta - pv$ $+\mu_w(1-S)$ $+\mu_s S$	$h = T\eta$ $+\mu_w(1-S)$ $+\mu_s S$	$f = -pv$ $+\mu_w(1-S)$ $+\mu_s S$	$g = \mu_w(1-S)$ $+\mu_s S$		
Additional variables		$\tilde{\kappa} := -\frac{1}{v}\frac{\partial v}{\partial p}$ $c^2 := \frac{1}{\rho\tilde{\kappa}}$ $\Gamma := \frac{\partial T}{\partial p}$ $\theta := T(p_0, \eta, S)$ $\rho_{pot} := \rho(p_0, \eta, S)$	$c_v := \frac{\partial e}{\partial T}$	$c_p := \frac{\partial h}{\partial T}$ $\kappa := -\frac{1}{v}\frac{\partial v}{\partial p}$ $\alpha := \frac{1}{v}\frac{\partial v}{\partial T}$ $\beta := -\frac{1}{v}\frac{\partial v}{\partial S}$ $a := \frac{\partial \Delta\mu}{\partial T}$	$\tilde{\alpha} := -\frac{1}{\rho}\frac{\partial \rho}{\partial \theta}$ $\tilde{\beta} := \frac{1}{\rho}\frac{\partial \rho}{\partial S}$	$\hat{\kappa} := \frac{1}{\rho}\frac{\partial \rho}{\partial p}$ $\hat{\alpha} := \frac{1}{\rho}\frac{\partial \rho}{\partial \rho_{pot}}$ $\hat{\beta} := \frac{1}{\rho}\frac{\partial \rho}{\partial S}$
Thermo-dynamic relations	$\frac{\partial p}{\partial v} = -\rho^2 c^2$ $\frac{\partial p}{\partial \eta} = \rho^2 c^2 \Gamma$ $\frac{\partial p}{\partial S} = -\rho c^2 \beta$ $+\rho^2 c^2 \Gamma a$ $\frac{\partial T}{\partial \eta} = \frac{T}{c_v}$ $\frac{\partial T}{\partial S} = -\rho c^2 \Gamma \beta$ $+\frac{Ta}{c_v}$	$\tilde{\kappa} = \kappa - \alpha\Gamma$ $\Gamma = \frac{\alpha v T}{c_p}$	$c_v = T\frac{\partial \eta}{\partial T}$ $c_v = c_p - \frac{\alpha^2 T v}{\kappa}$	$c_p = T\frac{\partial \eta}{\partial T}$ $\frac{\partial \theta}{\partial p} = -\Gamma\frac{\theta c_p}{T c_p^{pot}}$ $= -\Gamma\frac{\alpha}{\tilde{\alpha}}$ $\frac{\partial \theta}{\partial T} = \frac{\theta c_p}{T c_p^{pot}}$ $= \frac{\alpha}{\tilde{\alpha}}$ $\frac{\partial \theta}{\partial S} = \frac{\theta}{c_p^{pot}}(a_{pot} - a)$ $= \frac{\tilde{\beta} - \beta}{\tilde{\alpha}}$	$\tilde{\kappa} = \frac{1}{\rho}\frac{\partial \rho}{\partial p}$ $\tilde{\alpha} = \alpha\frac{T c_p^{pot}}{\theta c_p}$ $\tilde{\beta} = \beta + \rho\Gamma(a_{pot} - a)$ $\frac{\partial \rho_{pot}}{\partial p} = 0$ $\frac{\partial \rho_{pot}}{\partial \theta} = -\rho_{pot}\tilde{\alpha}_{pot}$ $\frac{\partial \rho_{pot}}{\partial S} = \rho_{pot}\tilde{\beta}_{pot}$	$\hat{\kappa} = \tilde{\kappa}$ $\hat{\alpha} = \frac{\tilde{\alpha}}{\rho_{pot}\tilde{\alpha}_{pot}}$ $\hat{\beta} = \tilde{\alpha}\left(\frac{\tilde{\beta}}{\tilde{\alpha}} - \frac{\tilde{\beta}_{pot}}{\tilde{\alpha}_{pot}}\right)$

a one-component system, θ is thus a (monotonic) function of η ($\partial\eta/\partial\theta = c_p^{\text{pot}}/\theta \geq 0$). The potential temperature θ, however, differs in one important aspect from the specific entropy η. Whereas the mixing of two water parcels does not decrease the entropy it might decrease the potential temperature (see Section 2.11). For a two-component system, like sea water, θ depends on η and S.

In the (p, T, S)-representation, the potential temperature is given by

$$\begin{aligned}\theta(p, T, S) &= T + \int_p^{p_0} \mathrm{d}p'\, \Gamma(p', \eta(p, T, S), S) \\ &= T + \int_p^{p_0} \mathrm{d}p'\, \Gamma(p', T', S)\end{aligned} \tag{2.12}$$

where the integration is along a path of constant η and S or constant θ and S. Equation (2.12) is thus an implicit formula for θ. The derivatives of θ with respect to (p, T, S) are listed in Table 2.1.

The adiabatic temperature gradient can be expressed as

$$\Gamma = \frac{\alpha v T}{c_p} \tag{2.13}$$

where α is the thermal expansion coefficient (see Section 2.4) and c_p is the specific heat at constant pressure (see Section 2.6). This is a *thermodynamic relation*, i.e., a relation that follows from transformation rules for derivatives and other mathematical identities (see Section A.4). The adiabatic temperature gradient and the potential temperature can thus be inferred from the equation of state and the specific heat, both of which are measured quantities. There exist numerical algorithms for the calculation of $\Gamma(p, T, S)$ and $\theta(p, T, S)$.

When θ is used as an independent variable instead of T, one arrives at the (p, θ, S)-representation. The thermodynamic potential for this representation has not been constructed.

Potential density If a fluid particle moves at constant η and S across a pressure surface its density changes at a rate given by the *adiabatic compressibility coefficient*

$$\tilde{\kappa} := \frac{1}{\rho}\left(\frac{\partial \rho}{\partial p}\right)_{\eta,S} \tag{2.14}$$

To remove this effect of pressure on density one introduces the *potential density*

$$\rho_{\text{pot}}(p, \eta, S; p_0) := \rho(p_0, \eta, S) \tag{2.15}$$

or

$$\rho_{\text{pot}}(p, \eta, S; p_0) = \rho(p, \eta, S) + \int_p^{p_0} \mathrm{d}p'\, c^{-2}(p', \theta, S) \tag{2.16}$$

where

$$c^2 := \left(\frac{\partial p}{\partial \rho}\right)_{\eta,S} \tag{2.17}$$

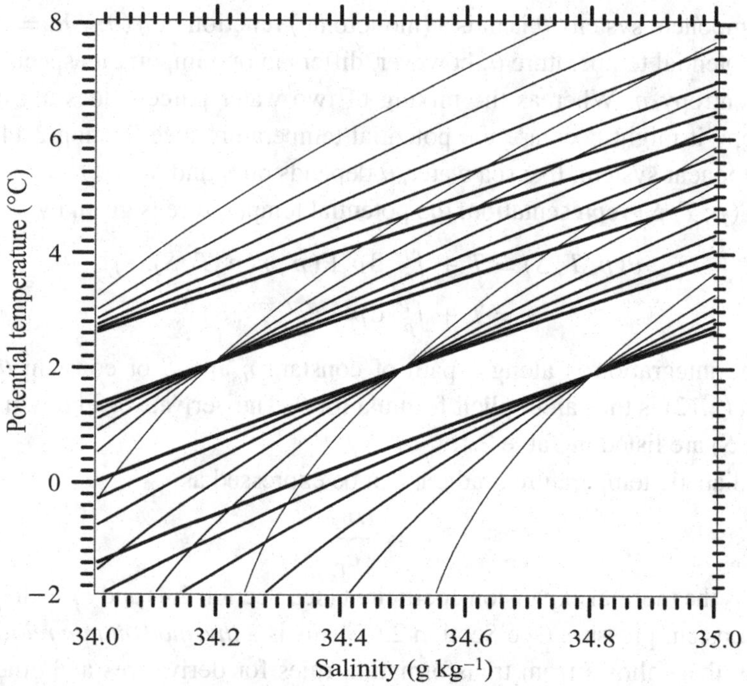

Figure 2.1. Equation of state for sea water. Contours of the density difference $\rho(p, \theta, S) - \rho(p, 2\,°\mathrm{C}, 34, 5\,\mathrm{psu})$ are shown in the (θ, S)-plane for different values of pressure corresponding to depths of 0 m (*thin lines*) to 5 km (*thick lines*) in 1 km intervals. The contour interval is $0.25\,\mathrm{kg\,m^{-3}}$. The equation of state is nonlinear. The contours (isopycnals) are *curved* and their slope *turns* with pressure. Courtesy of Ernst Maier-Reimer.

is the square of the sound speed. The potential density is the density a fluid parcel would attain if moved adiabatically to a reference pressure p_0. Again, the potential density depends on the choice of reference pressure. In the (p, θ, S)-representation the potential density is given by $\rho_{pot}(p, \theta, S; p_0) = \rho(p_0, \theta, S)$. The derivatives of ρ_{pot} with respect to (p, θ, S) are listed in Table 2.1.

The sub- or superscript "pot" is also applied to other variables. It always denotes the value of the variable at the same specific entropy η and salinity S but at a reference pressure p_0.

2.4 Equation of state

The density ρ expressed as a function of any set of independent thermodynamic variables is called *the equation of state*. The functions $\rho = \rho(p, T, S)$, $\rho = \rho(p, \theta, S)$, and $\rho = \rho(p, \rho_{pot}, S)$ are tabulated. As an example, Figure 2.1 shows contours of the density (isopycnals) in the (θ, S)-plane for different values of the pressure.

Since the density of sea water differs very little from 10^3 kg m^{-3} one often introduces

$$\begin{aligned} \sigma_{STp} &:= \rho(p, T, S)\frac{\text{m}^3}{\text{kg}} - 10^3 \\ \sigma_t &:= \rho(p_0, T, S)\frac{\text{m}^3}{\text{kg}} - 10^3 \\ \sigma_\theta &:= \rho(p_0, \theta, S)\frac{\text{m}^3}{\text{kg}} - 10^3 \end{aligned} \tag{2.18}$$

Note on notation In writing $\rho = \rho(p, T, S)$ we have used the same symbol "ρ" to denote a variable and a function. It would have been more proper to use different symbols and write $\rho = f(p, T, S)$, as in $y = f(x)$. Furthermore, in $\rho = \rho(p, \theta, S)$ the second ρ denotes a different function. One should introduce a different symbol and write $\rho = g(p, \theta, S)$. To limit the number of symbols we will always write $y = y(x)$ instead of $y = f(x)$. This convention also applies to the composition of functions. Thus instead of $y = f(x)$, $x = g(t)$, and $y = h(t)$, where $h = f \circ g$, we write $y = y(x)$, $x = x(t)$, and $y = y(t)$ and hope that the context always provides the proper interpretation.

First derivatives The dependence of density on pressure, temperature, and salinity is described by

$$\begin{aligned} \kappa &:= \frac{1}{\rho}\left(\frac{\partial \rho}{\partial p}\right)_{T,S} && \text{isothermal compressibility coefficient} \\ \alpha &:= -\frac{1}{\rho}\left(\frac{\partial \rho}{\partial T}\right)_{p,S} && \text{thermal expansion coefficient} \\ \beta &:= \frac{1}{\rho}\left(\frac{\partial \rho}{\partial S}\right)_{p,T} && \text{haline contraction coefficient} \end{aligned} \tag{2.19}$$

Table 2.1 lists the analogous coefficients for $\rho(p, \theta, S)$ (denoted by tildes) and for $\rho(p, \rho_{\text{pot}}, S)$ (denoted by hats) and the thermodynamic relations between these coefficients. Thermodynamic inequalities imply $0 \leq \tilde{\kappa} \leq \kappa$ and hence $\alpha\Gamma \geq 0$.

Nonlinearities The equation of state is nonlinear in the sense that the coefficients $\tilde{\kappa}$, $\tilde{\alpha}$, and $\tilde{\beta}$ are not constant but dependent on (p, θ, S). This dependence is described by the second derivatives. Among the six independent second derivatives two combinations are of particular importance. The first one is the *thermobaric coefficient*

$$\tilde{b} := \frac{1}{2}\left[\frac{1}{\tilde{\beta}}\left(\frac{\partial \tilde{\beta}}{\partial p}\right)_{\theta,S} - \frac{1}{\tilde{\alpha}}\left(\frac{\partial \tilde{\alpha}}{\partial p}\right)_{\theta,S}\right] \tag{2.20}$$

It describes the turning of the isopycnal slope (see Figure 2.1)

$$\tilde{\gamma} := \frac{\tilde{\beta}}{\tilde{\alpha}} \tag{2.21}$$

with pressure

$$\left(\frac{\partial \tilde{\gamma}}{\partial p}\right)_{\theta,S} = 2\tilde{b}\tilde{\gamma} \tag{2.22}$$

The coefficient $\tilde{b}$ is called the thermobaric coefficient because the largest contribution arises from the second term, the dependence of $\tilde{\alpha}$ on p. The thermobaric coefficient can also be expressed in terms of the adiabatic compressibility as

$$\tilde{b} = \frac{1}{2}\left[\frac{1}{\tilde{\beta}}\left(\frac{\partial \tilde{\kappa}}{\partial S}\right)_{p,\theta} + \frac{1}{\tilde{\alpha}}\left(\frac{\partial \tilde{\kappa}}{\partial \theta}\right)_{p,S}\right] \tag{2.23}$$

Again, the second term dominates.

The turning of the isopycnal slope with pressure has important dynamical consequences. Consider two fluid particles on a pressure surface that have the same density but different potential temperatures and salinities. When moved at constant θ and S to a different pressure surface they will have different densities (see Figure 2.1).

The second combination is the *cabbeling coefficient*

$$\tilde{d} := \frac{1}{\tilde{\alpha}\tilde{\beta}}\frac{\partial \tilde{\beta}}{\partial \theta} - \frac{1}{\tilde{\alpha}^2}\frac{\partial \tilde{\alpha}}{\partial \theta} + \frac{1}{\tilde{\beta}^2}\frac{\partial \tilde{\beta}}{\partial S} - \frac{1}{\tilde{\alpha}\tilde{\beta}}\frac{\partial \tilde{\alpha}}{\partial S} \tag{2.24}$$

which characterizes the curvature of the isopycnals (see Figure 2.1)

$$\frac{1}{\tilde{\alpha}}\left(\frac{\partial \tilde{\gamma}}{\partial \theta}\right)_{\rho,p} = \frac{1}{\tilde{\beta}}\left(\frac{\partial \tilde{\gamma}}{\partial S}\right)_{\rho,p} = \tilde{d}\tilde{\gamma} \tag{2.25}$$

The coefficient $\tilde{d}$ is called the cabbeling coefficient since it appears in expressions that describe cabbeling, i.e., the densification upon mixing (see Section 2.11).

Approximations The equation of state has been determined experimentally and algorithms are available to calculate it to various degrees of accuracy. The following approximations are often used in theoretical investigations:

1. Non-thermobaric fluid If one assumes $\tilde{b} = 0$ the fluid is called non-thermobaric. In this case the density is completely determined by pressure and potential density, $\rho = \rho(p, \rho_{\text{pot}})$ (see Table 2.1). The fluid behaves in many ways as a one-component system.

2. Incompressible fluid If one assumes $\tilde{\kappa} = 0$ the fluid is called incompressible and $\rho = \rho(\theta, S)$. This is a degenerate limit. The pressure cannot be determined from thermodynamic relations. It can only be an independent variable. The internal and free energy functions $e(v, \eta, S)$ and $f(v, \eta, S)$ do not exist. Thermodynamic relations imply $\kappa = \alpha\Gamma$.

3. Homentropic fluid If one assumes $\tilde{\alpha} = 0$ and $\tilde{\beta} = 0$ the fluid is called homentropic (or barotropic). The density depends on pressure only, $\rho = \rho(p)$. Thermodynamic relations imply $\Gamma = 0$ and the temperature equals the potential temperature.

4. Linear equation of state If one assumes $\tilde{\kappa}$, $\tilde{\alpha}$, and $\tilde{\beta}$ to be constant then the equation of state becomes linear

$$\rho(p, \theta, S) = \rho_0 \left[1 + \tilde{\kappa}_0(p - p_0) - \tilde{\alpha}_0(\theta - \theta_0) + \tilde{\beta}_0(S - S_0)\right] \qquad (2.26)$$

The quantities with subscript zero are constant reference values.

2.5 Spiciness

Consider the isolines of density at constant pressure in the (T, S) (or (θ, S)) plane. One can introduce a quantity $\chi(T, S)$ whose isolines intercept those of density. Such a quantity χ is often called *spiciness*. There are many different ways to define spiciness. Here we consider two.

The first one is to require that the isolines of χ are orthogonal to those of density ρ (Veronis, 1972)

$$\frac{\partial \rho}{\partial T}\frac{\partial \chi}{\partial T} + \frac{\partial \rho}{\partial S}\frac{\partial \chi}{\partial S} = 0 \qquad (2.27)$$

This condition is a first-order partial differential equation for χ. Its characteristics

$$\frac{\partial \rho}{\partial S}\mathrm{d}T = \frac{\partial \rho}{\partial T}\mathrm{d}S \qquad (2.28)$$

define the isolines of spiciness. Values of spiciness then need to be assigned to these isolines. The definition of orthogonality depends on the scales of temperature and salinity. The geometric orthogonality is lost when the scales are changed.

The second way is to require that the slopes of the isolines of density and spiciness are equal and of opposite sign (Flament, 2002; Figure 2.2). This requirement leads to the partial differential equation

$$\frac{\partial \rho}{\partial S}\frac{\partial \chi}{\partial T} + \frac{\partial \rho}{\partial T}\frac{\partial \chi}{\partial S} = 0 \qquad (2.29)$$

with characteristics

$$\frac{\partial \rho}{\partial T}\mathrm{d}T = \frac{\partial \rho}{\partial S}\mathrm{d}S \qquad (2.30)$$

This definition is independent of the scales of T and S. The labeling of the spiciness isolines can still be done in different ways.

Once $\chi(T, S)$ has been constructed for each pressure surface one can use (χ, ρ) instead of (T, S) as independent variables. For the scale-independent definition

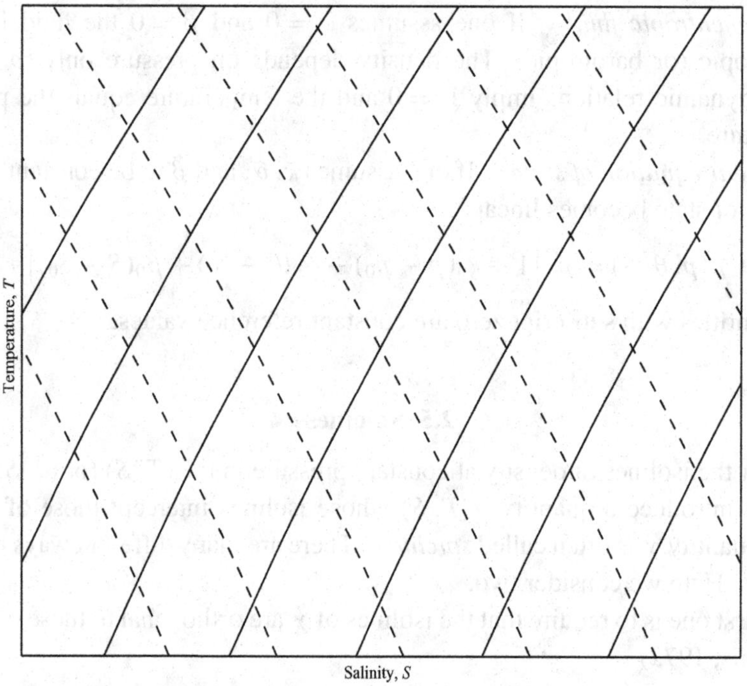

Figure 2.2. Sketch of isolines of density ρ (*solid lines*) and spiciness χ (*dashed lines*) in the (T, S)-plane for the scale-independent definition.

differentials transform according to

$$\begin{aligned} \mathrm{d}\rho &= -\rho\alpha \mathrm{d}T + \rho\beta \mathrm{d}S \\ \mathrm{d}\chi &= r(\rho\alpha \mathrm{d}T + \rho\beta \mathrm{d}S) \end{aligned} \tag{2.31}$$

$$\begin{aligned} \mathrm{d}T &= \frac{1}{2\rho\alpha}(r^{-1}\mathrm{d}\chi - \mathrm{d}\rho) \\ \mathrm{d}S &= \frac{1}{2\rho\beta}(r^{-1}\mathrm{d}\chi + \mathrm{d}\rho) \end{aligned} \tag{2.32}$$

where

$$r := \frac{\partial\chi/\partial S}{\partial\rho/\partial S} \tag{2.33}$$

For a linear equation of state $\rho = \rho_0[-\alpha_0(T - T_0) + \beta_0(S - S_0)]$ a particular convenient choice is

$$\chi = \rho_0[\alpha_0(T - T_0) + \beta_0(S - S_0)] \tag{2.34}$$

We will use spiciness and density as independent variables in Section 2.11 when we discuss cabbeling and in Sections 8.6 and 21.2 when we discuss the temperature–salinity mode.

2.6 Specific heat

The specific heat characterizes the amount of heat that is required to increase the temperature of a unit mass of sea water. Since this amount of heat depends on whether the heating takes place at constant pressure or at constant volume one distinguishes

$$\begin{aligned} c_p &:= \left(\frac{\partial h}{\partial T}\right)_{p,S} && \text{specific heat at constant pressure} \\ c_v &:= \left(\frac{\partial e}{\partial T}\right)_{v,S} && \text{specific heat at constant volume} \end{aligned} \tag{2.35}$$

Since $h = g + T\eta$ and $e = f + T\eta$ the specific heats are related to the specific entropy by

$$\begin{aligned} c_p &= T\left(\frac{\partial \eta}{\partial T}\right)_{p,S} \\ c_v &= T\left(\frac{\partial \eta}{\partial T}\right)_{v,S} \end{aligned} \tag{2.36}$$

Thermodynamic relations imply

$$\frac{c_p}{c_v} = \frac{\kappa}{\tilde{\kappa}} \tag{2.37}$$

$$c_v = c_p - \frac{T\alpha^2 v}{\kappa} \tag{2.38}$$

Thermodynamic inequalities imply $0 \leq c_v \leq c_p$. For an incompressible fluid c_v is not defined.

The specific heat c_p has been measured at atmospheric pressure p_a as a function of T and S. Its dependence on pressure can be inferred from the thermodynamic relation

$$\left(\frac{\partial c_p}{\partial p}\right)_{T,S} = -T\left(\frac{\partial^2 v}{\partial T^2}\right)_{p,S} \tag{2.39}$$

The inferred specific enthalpy h as a function of T for $S = 0$ and $p = p_a$ is shown in Figure 2.3.

2.7 Latent heat

The amount of heat required to vaporize (evaporate) a unit mass of liquid fresh water is called the specific heat of vaporization (or evaporation). It is given by

$$L_v := h_w^v - h_w^l \tag{2.40}$$

where $h_w^{v,l}$ are the specific enthalpies of water (subscript w) in its vapor phase (superscript v) and in its liquid phase (superscript l) (see Figure 2.3). Euler's identity

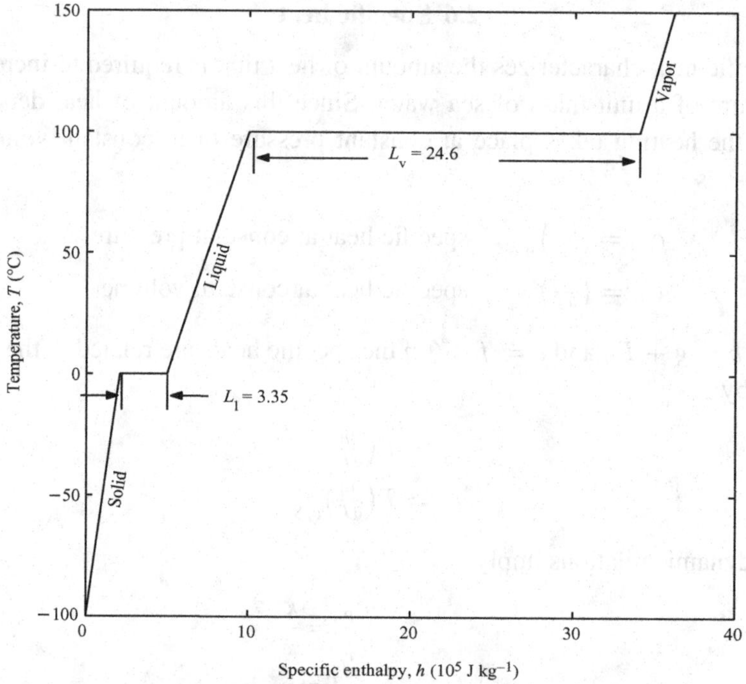

Figure 2.3. The specific enthalpy h as a function of temperature for $S = 0$ and $p = p_a$. The slope $\partial h/\partial T$ is the specific heat c_p. The steps Δh are the heats of liquification and vaporization L_l and L_v. Adapted from Apel (1987).

(2.8) implies

$$h_w^{v,l} = \mu_w^{v,l} + T\eta_w^{v,l} \tag{2.41}$$

for $S = 0$ where $\mu_w^{v,l}$ are the chemical potentials and $\eta_w^{v,l}$ the specific entropies of the respective phases. Since $\mu_w^v = \mu_w^l$ in phase equilibrium one obtains

$$L_v = T\left(\eta_w^v - \eta_w^l\right) \tag{2.42}$$

The analogous formulae for the heat of liquification (melting) of solid water (ice) are

$$L_l := h_w^l - h_w^s = T\left(\eta_w^l - \eta_w^s\right) \tag{2.43}$$

where the superscript s denotes the solid phase (see Figure 2.3). The specific heats of evaporation and liquification also give the amount of heat released when a unit mass of water vapor condenses or a unit mass of liquid water freezes. These specific heats are therefore also called latent heats.

The evaporation of water from sea water into air and the liquification of water from sea ice into sea water are also governed by the above formulae if $h_w^v(\eta_w^v)$ denote the partial enthalpy (entropy) of water vapor in air, $h_{\mathrm{w}}^{\mathrm{l}}(\eta_{\mathrm{w}}^{\mathrm{l}})$ the partial enthalpy (entropy) of water in sea water, and $h_{\mathrm{w}}^{\mathrm{s}}(\eta_{\mathrm{w}}^{\mathrm{s}})$ the partial enthalpy (entropy) of water in sea ice.[2]

Usually one assumes sea ice to consist of pure water, i.e., the salinity of sea ice is assumed to be zero. Then the partial enthalpy (entropy) of water in sea ice becomes the specific enthalpy (entropy). If the (small) amount of sea salt in sea ice is taken into account then L_{l} in (2.43) describes only the amount of heat required to transfer water from sea ice to sea water. An analogous expression is required to describe the amount of heat required to transfer sea salt from sea ice to sea water.

2.8 Boiling and freezing temperature

Different phases of water can coexist in thermal equilibrium. The simplest case of the phase equilibrium of pure water and water vapor is treated in Appendix A. Here we consider the phase equilibrium of sea water with water vapor, air, and sea ice.

Phase equilibrium of sea water and water vapor The temperature at which sea water coexists with water vapor is called the *boiling temperature* T_{b}. In thermodynamic equilibrium the pressure, temperature, and chemical potential of water must be equal in the two phases. The boiling temperature is thus determined by the relation

$$\mu_{\mathrm{w}}^{\mathrm{l}}(p, T_{\mathrm{b}}, S) = \mu_{\mathrm{w}}^{\mathrm{v}}(p, T_{\mathrm{b}}) \tag{2.44}$$

where $\mu_{\mathrm{w}}^{\mathrm{l}}$ is the chemical potential of water in sea water and $\mu_{\mathrm{w}}^{\mathrm{v}}$ is the chemical potential of water vapor. The boiling temperature is thus a function of p and S, $T_{\mathrm{b}} = T_{\mathrm{b}}(p, S)$. This result is consistent with Gibbs' phase rule, which states that an N-component, M-phase system is described by $f = N - M + 2$ variables. In the above case, $N = 2$, $M = 2$, and hence $f = 2$.

Differentiation of (2.44) with respect to p at constant S gives the *Clausius–Clapeyron relation*

$$\left(\frac{\partial T_{\mathrm{b}}}{\partial p}\right)_S = \frac{T_{\mathrm{b}}}{L_{\mathrm{v}}}\left(v_{\mathrm{w}}^{\mathrm{v}} - v_{\mathrm{w}}^{\mathrm{l}}\right) \tag{2.45}$$

where $v_{\mathrm{w}}^{\mathrm{v}}$ is the specific volume of water vapor, $v_{\mathrm{w}}^{\mathrm{l}}$ the partial specific volume of liquid water in sea water, and L_{v} the specific heat of vaporization. Since $L_{\mathrm{v}} > 0$ and $v_{\mathrm{w}}^{\mathrm{v}} > v_{\mathrm{w}}^{\mathrm{l}}$ the boiling temperature increases with pressure.

[2] Partial contributions z_j are the coefficients in the decomposition $z = \sum_{j=1}^{N} z_j c_j$ (see Section A.3).

Differentiation with respect to salinity at constant pressure gives

$$\left(\frac{\partial T_{\mathrm{b}}}{\partial S}\right)_p = -\frac{T_{\mathrm{b}}}{L_{\mathrm{v}}}\left(\frac{\partial \mu_{\mathrm{w}}^{1}}{\partial S}\right)_p = \frac{T_{\mathrm{b}}S}{L_{\mathrm{v}}}\left(\frac{\partial \Delta\mu}{\partial S}\right)_p \tag{2.46}$$

where we have used $\partial\mu_{\mathrm{w}}^{1}/\partial S = -S\partial\Delta\mu/\partial S$. The thermodynamic inequality $\partial\Delta\mu/\partial S \geq 0$ then implies that the boiling temperature increases with salinity. Equation (2.46) determines $\left(\frac{\partial\Delta\mu}{\partial S}\right)_{p,T}$ (p, T, S) at boiling temperature.

Phase equilibrium of sea water and air If air is considered a mixture of dry air and water vapor then the condition for phase equilibrium becomes

$$\mu_{\mathrm{w}}^{1}(p, T, S) = \mu_{\mathrm{w}}^{\mathrm{v}}(p, T, q) \tag{2.47}$$

where $q := \rho_{\mathrm{w}}^{\mathrm{v}}/\rho$ is the specific humidity. Gibbs' phase rule now implies that the boiling temperature is a function of three variables, $T_{\mathrm{b}} = T_{\mathrm{b}}(p, S, q)$. However, one usually solves (2.47) for the saturation humidity

$$q_{\mathrm{s}} = q_{\mathrm{s}}(p, T, S) \tag{2.48}$$

The Clausius–Clapeyron relations then take the form

$$\begin{aligned}
\left(\frac{\partial q_{\mathrm{s}}}{\partial T}\right)_{p,S} &= \frac{L_{\mathrm{v}}}{T\frac{\partial\mu_{\mathrm{w}}^{\mathrm{v}}}{\partial q}} \geq 0 \\
\left(\frac{\partial q_{\mathrm{s}}}{\partial p}\right)_{T,S} &= -\frac{v_{\mathrm{w}}^{\mathrm{v}} - v_{\mathrm{w}}^{1}}{\frac{\partial\mu_{\mathrm{w}}^{\mathrm{v}}}{\partial q}} \leq 0 \\
\left(\frac{\partial q_{\mathrm{s}}}{\partial S}\right)_{T,p} &= \frac{\frac{\partial\mu_{\mathrm{w}}^{1}}{\partial S}}{\frac{\partial\mu_{\mathrm{w}}^{\mathrm{v}}}{\partial q}} \leq 0
\end{aligned} \tag{2.49}$$

where the relations $\partial\mu_{\mathrm{w}}^{1}/\partial S = -S\partial\Delta\mu^{1}/\partial S \leq 0$ and $\partial\mu_{\mathrm{w}}^{\mathrm{v}}/\partial q = (1-q)$ $\partial\Delta\mu^{\mathrm{v}}/\partial q \geq 0$ (see Section 2.9) and $v_{\mathrm{w}}^{\mathrm{v}} > v_{\mathrm{w}}^{1}$ imply the inequalities.

If one further assumes that water vapor and dry air behave as ideal gases (see Section A.9) then one can introduce the saturation pressure by

$$p_{\mathrm{s}} = q_{\mathrm{s}}\frac{pR_{\mathrm{v}}}{R_{\mathrm{d}}\left[1 + q_{\mathrm{s}}\left(\frac{R_{\mathrm{v}}}{R_{\mathrm{d}}} - 1\right)\right]} \tag{2.50}$$

where R_{d} and R_{v} are the gas constants of dry air and water vapor. The saturation pressure depends on (p, T, S).

Phase equilibrium of sea water and sea ice The temperature at which sea water and sea ice coexist is called the *freezing temperature*. Phase equilibrium now

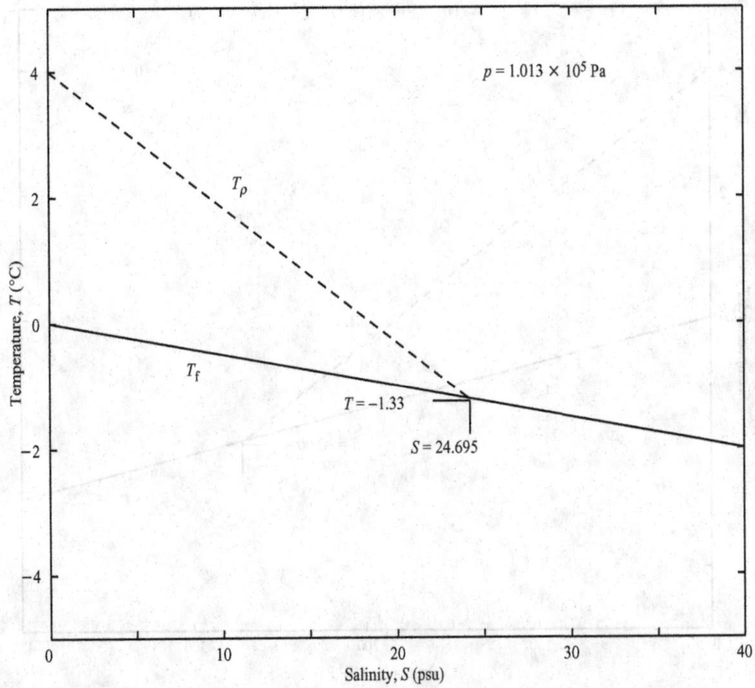

Figure 2.4. Freezing temperature T_f and maximal-density temperature T_ρ as a function of salinity S for $p = p_\mathrm{a}$. Adapted from Apel (1987).

requires

$$\begin{aligned} \mu_\mathrm{w}^\mathrm{l}(p, T_\mathrm{f}, S) &= \mu_\mathrm{w}^\mathrm{s}(p, T_\mathrm{f}, c_s^s) \\ \mu_\mathrm{s}^\mathrm{l}(p, T_\mathrm{f}, S) &= \mu_\mathrm{s}^\mathrm{s}(p, T_\mathrm{f}, c_s^s) \end{aligned} \tag{2.51}$$

where $\mu_\mathrm{w}^\mathrm{l,s}$ are the chemical potentials of water (subscript w) in sea water (superscript l) and in sea ice (superscript s), $\mu_\mathrm{s}^\mathrm{l,s}$ the chemical potentials of sea salt (subscript s) in sea water and sea ice, and c_s^s the salinity of sea ice. From these two relations one can determine $T_\mathrm{f} = T_\mathrm{f}(p, S)$ and $c_s^s = c_s^s(p, S)$, again in accordance with Gibbs' phase rule. If one assumes $c_s^s = 0$ then the Clausius–Clapeyron relations take the form

$$\begin{aligned} \left(\frac{\partial T_\mathrm{f}}{\partial p}\right)_S &= \frac{T_\mathrm{f}}{L_\mathrm{l}}\left(v_\mathrm{w}^\mathrm{l} - v_\mathrm{w}^\mathrm{s}\right) \leq 0 \\ \left(\frac{\partial T_\mathrm{f}}{\partial S}\right)_p &= -\frac{T_\mathrm{f} S}{L_\mathrm{l}}\left(\frac{\partial \Delta\mu}{\partial S}\right)_p \leq 0 \end{aligned} \tag{2.52}$$

where v_w^s is the specific volume of ice. Since $v_\mathrm{w}^\mathrm{s} > v_\mathrm{w}^\mathrm{l}$ and $\partial\Delta\mu/\partial S \geq 0$ the freezing temperature decreases with both p and S. The freezing temperature T_f as a function of S is shown in Figure 2.4 for $p = p_a$ and as a function of p in Figure 2.5 for $S = 0$.

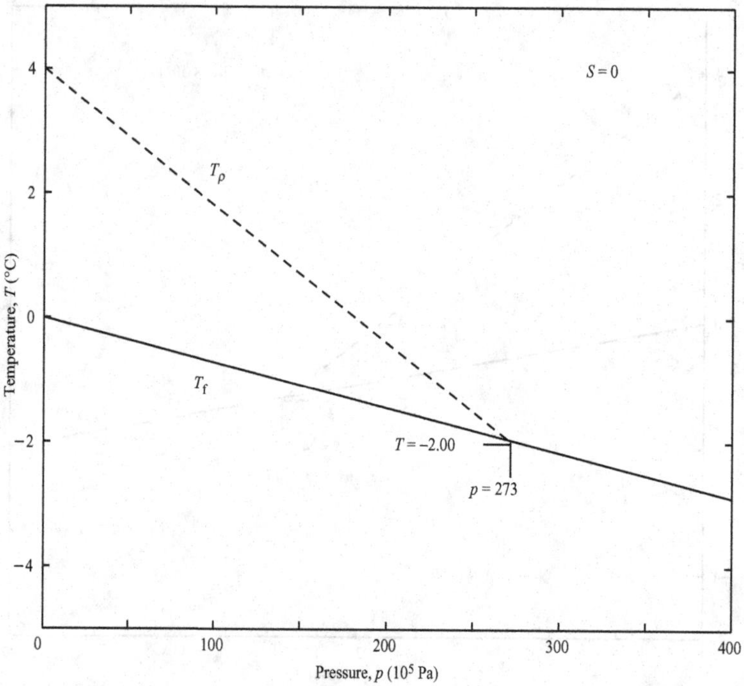

Figure 2.5. Freezing temperature T_f and maximal-density temperature T_ρ as a function of pressure p for $S = 0$. Adapted from Apel (1987).

Also shown in both figures is the temperature T_ρ where the density $\rho(p, T, S)$ is largest. The second relation in (2.52) determines $\left(\frac{\partial \Delta\mu}{\partial S}\right)_{p,T}(p, T, S)$ at freezing temperature.

Equation (2.45) and the first relation in (2.52) imply

$$L_{\text{v,l}} = \frac{T_{\text{b,f}}\left(v_{\text{w}}^{\text{v,l}} - v_{\text{w}}^{\text{l,s}}\right)}{\left(\frac{\partial T_{\text{b,f}}}{\partial p}\right)_S} \tag{2.53}$$

Since all quantities on the right-hand side are measured this formula determines the specific heats of vaporization and liquification.

2.9 Chemical potentials

The chemical potentials μ_{w} and μ_{s} of water and salt characterize the energy required to change the concentrations c_{w} and c_{s} of water and salt. Since the concentrations are not independent ($c_{\text{w}} = (1 - S)$, $c_{\text{s}} = S$) only the chemical potential difference

$$\Delta\mu := \mu_{\text{s}} - \mu_{\text{w}} \tag{2.54}$$

is determined by the specific free enthalpy

$$\Delta\mu = \left(\frac{\partial g}{\partial S}\right)_{p,T} \tag{2.55}$$

The individual potentials must be inferred from Euler's identity $g = \mu_w (1 - S) + \mu_s S$ and are given by

$$\begin{aligned} \mu_s &= g + (1 - S)\Delta\mu \\ \mu_w &= g - S\Delta\mu \end{aligned} \tag{2.56}$$

These relations imply

$$\begin{aligned} \left(\tfrac{\partial \mu_s}{\partial S}\right)_{p,T} &= (1 - S)\left(\tfrac{\partial \Delta\mu}{\partial S}\right)_{p,T} \geq 0 \\ \left(\tfrac{\partial \mu_w}{\partial S}\right)_{p,T} &= -S\left(\tfrac{\partial \Delta\mu}{\partial S}\right)_{p,T} \leq 0 \end{aligned} \tag{2.57}$$

The inequality signs follow from the thermodynamic inequality $\partial\Delta\mu/\partial S \geq 0$.

Euler's identity implies that the chemical potentials of water and salt are the partial contributions of water and salt to the specific free enthalpy g. We follow the traditional nomenclature that assigns the new symbols μ_w and μ_s to the partial contributions, rather than g_w and g_s.

The chemical potential difference $\Delta\mu$ can be constructed from measured quantities up to a function $a + bT$. The first relation in Table 2.2 implies

$$\Delta\mu(p, T, S) = \Delta\mu(p_0, T, S) + \int_{p_0}^{p} \mathrm{d}p' \, \frac{\partial v}{\partial S}(p', T, S) \tag{2.58}$$

One only needs to determine $\Delta\mu$ at a reference pressure p_0. The second relation in Table 2.2 implies

$$\frac{\partial \Delta\mu}{\partial T}(p, T, S) = \frac{\partial \Delta\mu}{\partial T}(p, T_0, S) - \int_{T_0}^{T} \mathrm{d}T' \, \frac{1}{T'}\frac{\partial c_p}{\partial S}(p, T', S) \tag{2.59}$$

The chemical potential difference at the reference pressure is thus given by

$$\begin{aligned} \Delta\mu(p_0, T, S) = {} & \Delta\mu(p_0, T_0, S) + \tfrac{\partial \Delta\mu}{\partial T}(p_0, T_0, S)(T - T_0) \\ & - \textstyle\int_{T_0}^{T} \mathrm{d}T' \int_{T_0}^{T'} \mathrm{d}T'' \, \tfrac{1}{T''}\tfrac{\partial c_p}{\partial S}(p_0, T'', S) \end{aligned} \tag{2.60}$$

and hence determined up to a function $a(S) + b(S)T$. The derivatives $\partial\Delta\mu/\partial S$ are given at boiling and freezing temperature (see (2.46) and (2.52)). These two relations can be solved for $\partial a/\partial S$ and $\partial b/\partial S$. Therefore $\Delta\mu$ is determined up to a function $a + bT$ where a and b are constants.

Because of this indeterminacy of $\Delta\mu$ one usually considers the specific enthalpy

$$h = h_w(1 - S) + h_s S \tag{2.61}$$

Table 2.2. *Derivatives of various quantities with respect to* (p, T, S) *and the functions up to which these quantities are determined by measurements*

The coefficients a, b, c, and d are constants

Chemical potential difference	$\frac{\partial \Delta\mu}{\partial p} = \frac{\partial v}{\partial S}$ $\frac{\partial^2 \Delta\mu}{\partial T^2} = -\frac{1}{T}\frac{\partial c_p}{\partial S}$ $\frac{\partial \Delta\mu}{\partial S} = \begin{cases} \frac{L_v}{T_b S}\frac{\partial T_b}{\partial S} & \text{at } T = T_b \\ -\frac{L_f}{T_f S}\frac{\partial T_f}{\partial S} & \text{at } T = T_f \end{cases}$	$a + bT$
Partial enthalpy difference	$\frac{\partial \Delta h}{\partial p} = \frac{\partial v}{\partial S} - T\frac{\partial^2 v}{\partial S \partial T}$ $\frac{\partial \Delta h}{\partial T} = \frac{\partial c_p}{\partial S}$ $\frac{\partial \Delta h}{\partial S} = \frac{\partial \Delta\mu}{\partial S} - T\frac{\partial^2 \Delta\mu}{\partial S \partial T}$	a
Specific entropy	$\frac{\partial \eta}{\partial p} = -\frac{\partial v}{\partial T}$ $\frac{\partial \eta}{\partial T} = \frac{c_p}{T}$ $\frac{\partial \eta}{\partial S} = -\frac{\partial \Delta\mu}{\partial T}$ (known up to b)	$-d - bS$
Specific internal energy	$\frac{\partial e}{\partial p} = -p\frac{\partial v}{\partial p} - T\frac{\partial v}{\partial T}$ $\frac{\partial e}{\partial T} = -p\frac{\partial v}{\partial T} + c_p$ $\frac{\partial e}{\partial S} = \Delta h - p\frac{\partial v}{\partial S}$ (known up to a)	$c + aS$
Specific free enthalpy	$\frac{\partial g}{\partial p} = v$ $\frac{\partial g}{\partial T} = -\eta$ (known up to $-d - bS$) $\frac{\partial g}{\partial S} = \Delta\mu$ (known up to $a + bT$)	$c + \mathrm{d}T + aS + bST$
Specific enthalpy	$\frac{\partial h}{\partial p} = v - T\frac{\partial v}{\partial T}$ $\frac{\partial h}{\partial T} = c_p$ $\frac{\partial h}{\partial S} = \Delta h$ (known up to a)	$c + aS$

with partial contributions h_w and h_s. The partial enthalpy difference

$$\Delta h := h_s - h_w = \left(\frac{\partial h}{\partial S}\right)_{p,T} \tag{2.62}$$

is related to $\Delta\mu$ by

$$\Delta h = \Delta\mu - T\left(\frac{\partial \Delta\mu}{\partial T}\right)_{p,S} \tag{2.63}$$

which follows from $h = g + T\eta$. The derivatives of Δh are listed in Table 2.2 and are all measured quantities. The partial enthalpy difference Δh is known up to a constant a.

2.10 Measured quantities

The following quantities have been measured directly:

- $v = v(p, T, S)$ equation of state (see Figure 2.1);
- $c_p = c_p(p_0, T, S)$ specific heat at atmospheric pressure (see Figure 2.3);
- $T_b = T_b(p, S)$ boiling temperature; and
- $T_f = T_f(p, S)$ freezing temperature (see Figures 2.4 and 2.5).

From these measured quantities one can determine:

- $\Gamma(p, T, S)$ adiabatic temperature gradient;
- $\theta(p, T, S)$ potential temperature;
- $c_p(p, T, S)$ specific heat;
- $L_{v,l}(p, S)$ specific heats of vaporization and liquification; and
- $\left(\frac{\partial \Delta\mu}{\partial S}\right)_{p,T}(p, T_{b,f}, S)$

as has been pointed out in the respective sections. Table 2.2 lists the derivatives of some other (theoretically relevant) quantities and the functions up to which these quantities are determined by measurements.

2.11 Mixing

Consider the mixing of two parcels of sea water labeled "A" and "B" resulting in a water parcel "C." If the mixing occurs at constant pressure then mass, salt, and enthalpy are additive

$$\begin{aligned} M_C &= M_A + M_B \\ S_C &= \frac{M_A}{M_C} S_A + \frac{M_B}{M_C} S_B \\ h_C &= \frac{M_A}{M_C} h_A + \frac{M_B}{M_C} h_B \end{aligned} \tag{2.64}$$

Salinity and specific enthalpy mix in mass proportion. Mixing occurs along a straight line in (h, S)-space (see Figure 2.6). (If mixing occurs without change of volume then the specific internal energy and salinity mix in mass proportion.) For any quantity $a(p, h, S)$ the deviation from such mixing in mass proportion is given by

$$\Delta a = a(p, h_C, S_C) - \left[\frac{M_A}{M_C} a(p, h_A, S_A) + \frac{M_B}{M_C} a(p, h_B, S_B)\right] \tag{2.65}$$

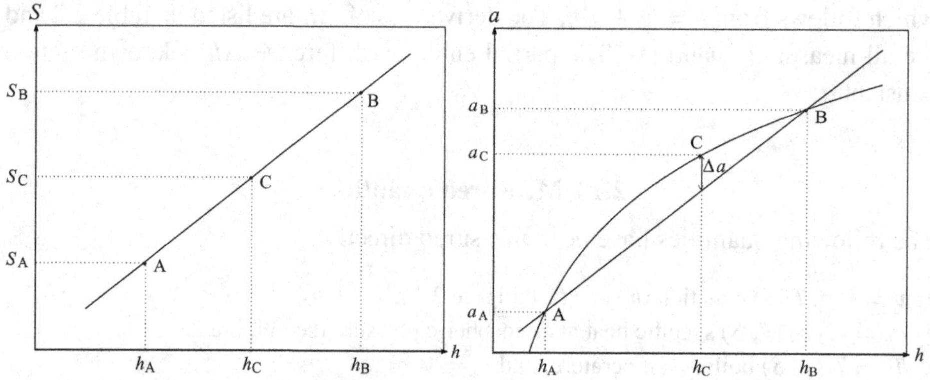

Figure 2.6. Mixing of two water masses. Salinity and specific enthalpy mix in mass proportion (*left*). Quantities a that depend nonlinearly on h (and S) deviate from mixing in mass proportion by Δa (*right*).

and is unequal to zero if a depends nonlinearly on h and S (see Figure 2.6). If the two initial states differ only slightly then a Taylor expansion yields

$$\Delta a = -\frac{1}{2}\frac{M_A M_B}{M_C^2}\left[\frac{\partial^2 a}{\partial h^2}(h_B - h_A)^2 + 2\frac{\partial^2 a}{\partial h \partial S}(h_B - h_A)(S_B - S_A) + \frac{\partial^2 a}{\partial S^2}(S_B - S_A)^2\right] \tag{2.66}$$

or

$$\begin{aligned}\Delta a = -\frac{1}{2}\frac{M_A M_B}{M_C^2}\Bigg\{&\frac{\partial^2 a}{\partial T^2}(T_B - T_A)^2 + 2\frac{\partial^2 a}{\partial T \partial S}(T_B - T_A)(S_B - S_A)\\ &+ \frac{\partial^2 a}{\partial S^2}(S_B - S_A)^2 - \frac{1}{c_p}\frac{\partial a}{\partial T}\Bigg[\frac{\partial c_p}{\partial T}(T_B - T_A)^2\\ &+ 2\frac{\partial c_p}{\partial S}(T_B - T_A)(S_B - S_A) + \frac{\partial \Delta h}{\partial S}(S_B - S_A)^2\Bigg]\Bigg\}\end{aligned} \tag{2.67}$$

in the (p, T, S)-representation.

Consider specifically the following quantities:

1. *Entropy* ($a = \eta$) Then

$$\Delta\eta = \frac{1}{2}\frac{M_A M_B}{M_C^2}\left[\frac{c_p}{T^2}(T_B - T_A)^2 + \frac{1}{T}\frac{\partial \Delta\mu}{\partial S}(S_B - S_A)^2\right] \tag{2.68}$$

Since $c_p \geq 0$, $T \geq 0$, and $\partial\Delta\mu/\partial S \geq 0$ for any thermodynamic system, $\Delta\eta \geq 0$ as required by the second law of thermodynamics.

2. *Temperature* ($a = T$) Then

$$\Delta T = \frac{1}{2}\frac{M_A M_B}{M_C^2}\frac{1}{c_p}\left[\frac{\partial c_p}{\partial T}(T_B - T_A)^2 + 2\frac{\partial c_p}{\partial S}(T_B - T_A)(S_B - S_A) + \frac{\partial \Delta h}{\partial S}(S_B - S_A)^2\right] \tag{2.69}$$

The temperature changes even if two parcels of the same temperature are mixed. This effect is called the *heat of mixing* and is governed by the coefficient $\partial \Delta h/\partial S$.

3. *Potential temperature* ($a = \theta$) Then

$$\Delta\theta(p,T) = \frac{1}{2}\frac{M_A M_B}{M_C^2}\frac{\theta c_p}{T^2 c_p^{\text{pot}}}\left(1 - \frac{c_p}{c_p^{\text{pot}}} + \frac{T}{c_p^{\text{pot}}}\frac{\partial c_p^{\text{pot}}}{\partial T}\right)(T_B - T_A)^2 \tag{2.70}$$

for fresh water ($S = 0$). At $p = p_0$ one has $T = \theta$ and $c_p = c_p^{\text{pot}}$ and hence

$$\Delta\theta(p_0,T) = \frac{T^2}{c_p^2}\frac{\partial c_p}{\partial T}\Delta\eta \tag{2.71}$$

The sign of $\Delta\theta$ thus depends on the sign of $\partial c_p/\partial T$. For water one finds $\partial c_p/\partial T < 0$ (e.g., Fofonoff, 1962). Mixing thus does not necessarily increase the potential temperature.

4. *Density* ($a = \rho$) Then

$$\begin{aligned}\Delta\rho = -\tfrac{1}{2}\tfrac{M_A M_B}{M_C^2}\Big[&\tfrac{\partial^2\rho}{\partial T^2}(T_B - T_A)^2 + 2\tfrac{\partial^2\rho}{\partial T\partial S}(T_B - T_A)(S_B - S_A)\\ &+ \tfrac{\partial^2\rho}{\partial S^2}(S_B - S_A)^2\Big] + \tfrac{\partial\rho}{\partial T}\Delta T\end{aligned} \tag{2.72}$$

If one introduces instead of the temperature and salinity differences the density and spiciness differences (see (2.31))

$$\begin{aligned}\rho_B - \rho_A &= -\rho\alpha(T_B - T_A) + \rho\beta(S_B - S_A)\\ \chi_B - \chi_A &= r[\rho\alpha(T_B - T_A) + \rho\beta(S_B - S_A)]\end{aligned} \tag{2.73}$$

then

$$\begin{aligned}\Delta\rho = -\frac{1}{2}\frac{M_A M_B}{M_C^2}\frac{1}{4}\frac{1}{\rho}&[c_1(\rho_B - \rho_A)^2 + 2c_2 r^{-1}(\rho_B - \rho_A)(\chi_B - \chi_A)\\ &+ c_3 r^{-2}(\chi_B - \chi_A)^2] + \frac{\partial\rho}{\partial T}\Delta T\end{aligned} \tag{2.74}$$

where

$$c_1 = \frac{1}{\rho}\left(\frac{1}{\alpha^2}\frac{\partial^2\rho}{\partial T^2} - 2\frac{1}{\alpha\beta}\frac{\partial^2\rho}{\partial T\partial S} + \frac{1}{\beta^2}\frac{\partial^2\rho}{\partial S^2}\right) \tag{2.75}$$

$$c_2 = \frac{1}{\rho}\left(-\frac{1}{\alpha^2}\frac{\partial^2\rho}{\partial T^2} + \frac{1}{\beta^2}\frac{\partial\rho}{\partial S^2}\right) \tag{2.76}$$

$$c_3 = \frac{1}{\rho}\left(\frac{1}{\alpha^2}\frac{\partial^2\rho}{\partial T^2} + 2\frac{1}{\alpha\beta}\frac{\partial^2\rho}{\partial T\partial S} + \frac{1}{\beta^2}\frac{\partial^2\rho}{\partial S^2}\right) \tag{2.77}$$

If two water parcels of the same density are mixed, the density changes owing to cabbeling, characterized by the cabbeling coefficient c_3, which corresponds to the cabbeling coefficient d defined in (2.24).

3

Balance equations

In this chapter we establish the equations that govern the evolution of a fluid that is not in thermodynamic, but in *local* thermodynamic equilibrium. These equations are the balance[1] or budget equations for water, salt, momentum, angular momentum, and energy within arbitrary volumes. We formulate these equations by making the continuum hypothesis. The balance equations then become partial differential equations in space and time. They all have the same general form. The rate of change of the amount within a volume is given by the fluxes through its enclosing surface and by the sources and sinks within the volume. This balance is self-evident. Specific physics enters when the fluxes and the sources and sinks are specified. If no sources and sinks are assumed then the balance equations become conservation equations. It will turn out that six balance equations, the ones for the mass of water, the mass of salt, the three components of the momentum vector, and energy completely determine the evolution of oceanic motions, given appropriate initial and boundary conditions. The vector and tensor calculus required for this chapter is briefly reviewed in Appendix B.

3.1 Continuum hypothesis

The ocean is assumed to be in *local thermodynamic equilibrium*. Each fluid parcel can be described by the usual thermodynamic variables such as temperature, pressure, etc., and the usual thermodynamic relations hold among these variables, but the values of these variables depend on the fluid particle. The ocean as a whole is not in thermodynamic equilibrium. Consider the *continuum limit* where the fluid particles are sufficiently small from a macroscopic point of view to be treated as points but still sufficiently large from a microscopic point of view to contain enough

[1] The term balance equations also denotes the equations that describe balanced, such as geostrophically balanced, flows. This is not the meaning here.

molecules for equilibrium thermodynamics to apply. In this limit the fluid particles can be labeled by their position $\mathbf{x}$ in space. Similarly, one assumes that a fluid parcel, when disturbed, reaches its new thermodynamic equilibrium so rapidly that the adjustment can be regarded as instantaneous from a macroscopic point of view. In this continuum limit, the ocean is described by intensive variables $p(\mathbf{x}, t)$, $T(\mathbf{x}, t)$, $\mu(\mathbf{x}, t)$, $v(\mathbf{x}, t)$, $\eta(\mathbf{x}, t)$, $e(\mathbf{x}, t)$, etc., which are (generally continuous) functions of position $\mathbf{x}$ and time t. Instead of specific quantities one can also use the densities $\rho(\mathbf{x}, t)$, $\tilde{\eta}(\mathbf{x}, t) = \rho(\mathbf{x}, t)\eta(\mathbf{x}, t)$, $\tilde{e}(\mathbf{x}, t) = \rho(\mathbf{x}, t)e(\mathbf{x}, t)$, etc.

In addition, one allows the fluid parcels to move with a macroscopic velocity $\mathbf{u}(\mathbf{x}, t)$ and to be exposed to an external potential $\phi(\mathbf{x}, t)$. These "external" variables must be included in the description of the system. The important point is, however, that the thermodynamic state is not affected by a (uniform) motion of the fluid parcel and by its exposure to an external potential. The specific internal energy does *not* depend on these external state variables. The usual thermodynamic relations hold at each point in space and time

$$e(\mathbf{x}, t) = e\left(v(\mathbf{x}, t), \eta(\mathbf{x}, t), S(\mathbf{x}, t)\right) \tag{3.1}$$

where the symbol "e" on the left-hand side denotes the internal energy and on the right-hand side the internal energy function (see the note on notation in Section 2.4).

3.2 Conservation equations

Consider a substance A with density $\rho_A(\mathbf{x}, t)$. The substance is *conserved* if the amount of the substance does not change within a fluid volume $V(t)$ that moves with the velocity $\mathbf{u}_A$ of the substance

$$\frac{\mathrm{d}}{\mathrm{d}t} \iiint_{V(t)} \mathrm{d}^3x\ \rho_A(\mathbf{x}, t) = 0 \tag{3.2}$$

This is the conservation equation in integral form. According to Reynolds' transport theorem (B.56) and Gauss' theorem (B.53), the time derivative of volume integrals is given by

$$\begin{aligned}\tfrac{\mathrm{d}}{\mathrm{d}t} \iiint_{V(t)} \mathrm{d}^3x\ \rho_A(\mathbf{x}, t) &= \iiint_{V(t)} \mathrm{d}^3x\ \partial_t \rho_A(\mathbf{x}, t) + \iint_{\delta V(t)} \mathrm{d}^2x\ \mathbf{n} \cdot \mathbf{u}_A\ \rho_A(\mathbf{x}, t) \\ &= \iiint_{V(t)} \mathrm{d}^3x\ \{\partial_t \rho_A(\mathbf{x}, t) + \nabla \cdot (\mathbf{u}_A\ \rho_A(\mathbf{x}, t))\}\end{aligned} \tag{3.3}$$

where $\delta V(t)$ is the surface enclosing the fluid volume with outward normal vector $\mathbf{n}$. The integral on the right-hand side is zero for all volumes that move with the velocity $\mathbf{u}_A$ of the substance. This can only be the case if

$$\partial_t \rho_A + \nabla \cdot (\rho_A \mathbf{u}_A) = 0 \tag{3.4}$$

This is the conservation equation in differential form. The vector

$$\mathbf{I}_A := \rho_A \mathbf{u}_A \tag{3.5}$$

is called the *flux vector* of substance A. It is the amount of the substance that is transported through a unit area in unit time.

3.3 Conservation of salt and water

Conservation of salt and water imply that the densities ρ_s and ρ_w of salt and water satisfy

$$\partial_t \rho_s + \nabla \cdot (\rho_s \mathbf{u}_s) = 0 \tag{3.6}$$

$$\partial_t \rho_w + \nabla \cdot (\rho_w \mathbf{u}_w) = 0 \tag{3.7}$$

where $\mathbf{u}_s$ and $\mathbf{u}_w$ are the velocities of salt and water.

Introduce the fluid velocity

$$\mathbf{u} := \frac{\rho_s \mathbf{u}_s + \rho_w \mathbf{u}_w}{\rho_s + \rho_w} \tag{3.8}$$

which is the velocity of the center of inertia or *barycentric* velocity. The two conservation equations then take the form

$$\partial_t \rho_s + \nabla \cdot \left(\mathbf{I}_s^{adv} + \mathbf{I}_s^{mol}\right) = 0 \tag{3.9}$$

$$\partial_t \rho_w + \nabla \cdot \left(\mathbf{I}_w^{adv} + \mathbf{I}_w^{mol}\right) = 0 \tag{3.10}$$

where

$$\mathbf{I}_s^{adv} := \rho_s \mathbf{u} \tag{3.11}$$

$$\mathbf{I}_w^{adv} := \rho_w \mathbf{u} \tag{3.12}$$

are the advective fluxes and

$$\mathbf{I}_s^{mol} := \rho_s(\mathbf{u}_s - \mathbf{u}) \tag{3.13}$$

$$\mathbf{I}_w^{mol} := \rho_w(\mathbf{u}_w - \mathbf{u}) \tag{3.14}$$

the molecular diffusive fluxes that represent the difference between the fluid velocity $\mathbf{u}$ and the individual velocities $\mathbf{u}_{s,w}$ of salt and water. They satisfy

$$\mathbf{I}_w^{mol} + \mathbf{I}_s^{mol} = 0 \tag{3.15}$$

The final form of the conservation equations is obtained by introducing the mass density

$$\rho := \rho_s + \rho_w \tag{3.16}$$

and salinity

$$S := \frac{\rho_s}{\rho} \tag{3.17}$$

For these one obtains the equations

$$\partial_t \rho + \nabla \cdot (\rho \mathbf{u}) = 0 \tag{3.18}$$

$$\partial_t (\rho S) + \nabla \cdot \left(\rho S \mathbf{u} + \mathbf{I}_s^{\mathrm{mol}}\right) = 0 \tag{3.19}$$

The first equation states the conservation of total mass, water plus sea salt. It is also called the *continuity equation*. Note that the total mass flux *does not* contain any diffusive flux component. The total mass moves with the fluid velocity, by definition.

The above conservation equations are written in their *flux form*. The local time rate of change of the density of a quantity is given by the divergence of its flux. An alternative form is the advective form

$$\frac{\mathrm{D}}{\mathrm{D}t} v = v \, \nabla \cdot \mathbf{u} \tag{3.20}$$

$$\rho \frac{\mathrm{D}}{\mathrm{D}t} S = -\nabla \cdot \mathbf{I}_s^{\mathrm{mol}} \tag{3.21}$$

where

$$\frac{\mathrm{D}}{\mathrm{D}t} := \partial_t + \mathbf{u} \cdot \nabla \tag{3.22}$$

is the *advective (Lagrangian, material, convective) derivative* and $v = \rho^{-1}$ the *specific volume*. The advective form describes changes along fluid trajectories.

The conservation laws state that the total amounts of salt and water are conserved. This global or integral conservation has to be distinguished from material conservation. A quantity $\psi(\mathbf{x}, t)$ is said to be *materially conserved* when

$$\frac{\mathrm{D}}{\mathrm{D}t} \psi = 0 \tag{3.23}$$

The value of ψ does not change following the motion of a fluid particle. Material conservation is a much stronger statement than global conservation. It applies to *all* fluid particles individually, not just to the fluid as a whole. Neither the specific volume v nor the salinity S are materially conserved. Since

$$\partial_t (\rho \psi) + \nabla \cdot (\rho \psi \mathbf{u}) = \rho \frac{\mathrm{D}}{\mathrm{D}t} \psi \tag{3.24}$$

global conservation of $\rho\psi$ implies material conservation of ψ only if the flux is purely advective and given by $\rho\psi\mathbf{u}$.

3.4 Momentum balance

Newton's second law states that the acceleration of a fluid volume is given by the applied forces

$$\frac{\mathrm{d}}{\mathrm{d}t}\iiint_{V(t)} \mathrm{d}^3x\ \rho\mathbf{u} = \iiint_{V(t)} \mathrm{d}^3x\ \rho\mathbf{F} + \iint_{\delta V(t)} \mathrm{d}^2x\ \mathbf{G} \tag{3.25}$$

Here $\mathbf{F}$ represents the *volume forces* acting within the fluid and $\mathbf{G}$ the *surface forces* acting on the fluid surface. We have further to distinguish between *internal* and *external* forces. An external force is a force that an external agent exerts on the fluid. An internal force is a force that the fluid exerts on itself. The major external force is the *gravitational force*

$$\mathbf{F}_g = -\boldsymbol{\nabla}\phi_g \tag{3.26}$$

which is given by the gradient of the gravitational potential ϕ_g. It is a volume force.

The internal forces in a fluid are all surface forces. They can be expressed by a stress tensor $\boldsymbol{\Pi}$, such that

$$G_i = \Pi_{ij} n_j \tag{3.27}$$

where $\mathbf{n}$ is the outward normal vector of the surface δV. The stress tensor can be further decomposed

$$\Pi_{ij} = -p\delta_{ij} + \sigma_{ij}^{\mathrm{mol}} \tag{3.28}$$

where p is the thermodynamic pressure and $\boldsymbol{\sigma}^{\mathrm{mol}}$ the molecular viscous stress tensor. The molecular viscous stress tensor accounts for the molecular diffusion of momentum as the molecular salt flux $\mathbf{I}_{\mathrm{s}}^{\mathrm{mol}}$ accounts for the molecular diffusion of salt. Molecular diffusion of momentum is also referred to as molecular friction.

In differential form the momentum balance takes the form

$$\partial_t(\rho u_i) + \partial_j(\rho u_i u_j) = -\partial_i p + \partial_j \sigma_{ij}^{\mathrm{mol}} - \rho\partial_i\phi_g \tag{3.29}$$

or

$$\rho\frac{\mathrm{D}}{\mathrm{D}t}\mathbf{u} = -\boldsymbol{\nabla}p + \boldsymbol{\nabla}\cdot\boldsymbol{\sigma}^{\mathrm{mol}} - \rho\boldsymbol{\nabla}\phi_g \tag{3.30}$$

where $\rho\mathbf{u}$ is the momentum density,

$$I_{ij}^{\mathbf{m}} := \rho u_i u_j + p\delta_{ij} - \sigma_{ij}^{\mathrm{mol}} \tag{3.31}$$

the *momentum flux tensor*, and $\rho\boldsymbol{\nabla}\phi_g$ the external force.

3.5 Momentum balance in a rotating frame of reference

In the previous section the momentum balance has been formulated in an inertial frame of reference. It can also be formulated in any other frame. If this frame moves with translational velocity $\mathbf{V}(t)$ and angular velocity $\mathbf{\Omega}(t)$ then the velocity $\mathbf{u}_i$ in the inertial frame is related to the velocity $\mathbf{u}_r$ relative to the moving frame by

$$\mathbf{u}_i = \mathbf{u}_r + \mathbf{V} + \mathbf{\Omega} \times \mathbf{x} \tag{3.32}$$

The time rates of change of any vector $\mathbf{B}$ in the inertial and moving frames are related by

$$\left(\frac{\mathrm{d}\mathbf{B}}{\mathrm{d}t}\right)_i = \left(\frac{\mathrm{d}\mathbf{B}}{\mathrm{d}t}\right)_r + \mathbf{\Omega} \times \mathbf{B} \tag{3.33}$$

Substituting $\mathbf{B} = \mathbf{u}_i$ one finds

$$\begin{aligned} \left(\tfrac{\mathrm{d}\mathbf{u}_i}{\mathrm{d}t}\right)_i &= \left(\tfrac{\mathrm{d}\mathbf{u}_i}{\mathrm{d}t}\right)_r + \mathbf{\Omega} \times \mathbf{u}_i \\ &= \left(\tfrac{\mathrm{d}\mathbf{u}_r}{\mathrm{d}t}\right)_r + \tfrac{\mathrm{d}\mathbf{V}}{\mathrm{d}t} + \tfrac{\mathrm{d}\mathbf{\Omega}}{\mathrm{d}t} \times \mathbf{x} + 2\mathbf{\Omega} \times \mathbf{u}_r + \mathbf{\Omega} \times \mathbf{V} + \mathbf{\Omega} \times (\mathbf{\Omega} \times \mathbf{x}) \end{aligned} \tag{3.34}$$

Relevant to geophysical fluid dynamics is the case $\mathbf{V} = 0$ and $\mathrm{d}\mathbf{\Omega}/\mathrm{d}t = 0$. Then

$$\left(\frac{\mathrm{d}\mathbf{u}_i}{\mathrm{d}t}\right)_i = \left(\frac{\mathrm{d}\mathbf{u}_r}{\mathrm{d}t}\right)_r + 2\mathbf{\Omega} \times \mathbf{u}_r + \mathbf{\Omega} \times (\mathbf{\Omega} \times \mathbf{x}) \tag{3.35}$$

The acceleration in the inertial frame is the relative acceleration plus the Coriolis acceleration $2\,\mathbf{\Omega} \times \mathbf{u}_r$ plus the centripetal acceleration $\mathbf{\Omega} \times (\mathbf{\Omega} \times \mathbf{x})$. When these terms are moved to the right-hand side of the equation they become additional apparent forces, the *Coriolis force* $-2\,\mathbf{\Omega} \times \mathbf{u}_r$, and the *centrifugal force* $-\mathbf{\Omega} \times (\mathbf{\Omega} \times \mathbf{x})$. The centrifugal force can be written as a gradient of a *centrifugal potential* ϕ_{c}

$$-\mathbf{\Omega} \times (\mathbf{\Omega} \times \mathbf{x}) = -\nabla\phi_{\mathrm{c}} \tag{3.36}$$

where

$$\phi_{\mathrm{c}} := -\frac{1}{2}(\mathbf{\Omega} \times \mathbf{x}) \cdot (\mathbf{\Omega} \times \mathbf{x}) = -\frac{1}{2}\Omega^2 a^2 \tag{3.37}$$

and where a is the distance perpendicular to the rotation axis. The centrifugal and gravitational potentials can be combined into the *geopotential*

$$\phi := \phi_g + \phi_{\mathrm{c}} \tag{3.38}$$

In a uniformly rotating frame of reference the momentum balance thus takes the form

$$\rho \frac{\mathrm{D}}{\mathrm{D}t}\mathbf{u} = -\nabla p + \nabla \cdot \sigma^{\mathrm{mol}} - \rho\, 2\boldsymbol{\Omega} \times \mathbf{u} - \rho \nabla \phi \tag{3.39}$$

where $\mathbf{u}$ is the relative velocity (the subscript "r" has been and will be dropped in the following). It describes the acceleration in the rotating frame due to the real pressure, viscous and gravitational forces, and the apparent Coriolis and centrifugal forces.

3.6 Angular momentum balance

Changes in the angular momentum are given by the torques exerted on the fluid. If it is assumed that the torques within the fluid arise only as the moments of the forces and that the angular momentum consists only of $\rho\mathbf{x} \times \mathbf{u}$ then

$$\frac{\mathrm{d}}{\mathrm{d}t}\iiint_{V(t)} \mathrm{d}^3x\ \rho\mathbf{x} \times \mathbf{u} = \iiint_{V(t)} \mathrm{d}^3x\ \ \rho\mathbf{x} \times \mathbf{F} + \iint_{\delta V(t)} \mathrm{d}^2x\ \mathbf{x} \times \mathbf{G} \tag{3.40}$$

or in differential form

$$\partial_t (\rho\mathbf{x} \times \mathbf{u}) + \nabla \cdot (\rho\mathbf{x} \times \mathbf{u}\mathbf{u}) = -\rho\mathbf{x} \times \nabla\phi - \rho\mathbf{x} \times (2\boldsymbol{\Omega} \times \mathbf{u}) + \nabla \cdot (\mathbf{x} \times \boldsymbol{\Pi}) \tag{3.41}$$

If one forms the vector product of $\mathbf{x}$ and the momentum balance (3.39) and subtracts the result from (3.41) one finds

$$\nabla \cdot \left(\mathbf{x} \times \sigma^{\mathrm{mol}}\right) - \mathbf{x} \times \left(\nabla \cdot \sigma^{\mathrm{mol}}\right) = 0 \tag{3.42}$$

which can only hold if the molecular viscous stress tensor is symmetric

$$\sigma_{ij}^{\mathrm{mol}} = \sigma_{ji}^{\mathrm{mol}} \tag{3.43}$$

Momentum and angular momentum balance require the viscous stress tensor to be symmetric. Conversely, momentum balance plus symmetry of the stress tensor imply the above angular momentum balance. It is thus sufficient to consider only the momentum balance with a symmetric viscous stress tensor. A non-symmetric stress tensor is required when one allows for an internal component of the angular momentum and for torques in addition to the moments of the forces, as is the case in other areas of continuum dynamics.

3.7 Energy balance

Consider a system whose parts move with macroscopic velocities $\mathbf{u}$. The first law of thermodynamics then states that the kinetic *plus* internal energy changes by the amount of mechanical work done by the volume and surface forces and by the amount of heat supplied to the system

$$\frac{d}{dt}\iiint_{V(t)} d^3x\,\rho(e_{\rm int}+e_{\rm kin}) = \iiint_{V(t)} d^3x\,\rho\mathbf{F}\cdot\mathbf{u} + \iint_{\delta V(t)} d^2x\,\mathbf{G}\cdot\mathbf{u} - \iint_{\delta V(t)} d^2x\,\mathbf{n}\cdot\tilde{\mathbf{q}}^{\rm mol} \tag{3.44}$$

Here $e_{\rm int}$ is the *specific internal energy*,[2] $e_{\rm kin} := \mathbf{u}\cdot\mathbf{u}/2$ the *specific kinetic energy*, and $\tilde{\mathbf{q}}^{\rm mol}$ the *molecular diffusive heat flux*. Molecular diffusion of heat is also referred to as molecular heat conduction. The differential form of the energy equation is

$$\partial_t\left[\rho(e_{\rm int}+e_{\rm kin})\right] + \partial_j\left[\rho(e_{\rm int}+e_{\rm kin})u_j + \tilde{q}_j^{\rm mol} - u_i\Pi_{ij}\right] = -\rho u_i\partial_i\phi \tag{3.45}$$

The Coriolis force does not appear because it does not do any work. The heat flux $\tilde{\mathbf{q}}^{\rm mol}$ constitutes all the energy transfers to the system that are not mechanical. Often, a chemical part

$$\mathbf{q}_{\rm chem}^{\rm mol} := h_{\rm s}\mathbf{I}_{\rm s}^{\rm mol} + h_{\rm w}\mathbf{I}_{\rm w}^{\rm mol} = \Delta h\,\mathbf{I}_{\rm s}^{\rm mol} \tag{3.46}$$

is subtracted out of the total heat flux and the heat flux

$$\mathbf{q}^{\rm mol} := \tilde{\mathbf{q}}^{\rm mol} - \mathbf{q}_{\rm chem}^{\rm mol} \tag{3.47}$$

is introduced. This heat flux is zero if mass is added to the system at constant pressure and temperature (see Section A.5). This definition also has the advantage that $\nabla\cdot\mathbf{q}^{\rm mol}$ is well defined despite the fact that $e_{\rm int}$ is only defined up to a function $c + aS$ and Δh only up to a constant a (see Table 2.2). The unknown terms cancel if the conservation equations for mass and salt are applied

$$\partial_t\left[\rho(c+aS)\right] + \partial_j\left[\rho(c+aS)u_j + aI_{{\rm s},j}^{\rm mol}\right] = 0 \tag{3.48}$$

An equation for the kinetic energy alone can be obtained by taking the scalar product of $\mathbf{u}$ and the momentum balance. The result is

$$\partial_t(\rho e_{\rm kin}) + \partial_j(\rho e_{\rm kin}u_j - u_i\Pi_{ij}) = -\Pi_{ij}D_{ij} - \rho u_i\partial_i\phi \tag{3.49}$$

where

$$D_{ij} := \frac{1}{2}\left(\frac{\partial u_i}{\partial x_j} + \frac{\partial u_j}{\partial x_i}\right) \tag{3.50}$$

is the *rate of deformation* (or rate of strain) tensor. Subtracting this equation from (3.45) gives the internal energy equation

$$\partial_t(\rho e_{\rm int}) + \partial_j\left[\rho e_{\rm int}u_j + q_j^{\rm mol} + \Delta h I_{{\rm s},j}^{\rm mol}\right] = \Pi_{ij}D_{ij} \tag{3.51}$$

[2] Now that we introduce additional forms of energy, the internal energy is characterized by the index "int."

The source term $\Pi_{ij} D_{ij}$ describes the transfer between kinetic and internal energy. The equations for potential energy $e_{\text{pot}} := \phi$, mechanical energy $e_{\text{mech}} := e_{\text{kin}} + e_{\text{pot}}$, and total energy $e_{\text{tot}} := e_{\text{int}} + e_{\text{kin}} + e_{\text{pot}}$ are listed in Table 3.1. Kinetic and potential energies are exchanged through the term $\rho \mathbf{u} \cdot \nabla \phi$. Total energy is conserved unless the geopotential is time dependent.

3.8 Radiation

Electromagnetic radiation is also a form of energy. Radiative energy is converted to internal energy when radiation is absorbed by matter. Internal energy is converted to radiative energy when matter emits radiation.

Electromagnetic radiation also obeys a balance equation, called the radiative balance equation. For natural unpolarized radiation it has the form

$$\partial_t \tilde{e}_{\text{rad}} + \nabla \cdot \mathbf{q}_{\text{rad}} = S_{\text{rad}} \tag{3.52}$$

where $\tilde{e}_{\text{rad}}$ is the energy density of the radiation, $\mathbf{q}_{\text{rad}}$ the radiative flux (the Poynting vector), and S_{rad} the source term describing the absorption and emission of radiation by matter. The term S_{rad} must be added to the internal energy equation with the opposite sign. Since radiation moves at the speed of light the storage term can safely be neglected for oceanographic applications and $S_{\text{rad}} = \nabla \mathbf{q}_{\text{rad}}$. Incorporation of radiative processes thus results in the internal energy equation

$$\partial_t (\rho e_{\text{int}}) + \partial_j \left[\rho e_{\text{int}} u_j + q_j^{\text{mol}} + q_j^{\text{rad}} + \Delta h I_{\text{s},j}^{\text{mol}} \right] = \Pi_{ij} D_{ij} \tag{3.53}$$

The balance equations for water, salt, momentum, and energy are all of the form

$$\partial_t (\rho \psi) + \nabla \cdot \mathbf{I}_\psi = S_\psi \tag{3.54}$$

where ψ, $\mathbf{I}_\psi$, and S_ψ are given in Table 3.1.

3.9 Continuity of fluxes

Consider an interface S. None of the conserved quantities (mass, salt, momentum, and total energy) can accumulate in the interface. Fluxes must be continuous across the interface. Formally, this condition can be derived by integrating the balance equations over a volume ΔV surrounding the interface. Application of Reynolds' transport and Gauss' theorems results in

$$\frac{\mathrm{d}}{\mathrm{d}t} \iiint_{\Delta V} \mathrm{d}^3x \, \rho \psi + \iint_{\delta \Delta V} \mathrm{d}^2x \, \mathbf{n} \cdot (\mathbf{I}_\psi - \rho \psi \mathbf{v}) = \iiint_{\Delta V} \mathrm{d}^3x \, S_\psi \tag{3.55}$$

Table 3.1. *The balance equations* $\partial_t(\rho\psi) + \boldsymbol{\nabla}\cdot\mathbf{I}_\psi = S_\psi$

Conserved quantity	ψ	$\mathbf{I}_\psi$	S_ψ
Mass	1	$\rho\mathbf{u}$	0
Salt	S	$\rho S\mathbf{u} + \mathbf{I}_s^{\text{mol}}$	0
Momentum	$\mathbf{u}$	$\rho\mathbf{u}\mathbf{u} - \boldsymbol{\Pi}$	$-\rho\, 2\boldsymbol{\Omega}\times\mathbf{u} - \rho\boldsymbol{\nabla}\phi$
Internal energy	e_{int}	$\rho e_{\text{int}}\mathbf{u} + \mathbf{q}^{\text{mol}} + \mathbf{q}^{\text{rad}} + \Delta h\mathbf{I}_s^{\text{mol}}$	$\boldsymbol{\Pi}:\mathbf{D}$
Kinetic energy	$e_{\text{kin}} = \frac{1}{2}\mathbf{u}\cdot\mathbf{u}$	$\rho e_{\text{kin}}\mathbf{u} - \mathbf{u}\cdot\boldsymbol{\Pi}$	$-\boldsymbol{\Pi}:\mathbf{D} - \rho\mathbf{u}\cdot\boldsymbol{\nabla}\phi$
Potential energy	$e_{\text{pot}} = \phi$	$\rho e_{\text{pot}}\mathbf{u}$	$\rho\mathbf{u}\cdot\boldsymbol{\nabla}\phi + \rho\partial_t\phi$
Mechanical energy	$e_{\text{mech}} = e_{\text{pot}} + e_{\text{kin}}$	$\rho e_{\text{mech}}\mathbf{u} - \mathbf{u}\cdot\boldsymbol{\Pi}$	$-\boldsymbol{\Pi}:\mathbf{D} + \rho\partial_t\phi$
Total energy	$e_{\text{tot}} = e_{\text{mech}} + e_{\text{int}}$	$\rho e_{\text{tot}}\mathbf{u} + \mathbf{q}^{\text{mol}} + \mathbf{q}^{\text{rad}} + \Delta h\mathbf{I}_s^{\text{mol}} - \mathbf{u}\cdot\boldsymbol{\Pi}$	$\rho\partial_t\phi$
Entropy	η	$\rho\eta\mathbf{u} + \frac{1}{T}\mathbf{q}^{\text{mol}} + \frac{1}{T}(\Delta h - \Delta\mu)\,\mathbf{I}_s^{\text{mol}}$	$\mathbf{q}^{\text{mol}}\cdot\boldsymbol{\nabla}\frac{1}{T} - \mathbf{I}_s^{\text{mol}}\cdot\left(\frac{1}{T}\boldsymbol{\nabla}\Delta\mu\right) + \frac{1}{T}\boldsymbol{\sigma}^{\text{mol}}:\mathbf{D} - \frac{1}{T}\boldsymbol{\nabla}\cdot\mathbf{q}^{\text{rad}}$

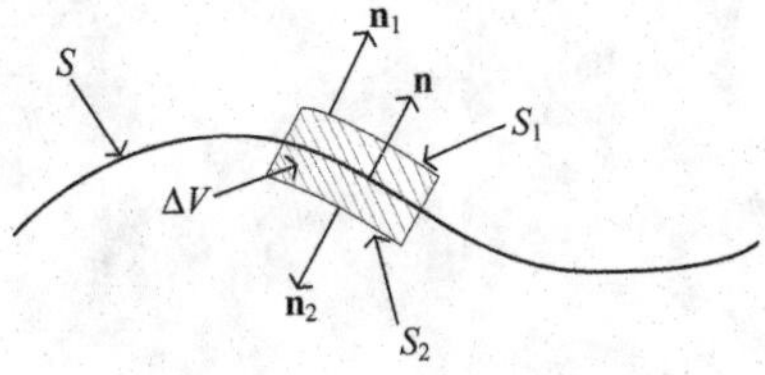

Figure 3.1. Volume ΔV surrounding the interface S.

where $\delta\Delta V$ is the surface enclosing the volume ΔV. In the limit $\Delta V \to 0$, the volume terms vanish and the surface integral becomes

$$\iint_{\delta\Delta V} \mathrm{d}^2x\, \mathbf{n}\cdot(\mathbf{I}_\psi - \rho\psi\mathbf{v}) \to \iint_{S_1} \mathrm{d}^2x\, \mathbf{n}_1\cdot(\mathbf{I}_\psi - \rho\psi\mathbf{v}_1) + \iint_{S_2} \mathrm{d}^2x\, \mathbf{n}_2\cdot(\mathbf{I}_\psi - \rho\psi\mathbf{v}_2) \tag{3.56}$$

where $S_{1,2}$ are the upper and lower surfaces (see Figure 3.1). In the same limit $\mathbf{n}_1 \to \mathbf{n}$, $\mathbf{n}_2 \to -\mathbf{n}$, and $\mathbf{v}_{1,2} \to \mathbf{v}$ where $\mathbf{n}$ is the outward normal vector and $\mathbf{v}$ the velocity of the surface S. Therefore the flux

$$\tilde{\mathbf{I}}_\psi\cdot\mathbf{n} := \mathbf{I}_\psi\cdot\mathbf{n} - \rho\psi\,\mathbf{v}\cdot\mathbf{n} \tag{3.57}$$

Table 3.2. *Continuity of fluxes*

Conserved quantity	$\tilde{\mathbf{I}}_\psi \cdot \mathbf{n}$
Mass	$\rho\,(\mathbf{u}-\mathbf{v})\cdot\mathbf{n} =: I_{\text{mass}}$
Salt	$I_{\text{mass}} S + \mathbf{I}_{\text{s}}^{\text{mol}} \cdot \mathbf{n}$
Momentum	$I_{\text{mass}} \mathbf{u} - \mathbf{\Pi} \cdot \mathbf{n}$
Total energy	$I_{\text{mass}} e_{\text{tot}} + \mathbf{q}^{\text{mol}} \cdot \mathbf{n} + \mathbf{q}^{\text{rad}} \cdot \mathbf{n} + \Delta h \mathbf{I}_{\text{s}}^{\text{mol}} \cdot \mathbf{n} - \mathbf{u} \cdot \mathbf{\Pi} \cdot \mathbf{n}$

must be continuous across the interface. The second term accounts for the fact that the interface moves with velocity $\mathbf{v}$. The explicit formulae for $\tilde{\mathbf{I}}_\psi$ are listed in Table 3.2. They are expressed in terms of the mass flux

$$I_{\text{mass}} := \rho(\mathbf{u}-\mathbf{v})\cdot\mathbf{n} \tag{3.58}$$

through the interface.

These continuity conditions will form part of the boundary conditions for the equations of oceanic motions.

4

Molecular flux laws

In this chapter we determine the salt flux $\mathbf{I}_s^{\text{mol}}$, the heat flux $\mathbf{q}^{\text{mol}}$, and the viscous stress tensor $\boldsymbol{\sigma}^{\text{mol}}$. These fluxes represent the molecular structure of sea water. Not all molecules move with the same (mean) velocity. The phenomenological theory of irreversible processes provides the basis for this determination. Its major principle is the second law of thermodynamics, which states that the entropy of a closed system cannot decrease. The starting point is thus the expression for the entropy production. The production term can be written as the sum of products of the fluxes and "thermodynamic forces." The basic assumption then is that the fluxes are proportional to the forces. The factors of proportionality are called phenomenological or molecular diffusion coefficients. Their number can be reduced by exploiting the symmetry of the system. Sea water can be assumed to be fully isotropic, as opposed to, say, crystals that only have very restrictive symmetries. A further reduction is due to *Onsager's law*, a phenomenological law that can be related to the time reversal invariance of the underlying microscopic equations. The final result is that the molecular fluxes are characterized by five molecular diffusion coefficients. They depend on the pressure, temperature, and salinity. They can in principle be derived from the microscopic structure of sea water, using methods from statistical mechanics. As with the equilibrium thermodynamic properties, this has not been achieved yet. The molecular diffusion coefficients must be and have been measured. The phenomenological theory for $\mathbf{q}^{\text{rad}}$ is not considered.

4.1 Entropy production

The evolution equation for the specific entropy η can be obtained by differentiating the internal energy function $e(v, \eta, S)$

$$\frac{\mathrm{D}}{\mathrm{D}t} e = T \frac{\mathrm{D}}{\mathrm{D}t} \eta - p \frac{\mathrm{D}}{\mathrm{D}t} v + \Delta\mu \frac{\mathrm{D}}{\mathrm{D}t} S \tag{4.1}$$

Substitution of the balance equations for e, v, and S then leads to

$$\rho \frac{\mathrm{D}}{\mathrm{D}t} \eta = -\frac{1}{T} \nabla \cdot \left(\mathbf{q}^{\mathrm{mol}} + \mathbf{q}^{\mathrm{rad}} + \Delta h \mathbf{I}_{\mathrm{s}}^{\mathrm{mol}}\right) + \frac{1}{T} \boldsymbol{\sigma}^{\mathrm{mol}} : \mathbf{D} + \frac{\Delta \mu}{T} \nabla \cdot \mathbf{I}_{\mathrm{s}}^{\mathrm{mol}} \tag{4.2}$$

where the double scalar product of two tensors is defined by $\mathbf{a} : \mathbf{b} = a_{ij} b_{ji}$. Despite the fact that η is only known up to a function $-d - bS$, Δh up to a constant a, and $\Delta \mu$ up to a function $a + bT$ (see Table 2.2), the expression for the evolution of the specific entropy is well defined because the unknown terms cancel

$$\rho \frac{\mathrm{D}}{\mathrm{D}t} (-d - bS) = -\frac{1}{T} \nabla \cdot \left(a \mathbf{I}_{\mathrm{s}}^{\mathrm{mol}}\right) + \frac{a + bT}{T} \nabla \cdot \mathbf{I}_{\mathrm{s}}^{\mathrm{mol}} \tag{4.3}$$

By simple regrouping, the entropy equation, (4.2), can be rewritten in the form

$$\rho \frac{\mathrm{D}}{\mathrm{D}t} \eta = -\nabla \cdot \mathbf{I}_{\eta}^{\mathrm{mol}} + S_{\eta}^{\mathrm{mol}} \tag{4.4}$$

where

$$\mathbf{I}_{\eta}^{\mathrm{mol}} := \frac{1}{T} \left[\mathbf{q}^{\mathrm{mol}} + (\Delta h - \Delta \mu) \mathbf{I}_{\mathrm{s}}^{\mathrm{mol}}\right] \tag{4.5}$$

is the entropy flux and

$$\begin{aligned} S_{\eta}^{\mathrm{mol}} &:= \left[\mathbf{q}^{\mathrm{mol}} + \mathbf{q}^{\mathrm{rad}} + (\Delta h - \Delta \mu) \mathbf{I}_{\mathrm{s}}^{\mathrm{mol}}\right] \cdot \nabla \frac{1}{T} - \frac{1}{T} \mathbf{I}_{\mathrm{s}}^{\mathrm{mol}} \cdot \nabla \Delta \mu + \frac{1}{T} \boldsymbol{\sigma}^{\mathrm{mol}} : \mathbf{D} \\ &= \mathbf{q}^{\mathrm{mol}} \cdot \nabla \frac{1}{T} - \mathbf{I}_{\mathrm{s}}^{\mathrm{mol}} \cdot \left(\frac{1}{T} \nabla \Delta \mu |_{T=\mathrm{const}}\right) + \boldsymbol{\sigma}^{\mathrm{mol}} : \frac{1}{T} \mathbf{D} - \frac{1}{T} \nabla \cdot \mathbf{q}^{\mathrm{rad}} \end{aligned} \tag{4.6}$$

is the entropy production rate. It is well defined since $\nabla \Delta \mu |_{T=\mathrm{const}}$ is unique despite the fact that $\Delta \mu$ is only known up to a function $a + bT$. The second law of thermodynamics requires $S_{\eta}^{\mathrm{mol}} \geq 0$.

4.2 Flux laws

Disregard the entropy production by the divergence of the radiative energy flux. Then all the terms in the expression for the entropy production rate are the product of a flux and a "force":

$$\begin{array}{ll} \text{flux} & \text{force} \\ \mathbf{q}^{\mathrm{mol}} & \nabla \frac{1}{T} \\ \mathbf{I}_{\mathrm{s}}^{\mathrm{mol}} & -\frac{1}{T} \nabla \Delta \mu |_{T=\mathrm{const}} \\ \boldsymbol{\sigma}^{\mathrm{mol}} & \frac{1}{T} \mathbf{D} \end{array} \tag{4.7}$$

In thermal equilibrium the forces are zero since $T =$ constant, $\Delta \mu =$ constant, and $\mathbf{D} = 0$ (see Section 6.5). It is the basic phenomenological law of non-equilibrium thermodynamics that the fluxes are linear functions of the forces for systems in local thermodynamic equilibrium

$$q_i^{\text{mol}} = A_{ij}^{(11)}\,\partial_j\left(\frac{1}{T}\right) + A_{ij}^{(12)}\left(-\frac{1}{T}\partial_j \Delta\mu|_{T=\text{const}}\right) + A_{ijk}^{(13)}\,\frac{1}{T}\,D_{jk} \tag{4.8}$$

$$I_{si}^{\text{mol}} = A_{ij}^{(21)}\,\partial_j\left(\frac{1}{T}\right) + A_{ij}^{(22)}\left(-\frac{1}{T}\partial_j \Delta\mu|_{T=\text{const}}\right) + A_{ijk}^{(23)}\,\frac{1}{T}\,D_{jk} \tag{4.9}$$

$$\sigma_{ij}^{\text{mol}} = A_{ijk}^{(31)}\,\partial_k\left(\frac{1}{T}\right) + A_{ijk}^{(32)}\left(-\frac{1}{T}\partial_k \Delta\mu|_{T=\text{const}}\right) + A_{ijkl}^{(33)}\,\frac{1}{T}\,D_{kl} \tag{4.10}$$

The 225 coefficients $A_{\dots}^{(\alpha\beta)}$ are called *phenomenological coefficients*. They depend on p, T, and S, but not on $\mathbf{u}$. The lower indices are tensor indices, the upper indices are not. In thermodynamic equilibrium the fluxes vanish.

Simplifications arise when the medium can be assumed to have certain symmetries. Sea water as a liquid can be assumed to be isotropic. In this case the phenomenological coefficients must be isotropic tensors (Curie's law). This implies (see Section B.6)

$$A_{ijk}^{(\alpha\beta)} = 0 \tag{4.11}$$

$$A_{ij}^{(\alpha\beta)} = a^{(\alpha\beta)}\,\delta_{ij} \tag{4.12}$$

$$A_{ijkl}^{(33)} = b^{(1)}\,\delta_{ij}\delta_{kl} + b^{(2)}\,(\delta_{ik}\delta_{jl} + \delta_{il}\delta_{jk}) \tag{4.13}$$

since A_{ijkl}^{33} is symmetric in the indices (i, j) and (k, l).

The phenomenological flux laws thus take the form

$$q_i^{\text{mol}} = a^{(11)}\,\partial_i\left(\frac{1}{T}\right) - a^{(12)}\,\frac{1}{T}\partial_i \Delta\mu|_{T=\text{const}} \tag{4.14}$$

$$I_{si}^{\text{mol}} = a^{(21)}\,\partial_i\left(\frac{1}{T}\right) - a^{(22)}\,\frac{1}{T}\partial_i \Delta\mu|_{T=\text{const}} \tag{4.15}$$

$$\sigma_{ij}^{\text{mol}} = b^{(1)}\,D_{kk}\,\delta_{ij} + 2b^{(2)}\,D_{ij} \tag{4.16}$$

They are called the *Onsager relations*. Finally, Onsager's law states that the matrix of phenomenological coefficients must be symmetric

$$a^{(21)} = a^{(12)} \tag{4.17}$$

Sea water is thus described by five phenomenological coefficients $a^{(11)}, a^{(12)}, a^{(22)}$, $b^{(1)}$, and $b^{(2)}$. Note that Onsager's law is not invariant under variable transformation and care must be taken to apply it to the right set of variables.

Since the phenomenological flux laws determine the molecular fluxes of heat, salt, and momentum they are also called the *molecular flux laws* and the coefficients are called the *molecular diffusion coefficients*. Often the word "diffusion" is reserved for the diffusion of salt only. The diffusion of momentum is then referred to as friction and the diffusion of heat as conduction.

4.3 Molecular diffusion coefficients

In oceanography, instead of $a^{(11)}, a^{(12)}, a^{(22)}, b^{(1)}$, and $b^{(2)}$, one introduces a different but equivalent set of coefficients.

Salt diffusion and heat conduction One first expresses the gradient of the chemical potential difference in (4.14) and (4.15) as

$$\begin{aligned} \nabla\Delta\mu|_{T=\text{const}} &= \frac{\partial\Delta\mu}{\partial p}\,\nabla p + \frac{\partial\Delta\mu}{\partial S}\,\nabla S \\ &= \frac{\partial\Delta\mu}{\partial S}\,(\nabla S - \gamma\nabla p) \end{aligned} \tag{4.18}$$

where

$$\gamma := -\frac{\partial\Delta\mu/\partial p}{\partial\Delta\mu/\partial S} \tag{4.19}$$

is the thermodynamic equilibrium gradient of salinity. The heat and salt fluxes then become

$$\mathbf{q}^{\text{mol}} = -\rho\,[\kappa_T\,\nabla T + \kappa_{TS}\,(\nabla S - \gamma\nabla p)] \tag{4.20}$$

$$\mathbf{I}_{\text{s}}^{\text{mol}} = -\rho\,[\kappa_S\,(\nabla S - \gamma\nabla p) + \kappa_{ST}\,\nabla T] \tag{4.21}$$

where

$$\begin{aligned} \kappa_T &:= \frac{a^{(11)}}{\rho T^2} \\ \kappa_S &:= \frac{a^{(22)}}{\rho T}\,\frac{\partial\Delta\mu}{\partial S} \\ \kappa_{TS} &:= \frac{a^{(12)}}{\rho T}\,\frac{\partial\Delta\mu}{\partial S} \\ \kappa_{ST} &:= \kappa_{TS}\,\frac{1}{T\,\partial\Delta\mu/\partial S} \end{aligned} \tag{4.22}$$

Molecular heat fluxes are thus caused by temperature gradients and by deviations of the salinity gradient from its thermodynamic equilibrium $\nabla S = \gamma\nabla p$. Salt fluxes are caused by the deviation of the salinity gradient from its equilibrium gradient and by temperature gradients.

The final expressions are obtained by isolating the heat flux in the absence of salt diffusion

$$\mathbf{q}^{\text{mol}} = -\rho c_p\lambda\nabla T + \frac{D'}{D}TS(1-S)\frac{\partial\Delta\mu}{\partial S}\,\mathbf{I}_{\text{s}}^{\text{mol}} \tag{4.23}$$

$$\mathbf{I}_{\text{s}}^{\text{mol}} = -\rho\left[D(\nabla S - \gamma\nabla p) + S(1-S)D'\nabla T\right] \tag{4.24}$$

where

$$\begin{array}{lll} \lambda := \frac{1}{c_p}\left(\kappa_T - \frac{\kappa_{TS}\kappa_{ST}}{\kappa_S}\right) & \text{thermal conduction coefficient} & \\ D := \kappa_S & \text{salt diffusion coefficient} & (4.25) \\ D' := \frac{\kappa_{ST}}{S(1-S)} & \text{thermal diffusion coefficient} & \end{array}$$

The ratio D/D' is often called the *Soret coefficient*.

Viscous friction The usual expressions for the viscous stress tensor are obtained by decomposing the rate of deformation tensor

$$D_{ij} = S_{ij} + N_{ij} \tag{4.26}$$

where

$$S_{ij} := D_{ij} - \frac{1}{3} D_{kk}\,\delta_{ij} \tag{4.27}$$

is the rate of shear deformation and

$$N_{ij} := \frac{1}{3} D_{kk}\,\delta_{ij} \tag{4.28}$$

the rate of normal deformation. Then

$$\sigma_{ij}^{\text{mol}} = 2\rho\nu S_{ij} + 3\rho\nu' N_{ij} \tag{4.29}$$

where

$$\nu = \frac{b^{(2)}}{\rho} \tag{4.30}$$

is the *shear viscosity* and

$$\nu' = \frac{3b^{(1)} + 2b^{(2)}}{3\rho} \tag{4.31}$$

the *expansion viscosity*. Often one introduces the *dynamic viscosities* $\mu = \rho\nu$ and $\mu' = \rho\nu'$ instead of kinematic viscosities ν and ν'.

The five phenomenological coefficients ($a^{(11)}, a^{(12)}, a^{(22)}, b^{(1)}, b^{(2)}$) are thus replaced by the five coefficients (λ, D, D', ν, ν'). For a one-component system one only has three coefficients (λ, ν, ν'). Often the assumption is made, originally due to Stokes, that $\nu' = 0$. Then $\sigma_{ii}^{\text{mol}} = 0$ and

$$p = -\frac{1}{3}\Pi_{ii} \tag{4.32}$$

The pressure is given by the trace of the stress tensor.

The molecular diffusion coefficients have been measured and tabulated (see, for example, Siedler and Peters, 1986). As an example, Figure 4.1 shows the coefficients ν, λ, and D as a function of temperature for $S = 35$ psu and $p = p_a$.

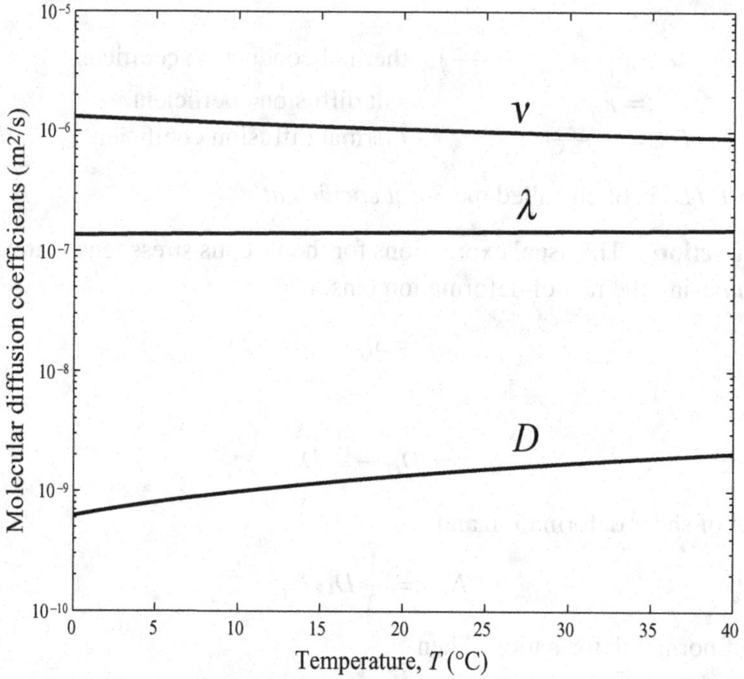

Figure 4.1. The molecular diffusion coefficients ν, λ, and D for sea water as a function of temperature T at $S = 35$ psu (32.9 psu for D) and $p = p_a$.

4.4 Entropy production and energy conversion

In terms of the new coefficients (λ, D, D', ν, ν'), the entropy production rate becomes

$$S_\eta^{\text{mol}} = \frac{1}{T}(2\rho\nu\mathbf{S} : \mathbf{S} + 3\rho\nu'\mathbf{N} : \mathbf{N}) + \frac{\rho c_p \lambda}{T^2}\boldsymbol{\nabla}T \cdot \boldsymbol{\nabla}T + \frac{\partial\Delta\mu/\partial S}{T\rho D}\mathbf{I}_s^{\text{mol}} \cdot \mathbf{I}_s^{\text{mol}} \tag{4.33}$$

The second law requires $S_\eta^{\text{mol}} \geq 0$. Hence $\nu \geq 0$, $\nu' \geq 0$, $\lambda \geq 0$, and $D \geq 0$ since $\partial\Delta\mu/\partial S \geq 0$. The coefficient D' can have an arbitrary sign. From the definition of the thermal conduction coefficient (the first equation in (4.25)) it follows that

$$\kappa_T \geq \frac{\kappa_{TS}\kappa_{ST}}{\kappa_S} \geq 0 \tag{4.34}$$

The conversion of kinetic to internal energy is given by

$$\boldsymbol{\Pi} : \mathbf{D} = -p\boldsymbol{\nabla} \cdot \mathbf{u} + 2\rho\nu\mathbf{S} : \mathbf{S} + 3\rho\nu'\mathbf{N} : \mathbf{N} \tag{4.35}$$

where the first term on the right-hand side describes the reversible exchange and the last two terms the irreversible conversion (dissipation) of kinetic energy into internal energy (heat).

4.5 Boundary conditions

The balance equations with the molecular fluxes as given in the previous section are partial differential equations in space and time. Their solution requires the specification of initial and boundary conditions. The number of boundary conditions required is determined by the highest-order spatial derivative, which is first order for the mass balance and second order for the salt, momentum, and energy balances. The balance equations thus require 11 boundary conditions. We first discuss the general set of boundary conditions at the interface between two fluids and then the boundary conditions for the air–sea, ice–sea, and sea–bottom interfaces.

General interface We only consider the situation where the two fluids (denoted by the superscripts "I" and "II") do not detach from each other. They have a common boundary that is characterized by its velocity $\mathbf{v}$ and a normal vector $\mathbf{n}$ which can be the outward normal vector for either fluid I or fluid II.

A first set of 6 conditions is the continuity of the fluxes of mass, salt, momentum, and total energy across the boundary

$$\tilde{\mathbf{I}}^{\mathrm{I}} \cdot \mathbf{n} = \tilde{\mathbf{I}}^{\mathrm{II}} \cdot \mathbf{n} \tag{4.36}$$

as discussed in Section 3.9. These fluxes obey phenomenological flux laws similar to the ones discussed, with gradients replaced by differences across the boundary. In oceanography one considers, however, the limit that the boundaries have no resistance. They are short-circuited. (This language comes from Ohm's law for resistors, which states that the electrical current is proportional to the electrical potential difference.) In this limit, the temperature, the chemical potentials of water and salt, and the tangential velocity components must be continuous across the boundary

$$\begin{aligned} T^{\mathrm{I}} &= T^{\mathrm{II}} \\ \mu_{\mathrm{w}}^{\mathrm{I}} &= \mu_{\mathrm{w}}^{\mathrm{II}} \\ \mu_{\mathrm{s}}^{\mathrm{I}} &= \mu_{\mathrm{s}}^{\mathrm{II}} \\ \mathbf{u}^{\mathrm{I}} \times \mathbf{n} &= \mathbf{u}^{\mathrm{II}} \times \mathbf{n} \end{aligned} \tag{4.37}$$

which constitutes a second set of 5 boundary conditions. The last condition of (4.37) implies that the fluids on the two sides do not slide across each other. Though written as a vector equation, it only constrains the two tangential velocity components to be continuous. The normal component might be discontinuous. These two sets of 11 continuity conditions constitute the 11 required boundary conditions for the balance equations.

The velocity $\mathbf{v}$ (and hence the position) of the boundary can be inferred from the relation $I_{\text{mass}} = \rho\,(\mathbf{u} - \mathbf{v}) \cdot \mathbf{n}$. If the boundary is specified as $G(\mathbf{x}, t) = 0$ then G obeys the *boundary equation*

$$(\partial_t + \mathbf{u} \cdot \nabla)\; G = \frac{I_{\text{mass}}}{\rho} |\nabla G| \tag{4.38}$$

If the mass flux is zero then the boundary is a material surface. The boundary moves with the fluid velocity. Fluid particles that are on the boundary remain on the boundary.

The air–sea and ice–sea boundary The air–sea and ice–sea boundaries are permeable for fresh water but are assumed to be impermeable for salt. One thus prescribes $\tilde{\mathbf{I}}^{\text{s}} \cdot \mathbf{n} = 0$ or

$$\mathbf{I}_{\text{s}}^{\text{mol}} \cdot \mathbf{n} = -I_{\text{mass}}\, S \tag{4.39}$$

at these boundaries where I_{mass} is now the freshwater flux, owing to evaporation and precipitation or freezing and melting. Then there is no condition for the chemical potential μ_{s} any more. The continuity conditions for the freshwater flux, momentum, and total energy can be rewritten

$$\left(\mathbf{u}^{\text{I}} - \mathbf{u}^{\text{II}}\right) \cdot \mathbf{n} = I_{\text{mass}}\, \Delta v \tag{4.40}$$

$$\mathbf{n} \cdot \left(\mathbf{\Pi}^{\text{I}} - \mathbf{\Pi}^{\text{II}}\right) \cdot \mathbf{n} = I_{\text{mass}}\, \Delta\mathbf{u} \cdot \mathbf{n} \tag{4.41}$$

$$\mathbf{n} \times \left(\mathbf{\Pi}^{\text{I}} - \mathbf{\Pi}^{\text{II}}\right) \cdot \mathbf{n} = 0 \tag{4.42}$$

$$\left(\mathbf{q}^{\text{I}} - \mathbf{q}^{\text{II}}\right) \cdot \mathbf{n} = I_{\text{mass}}\, \left(L_{v,l} + \Delta p \bar{v} + \Delta v\, \mathbf{n} \cdot \bar{\boldsymbol{\sigma}}^{\text{mol}} \cdot \mathbf{n}\right) \tag{4.43}$$

where $L_{v,l} = h_{\text{w}}^{\text{II}} - h_{\text{w}}^{\text{I}}$ is the latent heat of vaporization or liquification, $\mathbf{q} = \mathbf{q}^{\text{mol}} + \mathbf{q}^{\text{rad}}$ the total heat flux, $\bar{\psi} := (\psi^{\text{I}} + \psi^{\text{II}})/2$, and $\Delta\psi := \psi^{\text{I}} - \psi^{\text{II}}$. In deriving the heat flux condition use has been made of the thermodynamic identity $e_{\text{int}} - \Delta h S = h_{\text{w}} - pv$ and the analogous identities for air and sea ice. The continuity of the potential ϕ is also incorporated. The last two terms on the right-hand side of (4.43) account for the additional work that needs to be done in a non-equilibrium situation where there may be a pressure jump and a viscous stress. The latent heats are defined for the equilibrium situation where Δp and $\bar{\sigma}$ are zero. If there is a freshwater flux across the boundary then the normal components of the velocity, stress tensor, and total heat flux are discontinuous across the interface. The complete set of boundary conditions is listed in Table 4.1. The continuity condition (4.40) for the freshwater flux has been moved to the "momentum" row.

Table 4.1. *Boundary conditions at the air–sea or ice–sea boundary*

Boundary equation	$(\partial_t + \mathbf{u} \cdot \nabla) G = \frac{I_{mass}}{\rho} \|\nabla G\|$
Water	$\mu_w^I = \mu_w^{II}$
Salt	$\mathbf{I}_s^{mol} \cdot \mathbf{n} = -I_{mass}\, S$
Momentum	$\mathbf{u}^I \times \mathbf{n} = \mathbf{u}^{II} \times \mathbf{n}$ $\mathbf{u}^I \cdot \mathbf{n} = \mathbf{u}^{II} \cdot \mathbf{n} + I_{mass}\Delta v$ $\mathbf{n} \times \boldsymbol{\Pi}^I \cdot \mathbf{n} = \mathbf{n} \times \boldsymbol{\Pi}^{II} \cdot \mathbf{n}$ $\mathbf{n} \cdot \boldsymbol{\Pi}^I \cdot \mathbf{n} = \mathbf{n} \cdot \boldsymbol{\Pi}^{II} \cdot \mathbf{n} + I_{mass}\Delta\mathbf{u} \cdot \mathbf{n}$
Energy	$T^I = T^{II}$ $\mathbf{q}^I \cdot \mathbf{n} = \mathbf{q}^{II} \cdot \mathbf{n} + I_{mass}(L_{v,l} + \Delta p \bar{v} + \Delta v\, \mathbf{n} \cdot \bar{\boldsymbol{\sigma}}^{mol} \cdot \mathbf{n})$

Table 4.2. *Boundary conditions at the sea–bottom boundary*

Boundary equation	$(\partial_t + \mathbf{u} \cdot \nabla) G = 0$
Water	
Salt	$\mathbf{I}_s^{mol} \cdot \mathbf{n} = 0$
Momentum	$\mathbf{u}^I = \mathbf{u}^{II}$ $\boldsymbol{\Pi}^I \cdot \mathbf{n} = \boldsymbol{\Pi}^{II} \cdot \mathbf{n}$
Energy	$T^I = T^{II}$ $\mathbf{q}^I \cdot \mathbf{n} = \mathbf{q}^{II} \cdot \mathbf{n}$

The sea–bottom boundary At the sea–bottom boundary one assumes additionally that there is no freshwater flux, $I_{mass} = 0$. The boundary is then a material surface. There is no condition for the chemical potential of water any more. The velocity, stress, temperature, and molecular heat flux are continuous. The complete set of boundary conditions is listed in Table 4.2.

Comment It is emphasized that the original boundary conditions are continuity conditions. They couple the ocean to its surroundings. It is only by additional (closure) assumptions that they become prescriptions. There are two extreme limits. One limit is *isolation*. Generally, a system is called *closed* (as opposed to *open*) if there is no mass flux across its boundaries, $\tilde{\mathbf{I}}_{s,w} \cdot \mathbf{n} = 0$. We made this assumption for the salt flux at the air–sea and ice–sea boundary (Table 4.1), and for the salt and freshwater flux at the sea–bottom boundary (Table 4.2). A system is called *thermally isolated* if there is no heat flux across its boundaries, $\mathbf{q} \cdot \mathbf{n} = 0$. It is called *mechanically isolated* if there is no momentum flux across its boundaries, $\mathbf{\Pi} \cdot \mathbf{n} = 0$. The other limit is contact with an *infinite reservoir*. If a system is in contact with a system of infinite mechanical inertia then the boundary and its velocity become prescribed. If a system is in contact with a system of infinite thermal inertia then the temperature becomes prescribed. If the system is in contact with an infinite reservoir of salt then the salinity becomes prescribed. If the system is in contact with an infinite reservoir of water then the specific humidity becomes prescribed.

5
The gravitational potential

In this chapter we discuss the gravitational potential whose gradient enters the momentum equation. The potential is the sum of two terms, the potential of the Earth and the tidal potential caused by the Moon and Sun. For many problems, the gradient of the Earth's potential can be assumed to be a constant gravitational acceleration g_0. However, the actual equipotential surfaces (the geoid) have a fairly complicated shape. The gravitational potential is determined by the mass distribution, as the solution of a Poisson equation. For a prescribed mass distribution this solution can be expressed in terms of the Green's function of the Poisson equation. The determination of the gravitational potential then becomes a mere matter of integration, with well-known solutions for a sphere and other simple distributions. For a self-attracting rotating body, like the Earth, the mass distribution is not known a priori but needs to be determined simultaneously with the gravitational potential. For a fluid body of constant density, the solution to this implicit problem is MacLaurin's ellipsoid. For the Earth, the geoid needs to be measured.

The basic geometry of the geoid is an oblate ellipsoid. This suggests oblate spheroidal coordinates as the most convenient coordinate system. Since the eccentricity of the geoid is small one can approximate the metric coefficients of this coordinate system such that they look like the metric coefficients of spherical coordinates. This spherical approximation does not describe the ellipsoidal shape in spherical coordinates but maps the ellipsoidal shape onto a sphere. Surfaces of constant radius are equipotential surfaces. A second approximation utilizes the fact that the depth of the ocean is much smaller than the radius of the Earth. The radial coordinate is assumed to be constant in the metric coefficients. This leads to the pseudo-spherical coordinate system, which has metric coefficient that are independent of the vertical coordinate but is non-Euclidean. Basic formulae for these and other curvilinear coordinate systems are given in Appendix C.

The angular velocity of the Earth also enters the momentum equation, both through the centrifugal potential and the Coriolis force. Though the angular momentum of the Earth is conserved, the angular velocity changes when the Earth's moment of inertia changes. We do not consider these effects.

The construction of the tidal potential is straightforward, making appropriate geometric approximations. It is determined solely by the position of the Moon and Sun, and the best known forcing field of the ocean. Once these celestial bodies are included, conservation of angular momentum only holds for the complete Earth-Moon system.

5.1 Poisson equation

The gravitational potential ϕ_g satisfies the Poisson equation

$$\Delta\phi_g = 4\pi\, G\rho \tag{5.1}$$

where

$$G = 6.67 \cdot 10^{-11}\ \mathrm{m^3\,kg^{-1}\,s^{-2}} \tag{5.2}$$

is the gravitational constant and ρ the mass density. For a prescribed ρ the solution is given by

$$\phi_g(\mathbf{x}) = -G \iiint \mathrm{d}^3x' \frac{\rho(\mathbf{x}')}{|\mathbf{x}-\mathbf{x}'|} \tag{5.3}$$

Example 1: Sphere For a sphere of radius r_0, constant density ρ_0, and total mass $M = \frac{4\pi}{3} r_0^3 \rho_0$ the gravitational potential is

$$\begin{aligned} \phi_g^{\text{inside}} &= \frac{MG}{2r_0^3}\left(r^2 - 3r_0^2\right) && \text{for } r \le r_0 \\ \phi_g^{\text{outside}} &= -\frac{MG}{r} && \text{for } r_0 \le r \end{aligned} \tag{5.4}$$

and the gravitational acceleration is

$$g = |\nabla\phi_g| = \begin{cases} g_0 \frac{r}{r_0} & \text{for } r \le r_0 \\ g_0 \frac{r_0^2}{r^2} & \text{for } r_0 \le r \end{cases} \tag{5.5}$$

with a surface value

$$g_0 = MG\frac{1}{r_0^2} \tag{5.6}$$

Outside the sphere the gravitational potential is the same as that of a point source with the same mass as the sphere.

Example 2: Ellipsoids of rotation There are two types of ellipsoids of rotation (or revolution). *Oblate* ellipsoids are obtained by rotating an ellipse about its shorter axis. *Prolate* ellipsoids are obtained by rotation about the longer axis. In Cartesian coordinates (x, y, z) an oblate ellipsoid is given by

$$\frac{x^2+y^2}{r_\mathrm{e}^2}+\frac{z^2}{r_\mathrm{p}^2}=1 \tag{5.7}$$

where r_e is the equatorial and r_p the polar radius, and in spherical coordinates (r, θ, φ) it is given by

$$r(\theta)=\frac{r_0 \cos^{2/3}\mu}{(1-\sin^2\mu\cos^2\theta)^{1/2}} \tag{5.8}$$

where

$$\epsilon^2 := \sin^2\mu = \frac{r_\mathrm{e}^2 - r_\mathrm{p}^2}{r_\mathrm{e}^2} \tag{5.9}$$

is the eccentricity of the ellipsoid and

$$r_0 = \left(r_\mathrm{e}^2 r_\mathrm{p}\right)^{1/3} \tag{5.10}$$

the radius of a sphere that has the same volume as the ellipsoid. If $\mu = 0$ the ellipsoid becomes a sphere. If $\mu = \pi/2$ the ellipsoid becomes a disk. Additional definitions and relations are:

$$\begin{aligned}
r_\mathrm{p} &= r_0 \cos^{2/3}\mu && \text{polar radius}\\
r_\mathrm{e} &= r_0 \cos^{-1/3}\mu && \text{equatorial radius}\\
d^2 &:= r_\mathrm{e}^2 - r_\mathrm{p}^2 && \text{half-distance between foci}\\
e &:= \frac{r_\mathrm{e}-r_\mathrm{p}}{r_\mathrm{e}} = 1-\cos\mu && \text{ellipticity}
\end{aligned} \tag{5.11}$$

The gravitational potential inside of an ellipsoid with (r_0, μ), constant density ρ_0, and total mass $M = \frac{4\pi}{3} r_0^3 \rho_0$ is given by

$$\phi_g^{\mathrm{inside}}(r,\theta) = \frac{g_0}{2r_0} r^2 \left(1+2\nu-3\nu\cos^2\theta\right) \tag{5.12}$$

where $g_0 = MG/r_0^2$ and

$$\nu := \frac{3}{2}\frac{1-\mu\cos\mu}{\sin^2\mu} - \frac{1}{2} \tag{5.13}$$

or $\nu \approx \mu^2/5$ for $\mu \ll 1$. The equipotential surfaces are ellipsoids with a μ'-value given by

$$\sin^2\mu' = \frac{3\nu}{1+2\nu} \tag{5.14}$$

or $\mu'^2 \approx 3\mu^2/5$ for $\mu \ll 1$. In the limit $\mu = 0$ we recover the potential for a sphere.

5.2 The geoid

The geoid is the shape of a gravitationally self-attracting rotating body. The calculation of this shape for a fluid body has a long history, starting with Newton.

The fluid body is assumed to have constant density ρ_0, to be mechanically isolated, and to be in mechanical equilibrium. Mechanical equilibrium implies that the fluid rotates at constant angular velocity $\mathbf{\Omega}_0$ as a solid body and that the rate of deformation tensor is zero. Mechanical isolation then implies $p = 0$ at the surface since there are no viscous stresses. Mechanical equilibrium also implies the hydrostatic balance

$$\nabla p = -\rho_0 \nabla \phi \tag{5.15}$$

The surface must hence also be an equigeopotential surface. The geopotential is given by

$$\phi = \phi_g + \phi_c \tag{5.16}$$

where ϕ_g is the gravitational potential and ϕ_c is the centrifugal potential

$$\phi_c = -\frac{1}{2}\,\Omega_0^2 r^2 \cos^2\theta \tag{5.17}$$

The calculation of the shape is an implicit problem since the gravitational field is determined by the shape of the fluid body.

Huygens' ellipsoid If one assumes that the gravitational field is that of a sphere then the geopotential inside the ellipsoid is given by

$$\phi^{\text{inside}} = \frac{g_0}{2r_0}\,r^2 - \frac{1}{2}\,\Omega_0^2 r^2 \cos^2\theta \tag{5.18}$$

Its equipotential surfaces are ellipsoids with

$$\sin^2\mu = \frac{\Omega_0^2 r_0}{g_0} \tag{5.19}$$

For $\mu \ll 1$ one finds $\mu^2 \approx \Omega_0^2 r_0/g_0$ and hence

$$r_e - r_p \approx \frac{1}{2}\frac{\Omega_0^2 r_0^2}{g_0} \tag{5.20}$$

which is about 10 km, about a factor of 2 smaller than observed.

MacLaurin's ellipsoid To solve the implicit problem one anticipates the solution as an oblate ellipsoid rotating about the shorter axis. The geopotential is then given by

$$\phi^{\text{inside}}(r,\theta) = \frac{g_0}{2r_0} r^2 (1 + 2\nu - 3\nu \cos^2\theta) - \frac{1}{2}\Omega_0^2 r^2 \cos^2\theta \tag{5.21}$$

On the surface $r(\theta) = r_0 \cos^{2/3}\mu (1 - \sin^2\mu \cos^2\theta)^{-1/2}$ this potential is given by

$$\phi^{\text{surface}}(\theta) = \frac{g_0}{2r_0} \frac{r_0^2 \cos^{2/3}\mu}{1 - \sin^2\mu \cos^2\theta} \left[1 + 2\nu - \left(3\nu + \frac{\Omega_0^2 r_0}{g_0} \right) \cos^2\theta \right] \tag{5.22}$$

For the surface to be an equipotential surface this potential must be independent of θ, which is the case only if

$$\sin^2\mu = \frac{3\nu + \frac{\Omega_0^2 r_0}{g_0}}{1 + 2\nu} \tag{5.23}$$

or

$$\frac{3}{2}(1 - \mu \cos\mu)\left(2 - \frac{3}{\sin^2\mu}\right) + \frac{3}{2} = \frac{\Omega_0^2 r_0}{g_0} \tag{5.24}$$

if ν is substituted from (5.13). For the ellipsoid to be a solution, the parameters μ, Ω_0, g_0, and r_0 have to satisfy this equation. For the geoid problem Ω_0, g_0, and r_0 are prescribed. Then the above equation determines the eccentricity μ. In the limit $\mu \ll 1$

$$\mu^2 \approx \frac{5}{2} \frac{\Omega_0^2 r_0}{g_0} \tag{5.25}$$

which is Newton's result and leads to

$$r_e - r_p \approx \frac{5}{4} \frac{\Omega_0^2 r_0^2}{g_0} \tag{5.26}$$

which is about 25 km and close to the observed value.

The geopotential becomes

$$\phi^{\text{inside}}(r,\theta) = \frac{g_0}{2r_0} r^2 \,(1 + 2\nu)(1 - \sin^2\mu \cos^2\theta) \tag{5.27}$$

The surfaces of constant geopotential are ellipsoids with eccentricity $\sin^2\mu$. The surfaces of constant gravitational potential are ellipsoids with eccentricity $\sin^2\mu' = 3\nu/(1 + 2\nu)$.

5.3 The spherical approximation

Oblate spheroidal coordinates Since the equilibrium shape of the Earth is close to an oblate ellipsoid of rotation, the most appropriate coordinates are oblate spheroidal coordinates (q_1, q_2, q_3) that are related to Cartesian coordinates

(x, y, z) by

$$\begin{aligned} x &= d_0 \cosh q_1 \cos q_2 \cos q_3 \\ y &= d_0 \cosh q_1 \cos q_2 \sin q_3 \\ z &= d_0 \sinh q_1 \sin q_2 \end{aligned} \tag{5.28}$$

with $0 \le q_1 \le \infty$, $-\pi/2 \le q_2 \le \pi/2$, $0 \le q_3 \le 2\pi$, and $d_0 =$ constant. Since

$$\frac{x^2 + y^2}{d_0^2 \cosh^2 q_1} + \frac{z^2}{d_0^2 \sinh^2 q_1} = 1 \tag{5.29}$$

the surfaces $q_1 =$ constant are ellipsoids with foci at $d = \pm(d_0, 0, 0)$. Since

$$\frac{x^2 + y^2}{d_0^2 \cos^2 q_2} - \frac{z^2}{d_0^2 \sin^2 q_2} = 1 \tag{5.30}$$

the surfaces $q_2 =$ constant are hyperboloids with foci at $\pm(d_0, 0, 0)$. The surfaces $q_3 =$ constant are half-planes through the z-axis. The scale factors are given by

$$\begin{aligned} h_1^2 &= d_0^2 \left(\sinh^2 q_1 + \sin^2 q_2\right) \\ h_2^2 &= d_0^2 \left(\sinh^2 q_1 + \sin^2 q_2\right) \\ h_3^2 &= d_0^2 \cosh^2 q_2 \cos^2 q_2 \end{aligned} \tag{5.31}$$

The oblate spheroidal coordinates are optimal only near the surface since $d_0 =$ constant for the $q_1 =$ constant ellipsoids and $\mu =$ constant for the equigeopotential ellipsoids.

Modified oblate spheroidal coordinates An alternative set of coordinates is

$$\begin{aligned} p_1^2 &= d_0^2 \sinh^2 q_1 + \tfrac{d_0^2}{2} \\ p_2 &= q_2 \\ p_3 &= q_3 \end{aligned} \tag{5.32}$$

or

$$\begin{aligned} x &= \sqrt{p_1^2 + \tfrac{d_0^2}{2}} \cos p_2 \cos p_3 \\ y &= \sqrt{p_1^2 + \tfrac{d_0^2}{2}} \cos p_2 \sin p_3 \\ z &= \sqrt{p_1^2 - \tfrac{d_0^2}{2}} \sin p_2 \end{aligned} \tag{5.33}$$

with the scale factors

$$\begin{aligned} h_1^2 &= \frac{p_1^2\left(p_1^2 - \frac{d_0^2}{2} + d_0^2 \sin^2 p_2\right)}{\left(p_1^2 - \frac{d_0^2}{2}\right)\left(p_1^2 + \frac{d_0^2}{2}\right)} \\ h_2^2 &= p_1^2 - \tfrac{d_0^2}{2} + d_0^2 \sin^2 p_2 \\ h_3^2 &= \left(p_1^2 + \tfrac{d_0^2}{2}\right) \cos^2 p_2 \end{aligned} \tag{5.34}$$

The surfaces $p_1 = \text{constant}$, $p_2 = \text{constant}$, and $p_3 = \text{constant}$ remain ellipsoids, hyperboloids and half-planes.

Spherical coordinates At the surface of the Earth $d_0 = 521.854$ km and $d_0^2/r_0^2 \approx 6.7 \cdot 10^{-3} \ll 1$. In the limit $d_0^2 \ll r_0^2$ the scale factor of the (p_1, p_2, p_3) coordinate system can be approximated by

$$\begin{aligned} h_1^2 &= 1 \\ h_2^2 &= p_1^2 \\ h_3^2 &= p_1^2 \cos^2 p_2 \end{aligned} \tag{5.35}$$

In this limit the scale factors are identical to those of spherical coordinates

$$\begin{aligned} x &= r \cos\theta \cos\varphi \\ y &= r \cos\theta \sin\varphi \\ z &= r \sin\theta \end{aligned} \tag{5.36}$$

if we make the substitutions

$$\begin{aligned} p_1 &\to r = \text{radius} \\ p_2 &\to \theta = \text{latitude} \\ p_3 &\to \varphi = \text{longitude} \end{aligned} \tag{5.37}$$

The *spherical approximation* consists of making these substitutions. One thus assumes that the surfaces $r = \text{constant}$ are equigeopotential surfaces, i.e., $\phi = \phi(r)$ and thus $\nabla\phi \times \nabla r = 0$ exactly. These substitutions map the ellipsoidal shape of the Earth onto a sphere. One does not describe the ellipsoidal shape in spherical coordinates. We elaborate on this point in Section 5.4.

The spherical approximation generally also accounts for the fact that the ocean depth H is much smaller than the radius of the Earth. One introduces the local vertical coordinate z by

$$r = r_0 \left(1 + \frac{z}{r_0}\right) \tag{5.38}$$

and assumes $z/r_0 = O(H/r_0) \ll 1$. Then the scale factors can be approximated by

$$\begin{aligned} h_z^2 &= 1 \\ h_\theta^2 &= r_0^2 \\ h_\varphi^2 &= r_0^2 \cos^2\theta \end{aligned} \tag{5.39}$$

We refer to the coordinates (φ, θ, z) together with these metric coefficients as the *pseudo-spherical coordinate system*. It is a non-Euclidean geometry. The curvature tensor is non-zero. To the same degree of accuracy the geopotential can be approximated by

$$\phi = g_0 z \tag{5.40}$$

with $g_0 = 9.81 \text{ m s}^{-2}$.

We will employ further approximations to the scale factors when we consider the beta-plane and f-plane approximations in Chapter 18.

5.4 Particle motion in gravitational field

Here we consider particle motions in a (prescribed) gravitational field, i.e., solutions to the equation

$$\ddot{\mathbf{x}} = -\nabla\phi \tag{5.41}$$

This is a physics problem. In fluid dynamics, pressure and viscous forces also need to be considered. Specifically, we consider frictionless motions on rigid spheres and ellipsoids, rotating and non-rotating.

Motion on non-rotating sphere Consider particle motion on a frictionless non-rotating rigid sphere. In spherical coordinates the equations of motion take the form

$$\begin{aligned} a_\varphi &= r\cos\theta\ddot{\varphi} + 2\dot{\varphi}(\dot{r}\cos\theta - r\sin\theta\dot{\theta}) &&= 0 \\ a_\theta &= 2\dot{r}\dot{\theta} + r\ddot{\theta} + r\sin\theta\cos\theta\dot{\varphi}^2 &&= 0 \\ a_r &= \ddot{r} - r\dot{\theta}^2 - r\cos^2\theta\dot{\varphi}^2 &&= -g_0 \end{aligned} \tag{5.42}$$

If $r\dot{\theta}^2 + r\cos^2\theta\dot{\varphi}^2 \leq g_0$ and $\dot{r} = 0$ initially, then the particle stays on the surface $r = r_0 =$ constant and the equations reduce to

$$\begin{aligned} a_\varphi &= r_0\cos\theta\ddot{\varphi} - 2\dot{\varphi}r_0\sin\theta\dot{\theta} = 0 \\ a_\theta &= r_0\ddot{\theta} + r_0\sin\theta\cos\theta\dot{\varphi}^2 \quad = 0 \end{aligned} \tag{5.43}$$

Without loss of generality one can choose the orientation of (φ, θ) such that $\dot{\varphi} = 0$ initially. It is then obvious that solutions are motions with constant speed along great circles, including no motion, $\dot{\varphi} = 0$ and $\dot{\theta} = 0$.

Motion on a rotating sphere In an inertial frame, there are no new forces. Only gravity is acting. The particle still moves at constant speed along great circles. Transformation to a rotating frame, denoted by a tilde, is accomplished by

$$\begin{aligned} \tilde{\varphi} &= \varphi - \Omega_0 t \quad & \dot{\tilde{\varphi}} &= \dot{\varphi} - \Omega_0 \quad & \ddot{\tilde{\varphi}} &= \ddot{\varphi} \\ \tilde{\theta} &= \theta & \dot{\tilde{\theta}} &= \dot{\theta} \end{aligned} \tag{5.44}$$

Equations (5.43) then become

$$
\begin{aligned}
r_0 \cos\tilde{\theta}\ddot{\tilde{\varphi}} - 2\dot{\tilde{\varphi}} r_0 \sin\tilde{\theta}\dot{\tilde{\theta}} &= 2\Omega_0 r_0 \sin\tilde{\theta}\dot{\tilde{\theta}} \\
r_0\ddot{\tilde{\theta}} + r_0 \sin\tilde{\theta}\cos\tilde{\theta}\dot{\tilde{\varphi}}^2 &= -2\Omega_0 r_0 \sin\tilde{\theta}\cos\tilde{\theta}\dot{\tilde{\varphi}} - 2\Omega_0^2 r_0 \sin\tilde{\theta}\cos\tilde{\theta}
\end{aligned}
\tag{5.45}
$$

The new terms on the right-hand side are the Coriolis and centrifugal forces. Now, $\dot{\tilde{\varphi}} = 0$, $\dot{\tilde{\theta}} = 0$ is no longer a solution.

Motion on a rotating ellipsoid Next consider motion on an ellipsoid in the spherical approximation. Since the gravitational *plus* centrifugal forces act only in a radial direction ($\nabla\phi \times \nabla r = 0$), only the Coriolis force acts in the horizontal direction

$$
\begin{aligned}
r_0 \cos\tilde{\theta}\ddot{\tilde{\varphi}} - 2\dot{\tilde{\varphi}} r_0 \sin\tilde{\theta}\dot{\tilde{\theta}} &= 2\Omega_0 r_0 \sin\tilde{\theta}\dot{\tilde{\theta}} \\
r_0\ddot{\tilde{\theta}} + r_0 \sin\tilde{\theta}\cos\tilde{\theta}\dot{\tilde{\varphi}}^2 &= -2\Omega_0 r_0 \sin\tilde{\theta}\cos\tilde{\theta}\dot{\tilde{\varphi}}
\end{aligned}
\tag{5.46}
$$

Now, $\dot{\tilde{\varphi}} = 0$, $\dot{\tilde{\theta}} = 0$ is a solution again. In the inertial frame of reference

$$
\begin{aligned}
r_0 \cos\theta\ddot{\varphi} - 2\dot{\varphi} r_0 \sin\theta\dot{\theta} &= 0 \\
r_0\ddot{\theta} + r_0 \sin\theta\cos\theta\dot{\varphi}^2 &= \Omega^2 \sin\theta\cos\theta
\end{aligned}
\tag{5.47}
$$

where the term on the right-hand side is the gravitational force acting in the θ-direction. The surface of the ellipsoid is not a surface of constant gravitational potential, but a surface of constant geo-(= gravitational plus centrifugal) potential.[1]

5.5 The tidal potential

The gravitational forces exerted by the Moon and the Sun on the Earth give rise to tidal motions. If the celestial bodies are treated as spheres their gravitational potential at a point P on the Earth is given by

$$
\phi_{m,s} = -\frac{GM_{m,s}}{R} \tag{5.48}
$$

where $M_{m,s}$ is the total mass of the celestial body (m = Moon, s = Sun, see Table 5.1) and R the distance between the point P and the center of the celestial body (see Figure 5.1). The gravitational acceleration

$$
\mathbf{a}_{m,s} = -\nabla\phi_{m,s} \tag{5.49}
$$

[1] The results of this section can be summarized by the colloquialism that you can stand on a non-rotating ice-covered sphere and a rotating ice-covered ellipsoid but not on a rotating ice-covered sphere.

Table 5.1. *Values for the tidal problem*

	Moon	Sun	Earth
Mass	$M_\mathrm{m} = 7.33 \cdot 10^{22}$ kg	$M_\mathrm{s} = 1.98 \cdot 10^{30}$ kg	$M = 5.977 \cdot 10^{24}$ kg
Distance	$D_\mathrm{m} = 3.8 \cdot 10^8$ m	$D_\mathrm{s} = 1.49 \cdot 10^{11}$ m	
Radius			$r_0 = 6.37 \cdot 10^6$ m
Acceleration	$g_\mathrm{m} = \frac{GM_\mathrm{m} r_0}{D_\mathrm{m}^3} = 5.67 \cdot 10^{-7}\ \mathrm{m\,s^{-2}}$	$g_\mathrm{s} = \frac{GM_\mathrm{s} r_0}{D_\mathrm{s}^3} = 2.73 \cdot 10^{-7}\ \mathrm{m\,s^{-2}}$	$g_0 = \frac{GM}{r_0^2} = 9.81\ \mathrm{m\,s^{-2}}$

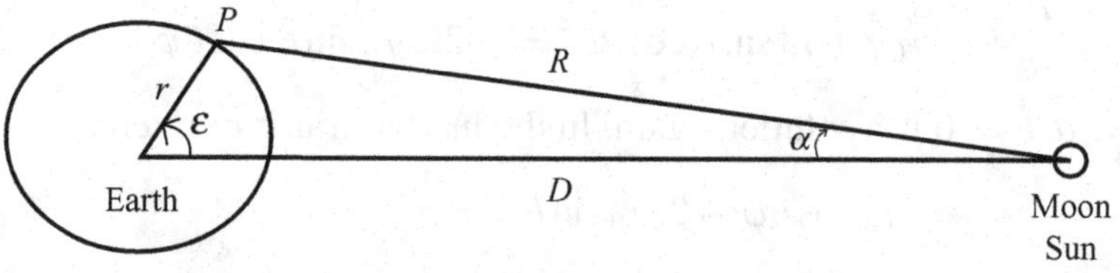

Figure 5.1. Geometry of the Earth–Moon/Sun system. Adapted from LeBlond and Mysak (1978).

is towards the celestial body and can be decomposed to

$$\mathbf{a}_{\mathrm{m,s}} = \frac{GM_{\mathrm{m,s}}}{R^2}\,(\cos(\epsilon + \alpha), \quad -\sin(\epsilon + \alpha)) \tag{5.50}$$

where the angles α and ϵ are defined in Figure 5.1. The first component is the acceleration perpendicular to the Earth's surface. The second component is the acceleration parallel to the Earth's surface, positive in the direction of increasing ϵ. The acceleration at the center of the Earth is

$$\mathbf{a}_\mathrm{c} = \frac{GM_{\mathrm{m,s}}}{D^2}\,(\cos\epsilon, \quad -\sin\epsilon) \tag{5.51}$$

where D is the distance between the center of the Earth and the center of the celestial body (see Figure 5.1). It determines the motion of the Earth as a whole relative to the celestial body. The difference between the two accelerations

$$\begin{aligned} \Delta\mathbf{a}_r &= \mathbf{a}_{\mathrm{m,s}} - \mathbf{a}_\mathrm{c} \\ &= \tfrac{GM_{\mathrm{m,s}}}{D^2}\left(\tfrac{D^2}{R^2}\cos(\epsilon+\alpha) - \cos\epsilon, -\tfrac{D^2}{R^2}\sin(\epsilon+\alpha) + \sin\epsilon\right) \end{aligned} \tag{5.52}$$

determines the relative motions on the Earth. Since $R^2 = D^2 - 2rD\cos\epsilon + r^2$ one finds to lowest order in $r/D \ll 1$

$$\frac{R^2}{D^2} = 1 - 2\frac{r}{D}\cos\epsilon \tag{5.53}$$

$$\cos\alpha = 1 \tag{5.54}$$

$$\sin\alpha = \frac{r}{D}\sin\epsilon \tag{5.55}$$

and hence

$$\Delta\mathbf{a}_r = \frac{GM_{\mathrm{m,s}}r}{D^3}(3\cos^2\epsilon - 1, -3\cos\epsilon\sin\epsilon) \tag{5.56}$$

correct to $O(r^2/D^2)$. This acceleration can be rewritten as

$$\Delta\mathbf{a}_r = -\left(\frac{\partial}{\partial r}, \frac{1}{r}\frac{\partial}{\partial\epsilon}\right)\phi_T \tag{5.57}$$

where

$$\phi_T = -\frac{GM_{\mathrm{m,s}}r^2}{D^3}\frac{1}{2}(3\cos^2\epsilon - 1) \tag{5.58}$$

is the *tidal potential*. Its equipotential surfaces are prolate ellipsoids of rotation with the rotational and major axes directed towards the celestial body. The tidal accelerations are much smaller than Earth's gravitational acceleration g_0 (see Table 5.1). One thus neglects the vertical component but not the horizontal component, which is unopposed by Earth's gravity. This approximation is achieved by replacing the radius r by the mean radius r_0 in the tidal potential, which then becomes independent of depth.

The tidal potential is often formulated in terms of an equilibrium tidal displacement defined by $\phi_T + g_0\eta_{equ} = 0$ or explicitly

$$\eta_{equ} = \eta_{equ}^{\max}\frac{1}{2}(3\cos^2\epsilon - 1) \tag{5.59}$$

where

$$\eta_{equ}^{\max} = \frac{GM_{\mathrm{m,s}}r_0^2}{g_0 D_{\mathrm{m,s}}^3} = \frac{g_{\mathrm{m,s}}}{g_0}r_0 = \begin{cases} 0.364\,\mathrm{m} & \text{Moon} \\ 0.168\,\mathrm{m} & \text{Sun} \end{cases} \tag{5.60}$$

and where $g_{\mathrm{m,s}}$ are defined in Table 5.1.

Because of the rotation of the Earth and the movement of the Moon and Sun relative to the Earth the ellipsoidal tidal potential moves around the Earth. In a coordinate system fixed to the rotating Earth, the angle ϵ depends on latitude θ and longitude φ. This dependence can be expanded in terms of spherical harmonics that form a complete set for describing functions on the surface of a sphere. The coefficients of this expansion and the distances $D_{\mathrm{m,s}}$ then depend on time with frequencies that are linear combinations of celestial frequencies. The most basic of these are

$$
\begin{aligned}
\Omega_{\mathrm{m}} &= \frac{2\pi}{27.321}\ \mathrm{d}^{-1} && \text{lunar month}\\
\Omega_{\mathrm{y}} &= \frac{2\pi}{365.242}\ \mathrm{d}^{-1} && \text{tropical year}\\
\Omega_{\mathrm{lp}} &= \frac{2\pi}{8.85}\ \mathrm{y}^{-1} && \text{lunar perigee}\\
\Omega_{\mathrm{ln}} &= \frac{2\pi}{18.61}\ \mathrm{y}^{-1} && \text{lunar node}\\
\Omega_{\mathrm{sp}} &= \frac{2\pi}{20{,}940}\ \mathrm{y}^{-1} && \text{solar perigee}\\
\Omega_{\mathrm{d}} &= 2\pi\ \mathrm{d}^{-1} && \text{solar day}\\
\Omega_{\mathrm{l}} &= \Omega_{\mathrm{d}} - \Omega_{\mathrm{m}} + \Omega_{\mathrm{y}} && \text{lunar day}
\end{aligned}
\tag{5.61}
$$

The tidal forces cause tides in the ocean, atmosphere, and solid Earth. These tidal motions redistribute mass and modify the gravitational potential. This *gravitational self-attraction* will be discussed in Chapter 17.

6
The basic equations

The balance equations describe changes of extensive quantities, the amounts of water, salt, momentum, and internal energy within a (infinitesimal) volume. They represent basic physical principles. Since the internal energy is not a directly measured quantity and since sea water can for many purposes be regarded as a nearly incompressible fluid it is convenient to introduce pressure and temperature as prognostic variables, instead of the density and specific internal energy. The basic equations for oceanic motions then consist of prognostic equations for the:

- pressure;
- velocity vector;
- temperature; and
- salinity.

There are six equations. They contain:

- "external" fields and parameters like the gravitational potential and earth's rotation rate, which need to be specified (or calculated);
- molecular diffusion coefficients, which need to be specified;
- thermodynamic coefficients, which need to be specified or are given by the equilibrium thermodynamic relations of Chapter 2. Most importantly, the density or equation of state must be specified.

These equations have to be augmented by appropriate boundary conditions. The equations, together with the boundary conditions, then determine the time evolution of any initial state. Nothing else is needed. It can, of course, be elucidating to study the evolution of other quantities, such as the circulation and vorticity. The evolution of such quantities is governed by theorems that are consequences of the basic equations. One important theorem is Ertel's potential vorticity theorem.

Exact solutions of the basic equations of oceanic motions are not known, except the thermodynamic equilibrium solution for a real fluid and the mechanical equilibrium solution for an ideal fluid.

The equations of oceanic motions describe the dynamic evolution of the fluid flow. The kinematic description and characterization of fluid flows are given in Appendix D.

6.1 The pressure and temperature equations

Instead of the specific volume v and specific energy e_{int} (or specific entropy η), it is common to introduce the pressure p and temperature T as prognostic variables. This transformation is accomplished by differentiating $p(v, \eta, S)$ and $T(v, \eta, S)$

$$\frac{\mathrm{D}}{\mathrm{D}t} p = \frac{\partial p}{\partial v}\frac{\mathrm{D}}{\mathrm{D}t} v + \frac{\partial p}{\partial \eta}\frac{\mathrm{D}}{\mathrm{D}t} \eta + \frac{\partial p}{\partial S}\frac{\mathrm{D}}{\mathrm{D}t} S \tag{6.1}$$

$$\frac{\mathrm{D}}{\mathrm{D}t} T = \frac{\partial T}{\partial v}\frac{\mathrm{D}}{\mathrm{D}t} v + \frac{\partial T}{\partial \eta}\frac{\mathrm{D}}{\mathrm{D}t} \eta + \frac{\partial T}{\partial S}\frac{\mathrm{D}}{\mathrm{D}t} S \tag{6.2}$$

By the substitution of the equations for v, η, and S and of the thermodynamic relations listed in Table 2.1 one finds

$$\frac{\mathrm{D}}{\mathrm{D}t} p = -\rho c^2 \boldsymbol{\nabla} \cdot \mathbf{u} + D_p^{\text{mol}} \tag{6.3}$$

$$\frac{\mathrm{D}}{\mathrm{D}t} T = -\rho c^2 \Gamma \boldsymbol{\nabla} \cdot \mathbf{u} + D_T^{\text{mol}} \tag{6.4}$$

where

$$D_p^{\text{mol}} := c^2 \beta \boldsymbol{\nabla} \cdot \mathbf{I}_{\text{s}}^{\text{mol}} + \rho c^2 \frac{\Gamma}{T} D_h^{\text{mol}} \tag{6.5}$$

$$D_T^{\text{mol}} := c^2 \beta \Gamma \boldsymbol{\nabla} \cdot \mathbf{I}_{\text{s}}^{\text{mol}} + \frac{1}{\rho c_v} D_h^{\text{mol}} \tag{6.6}$$

and where

$$D_h^{\text{mol}} := -\boldsymbol{\nabla} \cdot \mathbf{q}^{\text{mol}} + \boldsymbol{\sigma}^{\text{mol}} : \mathbf{D} - \mathbf{I}_{\text{s}}^{\text{mol}} \cdot \boldsymbol{\nabla} \Delta h \tag{6.7}$$

summarizes the molecular terms in the "heat equation" (as defined by (A.49)), consisting of the divergence of the molecular heat flux, the dissipation of mechanical energy into heat, and the "heat of mixing." Although Δh is only known up to a constant, the gradient of Δh in (6.7) is well defined. If penetrating solar radiation is included then a term $-\boldsymbol{\nabla} \cdot \mathbf{q}^{\text{rad}}$ must be added to D_h^{mol}.

The temperature equation can also be written

$$\frac{\mathrm{D}}{\mathrm{D}t} T = \Gamma \frac{\mathrm{D}}{\mathrm{D}t} p + \frac{1}{\rho c_p} D_h^{\text{mol}} \tag{6.8}$$

In this form, the "adiabatic" changes of temperature owing to changes in pressure are separated from the "diabatic" temperature changes owing to "heating." Note that the equations for p and T have been written in their advective forms and that new symbols $D_{p,T}^{\text{mol}}$ have been introduced to describe the effects of molecular diffusion.

One could write these equations in their flux form and define sources and sinks but these would have no real physical meaning since pressure and temperature are not extensive quantities.

Instead of temperature one also uses θ or ρ_{pot} as prognostic variables. Then

$$\frac{\mathrm{D}}{\mathrm{D}t}\theta = D_\theta^{\text{mol}} \tag{6.9}$$

$$\frac{\mathrm{D}}{\mathrm{D}t}\rho_{\text{pot}} = D_{\rho_{\text{pot}}}^{\text{mol}} \tag{6.10}$$

where

$$D_\theta^{\text{mol}} := \frac{\alpha}{\tilde{\alpha}}\frac{1}{\rho c_p}D_h^{\text{mol}} - \frac{\tilde{\beta}-\beta}{\tilde{\alpha}}\frac{1}{\rho}\nabla\cdot\mathbf{I}_s^{\text{mol}} \tag{6.11}$$

$$D_{\rho_{\text{pot}}}^{\text{mol}} := -\frac{\alpha}{\hat{\alpha}}\frac{1}{\rho c_p}D_h^{\text{mol}} + \frac{\hat{\beta}-\beta}{\hat{\alpha}}\frac{1}{\rho}\nabla\cdot\mathbf{I}_s^{\text{mol}} \tag{6.12}$$

Changes in potential temperature and potential density are only caused by molecular processes.

6.2 The complete set of basic equations

We now list the complete set of equations that describe oceanic motions. They consist of prognostic equations for $(p, \mathbf{u}, T, S)$, molecular flux laws and diffusion coefficients, thermodynamic specifications and relations, and boundary conditions.

Prognostic equations The prognostic equations are in their advective form

$$\frac{\mathrm{D}}{\mathrm{D}t}p = -\rho c^2\nabla\cdot\mathbf{u} + D_p^{\text{mol}} \tag{6.13}$$

$$\rho\frac{\mathrm{D}}{\mathrm{D}t}\mathbf{u} = -\nabla p - 2\rho\boldsymbol{\Omega}\times\mathbf{u} - \rho\nabla(\phi+\phi_T) + \mathbf{F}^{\text{mol}} \tag{6.14}$$

$$\frac{\mathrm{D}}{\mathrm{D}t}T = \Gamma\frac{\mathrm{D}}{\mathrm{D}t}p + \frac{1}{\rho c_p}D_h^{\text{mol}} \tag{6.15}$$

$$\frac{\mathrm{D}}{\mathrm{D}t}S = D_S^{\text{mol}} \tag{6.16}$$

where

$$D_p^{\text{mol}} := c^2\beta\nabla\cdot\mathbf{I}_s^{\text{mol}} + \rho c^2\frac{\Gamma}{T}D_h^{\text{mol}} \tag{6.17}$$

$$D_h^{\text{mol}} := -\nabla\cdot\mathbf{q}^{\text{mol}} + \boldsymbol{\sigma}^{\text{mol}}:\mathbf{D} - \mathbf{I}_s^{\text{mol}}\cdot\nabla\Delta h \tag{6.18}$$

$$D_S^{\text{mol}} := -\frac{1}{\rho}\nabla\cdot\mathbf{I}_s^{\text{mol}} \tag{6.19}$$

$$\mathbf{F}^{\text{mol}} := \nabla\cdot\boldsymbol{\sigma}^{\text{mol}} \tag{6.20}$$

The Earth's rotation rate $\mathbf{\Omega}$, the geopotential ϕ, and the tidal potential ϕ_T are assumed to be prescribed parameters or fields. Tidal forcing will not be considered in the following chapters except in the chapter on tidal equations (Chapter 17).

Molecular flux laws and diffusion coefficients The phenomenological flux laws are

$$\mathbf{q}^{\mathrm{mol}} = -\rho c_p \lambda \nabla T + \frac{D'}{D} T S(1-S) \frac{\partial \Delta\mu}{\partial S} \mathbf{I}_{\mathrm{s}}^{\mathrm{mol}} \tag{6.21}$$

$$\mathbf{I}_{\mathrm{s}}^{\mathrm{mol}} = -\rho \left[D(\nabla S - \gamma \nabla p) + S(1-S) D' \nabla T \right] \tag{6.22}$$

$$\boldsymbol{\sigma}^{\mathrm{mol}} = 2\rho\nu \mathbf{S} + 3\rho\nu' \mathbf{N} \tag{6.23}$$

with

$$\gamma := -\frac{\partial \Delta\mu / \partial p}{\partial \Delta\mu / \partial S} \tag{6.24}$$

and the five molecular diffusion coefficients λ, D, D', ν, and ν', which need to be specified as a function of (p, T, S).

Thermodynamic specifications and relations The prognostic equations and molecular flux laws contain thermodynamic coefficients that need to be specified as a function of (p, T, S). These are

$$\rho = \rho(p, T, S) \tag{6.25}$$

$$c_p = c_p(p, T, S) \tag{6.26}$$

$$\Delta h = \Delta h(p, T, S) \tag{6.27}$$

$$\frac{\partial \Delta\mu}{\partial S} = \frac{\partial \Delta\mu}{\partial S}(p, T, S) \tag{6.28}$$

The other thermodynamic coefficients are defined by or can be inferred from the thermodynamic relations

$$\kappa := \frac{1}{\rho} \frac{\partial \rho}{\partial p} \tag{6.29}$$

$$\alpha := -\frac{1}{\rho} \frac{\partial \rho}{\partial T} \tag{6.30}$$

$$\beta := \frac{1}{\rho} \frac{\partial \rho}{\partial S} \tag{6.31}$$

$$\tilde{\kappa} = \kappa - \frac{\alpha^2 v T}{c_p} \tag{6.32}$$

$$\Gamma = \frac{\alpha v T}{c_p} \tag{6.33}$$

$$\frac{\partial \Delta\mu}{\partial p} = -\frac{\beta}{\rho} \tag{6.34}$$

$$c^2 := \frac{1}{\rho \tilde{\kappa}} \tag{6.35}$$

Boundary conditions The boundary condition at the air–sea and sea–ice interface, and at the sea–bottom interface are given in Tables 4.1 and 4.2.

6.3 Tracers

Balance equations may be formulated for other substances as well. They are also of the form

$$\partial_t(\rho_A) + \nabla \cdot (\rho_A \mathbf{u} + \mathbf{I}_A) = S_A \tag{6.36}$$

or

$$\rho \frac{D}{Dt} c_A = -\nabla \cdot \mathbf{I}_A + S_A \tag{6.37}$$

where $\rho_A = \rho c_A$ is the density and c_A the concentration of the substance "A." The flux $\mathbf{I}_A$ accounts for the effect that the substance moves with a velocity other than the fluid velocity. The source term S_A accounts for sources and sinks (chemical reactions, radioactive decay, etc.)

If $\mathbf{I}_A = 0$, then the substance moves with the fluid velocity and the substance is called a *tracer*. If the tracer does not affect the dynamic evolution, it is called a *passive* tracer. If $S_A = 0$, the tracer is called a *conservative* tracer. A conservative tracer obeys

$$\frac{D}{Dt} c_A = 0 \tag{6.38}$$

The concentration is materially conserved.

6.4 Theorems

Theorems are consequences of the equations of oceanic motions. Since the equations for oceanic motions are prognostic equations for $(p, T, S, \mathbf{u})$, the density, energy, and entropy equations listed in Table 3.1 become theorems. Here we discuss vorticity and circulation theorems.

Vorticity One has to distinguish three vorticity vectors

$$\begin{aligned} \omega &= \nabla \times \mathbf{u} && \text{relative vorticity} \\ 2\Omega &= \nabla \times \mathbf{U} && \text{planetary vorticity} \\ \omega^a &= \omega + 2\Omega && \text{absolute vorticity} \end{aligned} \tag{6.39}$$

where $\mathbf{U} = \Omega \times \mathbf{x}$ is the velocity of the rotating frame. The planetary vorticity is assumed to be constant.

The vorticity theorem is obtained by taking the curl of the momentum balance. Making use of the vector identity $\mathbf{u} \cdot \nabla \mathbf{u} = \omega \times \mathbf{u} + \nabla(\mathbf{u} \cdot \mathbf{u}/2)$ and the vector identities of Section B.2 one obtains

$$\frac{\mathrm{D}}{\mathrm{D}t}\omega = (\omega + 2\Omega)\cdot\nabla\mathbf{u} - (\omega + 2\Omega)\,\nabla\cdot\mathbf{u} + \frac{1}{\rho^2}\nabla\rho\times\nabla p + \nabla\times(\rho^{-1}\mathbf{F}^{\mathrm{mol}}) \tag{6.40}$$

Changes in the relative vorticity are due to vortex stretching and twisting (first term), volume changes (second term), the fluid being non-homentropic or non-barotropic (third term), and viscosity (fourth term). A fluid is called homentropic (or barotropic) if $\rho = \rho(p)$ and hence $\nabla\rho\times\nabla p = 0$ and non-homentropic (or non-barotropic = baroclinic) otherwise. Planetary vorticity generates relative vorticity through the terms $2\Omega\cdot\nabla\mathbf{u}$ and $2\Omega\nabla\cdot\mathbf{u}$. The volume term can be eliminated by considering

$$\rho\frac{\mathrm{D}}{\mathrm{D}t}\left(\frac{\omega + 2\Omega}{\rho}\right) = (\omega + 2\Omega)\cdot\nabla\mathbf{u} + \frac{1}{\rho^2}\nabla\rho\times\nabla p + \nabla\times(\rho^{-1}\mathbf{F}^{\mathrm{mol}}) \tag{6.41}$$

The evolution of the velocity gradient tensor $\mathbf{G}$ with components

$$G_{ij} := \partial_j u_i \tag{6.42}$$

is governed by

$$\begin{aligned}\frac{\mathrm{D}}{\mathrm{D}t}G_{ij} + G_{ik}G_{kj} = &-\frac{1}{\rho}\partial_i\partial_j p + \frac{1}{\rho^2}\partial_j\rho\partial_i p - \epsilon_{ikl}2\Omega_k G_{lj}\\ &- \partial_j\partial_i\phi + \partial_j\left(\rho^{-1}F_i^{\mathrm{mol}}\right)\end{aligned} \tag{6.43}$$

The symmetric part of the velocity gradient tensor $\mathbf{G}$ is the rate of deformation tensor $\mathbf{D}$. The antisymmetric part

$$V_{ij} := \frac{1}{2}(G_{ij} - G_{ji}) \tag{6.44}$$

is the vorticity tensor. The relative vorticity vector is related to the vorticity tensor by

$$\omega_i = -\epsilon_{ijk}V_{jk} \tag{6.45}$$

The vorticity theorem and theorems for other parts of the velocity gradient tensor, such as the divergence $\nabla\cdot\mathbf{u} = G_{ii}$, can be inferred from (6.43).

Circulation The circulation is a quantity related to the vorticity. Again, one has to distinguish between

$$\begin{aligned}\Gamma &:= \oint_C \mathbf{dx}\cdot\mathbf{u} && \text{relative circulation}\\ \Gamma^{\mathrm{p}} &:= \oint_C \mathbf{dx}\cdot\mathbf{U} && \text{planetary circulation}\\ \Gamma^{\mathrm{a}} &:= \Gamma^{\mathrm{p}} + \Gamma && \text{absolute circulation}\end{aligned} \tag{6.46}$$

where C is a closed curve. If S is a cap of C (see Section B.4) then Stokes' theorem (B.55) implies

$$\Gamma = \iint_S \mathrm{d}^2x\ \mathbf{n}\cdot\boldsymbol{\omega} \tag{6.47}$$

$$\Gamma^{\mathrm{p}} = \iint_S \mathrm{d}^2x\ \mathbf{n}\cdot 2\boldsymbol{\Omega} = 2\Omega S_\perp \tag{6.48}$$

where $S_\perp$ is the projection of S onto the plane perpendicular to $\boldsymbol{\Omega}$.

If $C(t)$ is a material line, then (D.42) for the differentiation of a line integral and Stokes' theorem imply

$$\frac{\mathrm{d}}{dt}\Gamma = \iint_S \mathrm{d}^2x \frac{1}{\rho^2}\mathbf{n}\cdot(\boldsymbol{\nabla}\rho\times\boldsymbol{\nabla}p) - 2\Omega\frac{\mathrm{d}}{\mathrm{d}t}S_\perp + \iint_S \mathrm{d}^2x\ \mathbf{n}\cdot[\boldsymbol{\nabla}\times(\rho^{-1}\mathbf{F}^{\mathrm{mol}})] \tag{6.49}$$

$$\frac{\mathrm{d}}{\mathrm{d}t}\Gamma^{\mathrm{p}} = 2\Omega\frac{\mathrm{d}}{\mathrm{d}t}S_\perp \tag{6.50}$$

$$\frac{\mathrm{d}}{\mathrm{d}t}\Gamma^{\mathrm{a}} = \iint_S \mathrm{d}^2x \frac{1}{\rho^2}\mathbf{n}\cdot(\boldsymbol{\nabla}\rho\times\boldsymbol{\nabla}p) + \iint_S \mathrm{d}^2x\ \mathbf{n}\cdot[\boldsymbol{\nabla}\times(\rho^{-1}\mathbf{F}^{\mathrm{mol}})] \tag{6.51}$$

The absolute circulation changes only when the fluid is non-homentropic and when molecular viscosity is present. The term describing the exchange between planetary and relative circulation can alternatively be written as

$$2\Omega\frac{\mathrm{d}}{\mathrm{d}t}S_\perp = \oint_C \mathbf{dx}\cdot(2\boldsymbol{\Omega}\times\mathbf{u}) \tag{6.52}$$

Ertel's potential vorticity For any quantity $\psi(\mathbf{x}, t)$ *Ertel's potential vorticity* is defined by

$$q_\psi := \frac{(\boldsymbol{\omega}+2\boldsymbol{\Omega})\cdot\boldsymbol{\nabla}\psi}{\rho} \tag{6.53}$$

The quantity ψ does not necessarily need to be the concentration of a substance. The vorticity equation (6.41) then implies

$$\partial_t(\rho q_\psi) + \boldsymbol{\nabla}\cdot(\rho q_\psi\mathbf{u} + \mathbf{I}_q) = 0 \tag{6.54}$$

where

$$\mathbf{I}_q := \frac{1}{\rho}\boldsymbol{\nabla}p\times\boldsymbol{\nabla}\psi - \frac{1}{\rho}\mathbf{F}^{\mathrm{mol}}\times\boldsymbol{\nabla}\psi - (\boldsymbol{\omega}+2\boldsymbol{\Omega})\dot{\psi} \tag{6.55}$$

and where $\dot{\psi} := D\psi/Dt$. Ertel's potential vorticity is globally (integrally) conserved. The amount of a potential vorticity substance of which q is the amount per unit mass and ρq_ψ the amount per unit volume does not change in a volume that moves with velocity

$$\mathbf{u}_q := \mathbf{u} + \mathbf{I}_q/\rho q_\psi \tag{6.56}$$

Its component in the direction of $\boldsymbol{\nabla}\psi$ is given by $\mathbf{u}_q \cdot \boldsymbol{\nabla}\psi = \mathbf{v} \cdot \boldsymbol{\nabla}\psi$ where $\mathbf{v} = -\partial_t \psi \boldsymbol{\nabla}\psi/|\boldsymbol{\nabla}\psi|^2$ is the velocity of the ψ-surface. The substance cannot cross ψ-surfaces.

The global conservation is solely a consequence of the fact that ρq_ψ can be written as a divergence

$$\rho q_\psi = (\boldsymbol{\nabla} \times (\mathbf{u} + \mathbf{U})) \cdot \boldsymbol{\nabla}\psi = \boldsymbol{\nabla} \cdot \mathbf{G} \tag{6.57}$$

where

$$\mathbf{G} := (\mathbf{u} + \mathbf{U}) \times \boldsymbol{\nabla}\psi \tag{6.58}$$

Hence

$$\partial_t(\rho q_\psi) = \boldsymbol{\nabla} \cdot \partial_t \mathbf{G} \tag{6.59}$$

The global conservation of potential vorticity can be rewritten as

$$\rho \frac{\mathrm{D}}{\mathrm{D}t} q_\psi = -\boldsymbol{\nabla} \cdot \mathbf{I}_q \tag{6.60}$$

Ertel's potential vorticity is materially conserved if the divergence of the flux $\mathbf{I}_q$ vanishes. The first term of the divergence can be written as

$$\boldsymbol{\nabla} \cdot \left(\frac{1}{\rho} \boldsymbol{\nabla} p \times \boldsymbol{\nabla}\psi \right) = -\frac{1}{\rho^2} (\boldsymbol{\nabla}\rho \times \boldsymbol{\nabla} p) \cdot \boldsymbol{\nabla}\psi \tag{6.61}$$

and vanishes if $\psi = \psi(p, \rho)$.

Ertel's potential vorticity theorem is the most general vorticity theorem. By specifying ψ to be the components of the position vector $\mathbf{x}$ one recovers the vorticity equation (6.41).

6.5 Thermodynamic equilibrium

The conditions for thermodynamic equilibrium of a system can be derived from the condition that the entropy S of the system has to be a maximum. For a one-component system of fixed volume V the condition

$$S = \iiint_V \mathrm{d}^3x \, \rho\eta(\rho, \rho e_{\text{int}}) = \text{maximum} \tag{6.62}$$

has to be subjected to the constraints that

$$\begin{aligned}\int\int\int_V \mathrm{d}^3x\ \rho &= \text{total mass}\\ \int\int\int_V \mathrm{d}^3x\ \rho\mathbf{u} &= \text{total momentum}\\ \int\int\int_V \mathrm{d}^3x\ \rho\mathbf{x}\times\mathbf{u} &= \text{total angular momentum}\\ \int\int\int_V \mathrm{d}^3x\ \rho\left(e_{\mathrm{int}}+\phi_g+\tfrac{1}{2}\mathbf{u}\cdot\mathbf{u}\right) &= \text{total energy}\end{aligned} \tag{6.63}$$

remain constant. The total energy is the sum of the internal, potential, and kinetic energy. Only the gravitational potential energy appears since the problem is formulated in an inertial frame of reference. The important point is that the entropy density $\rho\eta$ depends on the internal and not on the total energy density. The necessary conditions for thermodynamic equilibrium can be obtained by $\delta S = 0$, i.e., by setting to zero the variation of the entropy functional with respect to ρ, ρe_{tot}, and $\rho\mathbf{u}$. The constraints can be accounted for by the method of Lagrange multipliers. The resulting conditions are

$$T = c_0 \tag{6.64}$$

$$\mu + \phi_g - \frac{1}{2}\mathbf{u}\cdot\mathbf{u} = c_1 \tag{6.65}$$

$$\mathbf{u} = \mathbf{V} + \boldsymbol{\Omega}\times\mathbf{x} \tag{6.66}$$

where c_0, c_1, $\mathbf{V}$, and $\boldsymbol{\Omega}$ are constants.

The third condition states that the fluid moves as a rigid body with a constant translation velocity $\mathbf{V}$ and a constant angular velocity $\boldsymbol{\Omega}$. The rate of deformation tensor $\mathbf{D}$ is hence identically zero. We assume $\mathbf{V} = 0$ for simplicity.

The equilibrium condition for the pressure can be derived from the Gibbs–Durham relation $v\nabla p - \eta\nabla T = \nabla\mu$. In thermal equilibrium $\nabla T = 0$ and $\nabla\mu = -\nabla(\phi_g + \phi_c)$ where $\phi_c = -1/2(\boldsymbol{\Omega}\times\mathbf{x})\cdot(\boldsymbol{\Omega}\times\mathbf{x})$ is the centrifugal potential. Hence

$$\nabla p = -\rho\nabla\phi \tag{6.67}$$

where $\phi = \phi_g + \phi_c$ is the geopotential. The pressure is in hydrostatic balance. Similarly, $\nabla\rho = \left(\frac{\partial\rho}{\partial p}\right)_T \nabla p + \left(\frac{\partial\rho}{\partial T}\right)_p \nabla T$ implies

$$\nabla\rho = \rho\kappa\nabla p \tag{6.68}$$

If $\phi_g = 0$ and $\boldsymbol{\Omega} = 0$ then $p =$ constant and $\rho =$ constant.

For a two-component system one finds

$$\mu_1 + \phi_g - \frac{1}{2}\mathbf{u}\cdot\mathbf{u} = c_1 \tag{6.69}$$

$$\mu_2 + \phi_g - \frac{1}{2}\mathbf{u}\cdot\mathbf{u} = c_2 \tag{6.70}$$

and the necessary condition for thermodynamic equilibrium can be stated as

$$T = \text{constant}, \quad \Delta\mu = \text{constant}, \quad \mathbf{D} = 0 \tag{6.71}$$

If tidal and other forcing mechanisms are neglected the basic equations of oceanic motions have the time-independent solution

$$\mathbf{u} = 0, \quad T = \text{constant} \tag{6.72}$$

$$\nabla p = -\rho \nabla \phi \tag{6.73}$$

$$\nabla S = \gamma \nabla p \tag{6.74}$$

where $\gamma := -\frac{\partial \Delta\mu/\partial p}{\partial \Delta\mu/\partial S}$. In this case all the molecular fluxes $\mathbf{q}^{\text{mol}}$, $\mathbf{I}_{\text{s}}^{\text{mol}}$, and σ^{mol} vanish. This is of course the thermodynamic equilibrium solution $T =$ constant, $\Delta\mu =$ constant, and $\mathbf{D} = 0$.

6.6 Mechanical equilibrium

If heat and salt diffusion and tidal forcing are neglected ($\mathbf{q}^{\text{mol}} = \mathbf{I}_{\text{s}}^{\text{mol}} = 0$, $\phi_T = 0$) then the basic equations of oceanic motions have the solution $\tilde{\mathbf{u}} = 0$, $\tilde{p} = \tilde{p}(z)$, $\tilde{\theta} = \tilde{\theta}(z)$, $\tilde{S} = \tilde{S}(z)$, and $\tilde{\rho} = \tilde{\rho}(z)$ where $\mathrm{d}z = g_0^{-1}\, \mathrm{d}\phi$. The fluid is motionless and stratified. The momentum equation reduces to the hydrostatic balance

$$\nabla p = -\rho \nabla \phi \tag{6.75}$$

If a fluid particle is displaced at constant θ and S from a position z to a position $z + \delta z$ its density will be

$$\rho(z + \delta z) = \rho(\tilde{p}(z + \delta z), \theta, S) = \tilde{\rho}(z) + \left(\frac{\partial \rho}{\partial p}\right)_{\theta,\text{s}} \frac{\mathrm{d}\tilde{p}}{\mathrm{d}z}\, \delta z + \ldots \tag{6.76}$$

The density of the surrounding fluid is

$$\tilde{\rho}(z + \delta z) = \tilde{\rho}(z) + \frac{\mathrm{d}\tilde{\rho}}{\mathrm{d}z}\, \delta z + \ldots \tag{6.77}$$

For infinitesimal displacements the particle thus experiences a buoyancy force

$$\frac{[\rho(z + \delta z) - \tilde{\rho}(z + \delta z)]\, g_0}{\tilde{\rho}(z)} = N^2\, \delta z \tag{6.78}$$

where

$$N^2 := -\frac{g_0}{\tilde{\rho}} \left[\frac{\mathrm{d}\tilde{\rho}}{\mathrm{d}z} - \frac{1}{\tilde{c}^2} \frac{\mathrm{d}\tilde{p}}{\mathrm{d}z} \right] \tag{6.79}$$

is the square of the buoyancy (or Brunt–Väisälä) frequency. If $N^2 > 0$ the particle is forced back to its original position. The water column is said to be *stably stratified* or in *stable mechanical equilibrium*. Alternative expressions for N^2 are

$$N^2 = g_0 \tilde{\alpha} \frac{\mathrm{d}\tilde{\theta}}{\mathrm{d}z} - g_0 \tilde{\beta} \frac{\mathrm{d}\tilde{S}}{\mathrm{d}z} \tag{6.80}$$

$$N^2 = g_0 \alpha \frac{\mathrm{d}\tilde{T}}{\mathrm{d}z} + \rho g_0^2 \alpha \Gamma - g_0 \beta \frac{\mathrm{d}\tilde{S}}{\mathrm{d}z} \tag{6.81}$$

$$N^2 = -\frac{g_0}{\tilde{\rho}} \frac{\mathrm{d}\tilde{\rho}}{\mathrm{d}z} - \frac{g_0^2}{\tilde{c}^2} \tag{6.82}$$

where the hydrostatic balance has been employed.

If the pressure is used as an independent variable one finds

$$\begin{aligned} N^2 &= g_0^2 \left[\frac{\mathrm{d}\tilde{\rho}}{\mathrm{d}p} - \frac{1}{\tilde{c}^2} \right] \\ &= g_0^2 \rho \left[\alpha \left(\Gamma - \frac{\mathrm{d}\tilde{T}}{\mathrm{d}p} \right) + \beta \frac{\mathrm{d}\tilde{S}}{\mathrm{d}p} \right] \end{aligned} \tag{6.83}$$

The last formula is important since it allows the calculation of the buoyancy frequency from the thermodynamic functions $\rho(p, T, S)$ and $\Gamma(p, T, S)$, and from the gradient $\mathrm{d}\tilde{T}/\mathrm{d}p$ and $\mathrm{d}\tilde{S}/\mathrm{d}p$ that are measured with a CTD (conductivity, temperature, and depth) recorder.

The buoyancy frequency of a water column in thermodynamic equilibrium is given by

$$N^2 = \tilde{\rho} g_0 (\kappa - \tilde{\kappa}) + \frac{g_0^2 \beta^2}{\partial \Delta\mu / \partial S} \geq 0 \tag{6.84}$$

It is non-negative since $\kappa \geq \tilde{\kappa}$ and $\partial \Delta\mu / \partial S \geq 0$ for any thermodynamic system.

In general, the gradient of the density profile is given by

$$\frac{\mathrm{d}\tilde{\rho}}{\mathrm{d}z} = -\rho \tilde{\alpha} \frac{\mathrm{d}\tilde{\theta}}{\mathrm{d}z} + \rho \tilde{\beta} \frac{\mathrm{d}\tilde{S}}{\mathrm{d}z} + \frac{1}{\tilde{c}^2} \frac{\mathrm{d}\tilde{p}}{\mathrm{d}z} \tag{6.85}$$

The last term is the *adiabatic gradient* and the first two terms constitute the *diabatic gradient*. The buoyancy frequency is thus proportional to the diabatic gradient of density. This diabatic gradient should be distinguished from the gradient of potential density $\tilde{\rho}_{\mathrm{pot}}(z) = \rho(p_0, \tilde{\theta}(z), \tilde{S}(z))$ that is given by

$$\frac{\mathrm{d}\tilde{\rho}_{\mathrm{pot}}}{\mathrm{d}z} = \rho_{\mathrm{pot}} \tilde{\alpha}_{\mathrm{pot}} \frac{\mathrm{d}\tilde{\theta}}{\mathrm{d}z} + \rho_{\mathrm{pot}} \tilde{\beta}_{\mathrm{pot}} \frac{\mathrm{d}\tilde{S}}{\mathrm{d}z} \tag{6.86}$$

and differs from the diabatic gradient in that the coefficients $\tilde{\alpha}$ and $\tilde{\beta}$ are evaluated at the reference pressure p_0 and not at the *in situ* pressure p.

The density stratification is characterized by the two depth scales

$$\tilde{D}_\mathrm{a} := \frac{\tilde{c}^2}{g} = \text{adiabatic depth scale} \tag{6.87}$$

$$\tilde{D}_\mathrm{d} := \frac{g}{N^2} = \text{diabatic depth scale} \tag{6.88}$$

These depth scales are typically of the order of a few hundred kilometers and thus much larger than the ocean depth H or the depth scale D of oceanic motions. This fact becomes part of various approximations to the equations of oceanic motions. It constitutes one of the major differences between oceanic and atmospheric equations since the corresponding height scales in the atmosphere are comparable to the height of the atmosphere.

6.7 Neutral directions

When a fluid particle is displaced at constant θ and S by $\mathbf{dx}$ then its density changes by

$$\mathrm{d}\rho = \frac{\partial \rho}{\partial p} \boldsymbol{\nabla} p \cdot \mathbf{dx} \tag{6.89}$$

The density of the surrounding fluid at that point is given by

$$\mathrm{d}\rho = \frac{\partial \rho}{\partial p} \boldsymbol{\nabla} p \cdot \mathbf{dx} + \frac{\partial \rho}{\partial \theta} \boldsymbol{\nabla} \theta \cdot \mathbf{dx} + \frac{\partial \rho}{\partial S} \boldsymbol{\nabla} S \cdot \mathbf{dx} \tag{6.90}$$

The particle does not experience any buoyancy force if these two densities are the same. This is the case if the particle is displaced in directions that are perpendicular to the vector

$$\mathbf{d} = \frac{\partial \rho}{\partial \theta} \boldsymbol{\nabla} \theta + \frac{\partial \rho}{\partial S} \boldsymbol{\nabla} S \tag{6.91}$$

These directions are called neutral directions. For a nonlinear equation of state $\boldsymbol{\nabla} \times \mathbf{d} \neq 0$ in general, so that in general there exist no neutral surfaces $G(\mathbf{x}, t) = 0$ such that $\mathbf{d} = \boldsymbol{\nabla} G$. For a non-thermobaric fluid the gradient

$$\boldsymbol{\nabla} \rho_\mathrm{pot} = \frac{\partial \rho_\mathrm{pot}}{\partial \theta} \boldsymbol{\nabla} \theta + \frac{\partial \rho_\mathrm{pot}}{\partial S} \boldsymbol{\nabla} S \tag{6.92}$$

is parallel to $\mathbf{d}$ since the ratio $\frac{\partial \rho / \partial \theta}{\partial \rho / \partial S}$ does not depend on pressure. In this case, surfaces $\rho_\mathrm{pot} = \text{constant}$ are neutral surfaces.

7

Dynamic impact of the equation of state

In this chapter we look at the impact of different forms of the equation of state on the dynamic evolution. We first consider the general case of a two-component fluid and then the limiting cases of a one-component, a homentropic (or barotropic), an incompressible, and a homogeneous fluid. In these limiting cases the fluid flow becomes constrained, as can be most clearly seen in the vorticity and circulation theorems. Of course, sea water is a two-component fluid, but in the following chapters we will encounter approximations to the equation of state that mimic these limiting cases.

To simplify the discussion we consider these limiting cases for an ideal non-rotating fluid. A fluid is called ideal or perfect when all molecular diffusion processes are neglected. The molecular diffusion coefficients λ, D, D', ν, and ν' are set to zero and the molecular fluxes $\mathbf{I}_S^{\text{mol}}$, $\mathbf{q}^{\text{mol}}$, and $\boldsymbol{\sigma}^{\text{mol}}$ vanish. Ideal fluid flows show no evolution towards thermodynamic equilibrium. They are reversible. At the surface they can only be forced by an applied pressure. They do not satisfy all of the boundary conditions. For these reasons solutions of the diffusive equations of oceanic motions do not converge uniformly to solutions of the ideal fluid equations in the limit of vanishing molecular diffusion coefficients. Ideal fluid theory is of limited practical value in oceanography. In some cases, ideal fluid behavior is assumed in the interior of the ocean. Diffusion is then taken into account only in boundary layers, to satisfy the boundary conditions. Ideal fluid theory is, however, useful in understanding fundamental aspects of fluid dynamics.

7.1 Two-component fluids

The equations for an ideal two-component, non-rotating fluid are

$$\frac{\mathrm{D}}{\mathrm{D}t}\mathbf{u} = -\frac{1}{\rho}\nabla p - \nabla\phi \tag{7.1}$$

$$\frac{\mathrm{D}}{\mathrm{D}t}\eta = 0 \tag{7.2}$$

$$\frac{\mathrm{D}}{\mathrm{D}t}c_2 = 0 \tag{7.3}$$

$$\frac{\mathrm{D}}{\mathrm{D}t}p = -\rho c^2 \boldsymbol{\nabla}\cdot\mathbf{u} \tag{7.4}$$

$$\rho = \rho(p, \eta, c_2) \tag{7.5}$$

If one introduces ρ as a prognostic variable and p as a diagnostic variable then

$$\frac{\mathrm{D}}{\mathrm{D}t}\rho = -\rho\boldsymbol{\nabla}\cdot\mathbf{u} \tag{7.6}$$

$$p = p(\rho, \eta, c_2) \tag{7.7}$$

instead of (7.4) and (7.5).

The boundary conditions are

$$\mathbf{u}\cdot\mathbf{n} = \mathbf{v}\cdot\mathbf{n}, \quad p = 0 \quad \text{at free surfaces} \tag{7.8}$$

$$\mathbf{u}^{\mathrm{I}}\cdot\mathbf{n} = \mathbf{v}\cdot\mathbf{n}, \quad \mathbf{u}^{\mathrm{I}}\cdot\mathbf{n} = \mathbf{u}^{\mathrm{II}}\cdot\mathbf{n}, \quad p^{\mathrm{I}} = p^{\mathrm{II}} \quad \text{at interfaces} \tag{7.9}$$

$$\mathbf{u}\cdot\mathbf{n} = 0 \quad \text{at rigid surfaces} \tag{7.10}$$

The boundaries are material surfaces. There are no conditions for the tangential velocity components.

Theorems The vorticity, circulation, and potential vorticity theorems are given by

$$\frac{\mathrm{D}}{\mathrm{D}t}\hat{\omega} = \hat{\omega}\cdot\boldsymbol{\nabla}\boldsymbol{u} + \frac{1}{\rho^3}\boldsymbol{\nabla}\rho\times\boldsymbol{\nabla}p \tag{7.11}$$

$$\frac{\mathrm{d}}{\mathrm{d}t}\Gamma = \iint \mathrm{d}^2x\, \frac{\mathbf{n}\cdot(\boldsymbol{\nabla}\rho\times\boldsymbol{\nabla}p)}{\rho^2} \tag{7.12}$$

$$\frac{\mathrm{D}}{\mathrm{D}t}q_\eta = \frac{1}{\rho^3}(\boldsymbol{\nabla}\rho\times\boldsymbol{\nabla}p)\cdot\boldsymbol{\nabla}\eta \tag{7.13}$$

where $\hat{\omega} := \omega/\rho$. Neither the vorticity, nor the circulation, nor the potential vorticity are materially conserved.

7.2 One-component fluids

For a one-component fluid $\rho = \rho(p, \eta)$. This does not affect the vorticity equation but affects the circulation and potential vorticity equations. The circulation on an isentropic surface now does not change

$$\frac{\mathrm{d}}{\mathrm{d}t}\Gamma_\eta = 0 \tag{7.14}$$

since $\mathbf{n} = \boldsymbol{\nabla}\eta/|\boldsymbol{\nabla}\eta|$ and $\boldsymbol{\nabla}\eta \cdot (\boldsymbol{\nabla}\rho \times \boldsymbol{\nabla} p) = 0$ for $\rho = \rho(p, \eta)$. For the same reason Ertel's theorem becomes

$$\frac{\mathrm{D}}{\mathrm{D}t} q_\eta = 0 \tag{7.15}$$

The material conservation of η, a thermodynamic variable, implies the material conservation of Ertel's potential vorticity q_η, a *dynamic variable*, dynamic in the sense that q_η involves the fluid velocity.

The conservation of potential vorticity implies that there exist motions with $q_\eta \equiv 0$, for appropriate boundary conditions. These motions do not carry any potential vorticity. Their vorticity component parallel to $\boldsymbol{\nabla}\eta$ is zero.

7.3 Homentropic fluids

A fluid is called homentropic (or barotropic) if $\eta = \eta_0 =$ constant. Then $\rho = \rho(p)$ and $\boldsymbol{\nabla}\rho \times \boldsymbol{\nabla} p = 0$. Isopycnal surfaces are isobaric surfaces. This has important consequences for the momentum equation. The pressure term can be written as a gradient

$$\frac{1}{\rho}\boldsymbol{\nabla} p = \boldsymbol{\nabla}\phi_p \tag{7.16}$$

where

$$\phi_p(\mathbf{x}, t) := \int_{p_0}^{p(\mathbf{x},t)} \mathrm{d}p' \, \frac{1}{\rho(p')} \tag{7.17}$$

The equations of motion thus take the form

$$\frac{\mathrm{D}}{\mathrm{D}t}\mathbf{u} = -\boldsymbol{\nabla}(\phi_p + \phi) \tag{7.18}$$

$$\frac{\mathrm{D}}{\mathrm{D}t} p = -\rho c^2 \boldsymbol{\nabla} \cdot \mathbf{u} \tag{7.19}$$

$$\rho = \rho(p) \tag{7.20}$$

The acceleration in the momentum balance is a gradient.

If one applies the vector identity $\mathbf{u} \cdot \boldsymbol{\nabla}\mathbf{u} = \boldsymbol{\omega} \times \mathbf{u} + \boldsymbol{\nabla}(\frac{1}{2}\mathbf{u} \cdot \mathbf{u})$ then the momentum balance takes the form

$$\partial_t \mathbf{u} + \boldsymbol{\omega} \times \mathbf{u} = -\boldsymbol{\nabla} B \tag{7.21}$$

where

$$B := \phi_p + \phi + \frac{1}{2}\mathbf{u}\cdot\mathbf{u} \tag{7.22}$$

is called the Bernoulli function. For steady flows ($\partial_t \mathbf{u} = 0$) ∇B is therefore perpendicular to $\boldsymbol{\omega}$ and $\mathbf{u}$. The Bernoulli function is constant along streamlines and vortex lines. Streamlines and vortex lines lie in surfaces $B =$ constant.

For homentropic fluids the vorticity, circulation, and potential vorticity equations take the form

$$\frac{\mathrm{D}}{\mathrm{D}t}\hat{\omega} = \hat{\omega}\cdot\nabla\mathbf{u} \tag{7.23}$$

$$\frac{\mathrm{d}}{\mathrm{d}t}\Gamma = 0 \tag{7.24}$$

$$\frac{\mathrm{D}}{\mathrm{D}t}q_\rho = -\nabla\cdot(\hat{\omega}\rho^2\nabla\cdot\boldsymbol{u}) \tag{7.25}$$

The vorticity equation is solved explicitly by

$$\hat{\omega}_i(\mathbf{r},t) = \frac{\partial x_i}{\partial r_j}\,\hat{\omega}_j(\mathbf{r},t=0) \tag{7.26}$$

where $\mathbf{r}$ is the initial position and $\mathbf{x}(\mathbf{r},t)$ the actual position of the fluid particle. The vorticity vector transforms like infinitesimal line elements. Vortex lines are material lines. One says "vortex lines are frozen into the fluid." The circulation does not change in time. The potential vorticity q_ρ is not materially conserved.

Irrotational flow If $\hat{\omega} \equiv 0$ initially then $\hat{\omega} \equiv 0$ for all times, for appropriate boundary conditions. The homentropic ideal fluid equations thus have irrotational flows $\boldsymbol{\omega} \equiv 0$ as solutions. Their fluid velocity can be represented as

$$\mathbf{u} = \nabla\varphi \tag{7.27}$$

where φ is the velocity potential. The velocity potential is not uniquely determined. A function of time can be added to φ without affecting $\mathbf{u}$. For irrotational flows the momentum balance takes the form

$$\nabla(\partial_t\varphi + B) = 0 \tag{7.28}$$

or $\partial_t\varphi + B = c(t)$. The function $c(t)$ can be absorbed into φ. Irrotational homentropic flow is thus governed by

$$\partial_t\varphi = -B \tag{7.29}$$

$$\frac{\mathrm{D}}{\mathrm{D}t}p = -\rho c^2\Delta\varphi \tag{7.30}$$

$$\rho = \rho(p) \tag{7.31}$$

These equations describe the acoustic mode of motion. Acoustic (or sound) waves in the ocean are discussed in Chapter 22.

7.4 Incompressible fluids

If the adiabatic compressibility coefficient is assumed to be zero,

$$\tilde{\kappa} := \frac{1}{\rho}\left(\frac{\partial \rho}{\partial p}\right)_\eta = 0 \tag{7.32}$$

(or equivalently $c^2 := 1/\rho\tilde{\kappa} = \infty$) then $\rho = \rho(\eta)$ and the fluid is called *incompressible*. The equations of motion then reduce to

$$\frac{\mathrm{D}}{\mathrm{D}t}\mathbf{u} = -\frac{1}{\rho}\boldsymbol{\nabla} p - \boldsymbol{\nabla}\phi \tag{7.33}$$

$$\boldsymbol{\nabla}\cdot\mathbf{u} = 0 \tag{7.34}$$

$$\frac{\mathrm{D}}{\mathrm{D}t}\rho = 0 \tag{7.35}$$

The velocity field must be non-divergent or solenoidal. This follows from the pressure equation (7.4). The density is materially conserved. This follows from the entropy equation (7.2). No equation of state is needed. The pressure must be determined dynamically from the constraint on the velocity field

$$\boldsymbol{\nabla}\left(\frac{1}{\rho}\boldsymbol{\nabla} p\right) = -\boldsymbol{\nabla}\cdot(\mathbf{u}\cdot\boldsymbol{\nabla}\mathbf{u}) \tag{7.36}$$

The vorticity, circulation, and potential vorticity equations reduce to

$$\frac{\mathrm{D}}{\mathrm{D}t}\hat{\omega} = \hat{\omega}\cdot\boldsymbol{\nabla}\mathbf{u} + \frac{1}{\rho^3}\boldsymbol{\nabla}\rho\times\boldsymbol{\nabla} p \tag{7.37}$$

$$\frac{\mathrm{d}}{\mathrm{d}t}\Gamma_\rho = 0 \tag{7.38}$$

$$\frac{\mathrm{D}}{\mathrm{D}t}q_\rho = 0 \tag{7.39}$$

The potential vorticity q_ρ is materially conserved. There are thus solutions for which $q_\rho \equiv 0$ or $\omega\cdot\boldsymbol{\nabla}\rho \equiv 0$. Their velocity component within the isopycnal surface can be obtained by the gradient of a velocity potential. This fact is exploited in Section 20.3 when we consider nonlinear internal gravity waves.

7.5 Homogeneous fluids

For both homentropic and incompressible fluids one can consider the limit of a homogeneous fluid for which $\rho = \rho_0 =$ constant. Flows in ideal homogeneous

Table 7.1. *Vorticity, circulation, and potential vorticity equations for various ideal fluids*

	Vorticity $\hat{\omega} = \frac{\omega}{\rho}$	Circulation $\Gamma = \oint_C \mathrm{d}x \cdot \mathbf{u}$	Potential vorticity $q_\psi = \frac{\omega \cdot \boldsymbol{\nabla}\psi}{\rho}$
Two-component $\rho = \rho(p, \eta, c_2)$	$\frac{\mathrm{D}}{\mathrm{D}t}\hat{\omega} = \hat{\omega} \cdot \boldsymbol{\nabla} u + \frac{1}{\rho^3}\boldsymbol{\nabla}\rho \times \boldsymbol{\nabla} p$	$\frac{\mathrm{d}}{\mathrm{d}t}\Gamma = \iint_{\mathrm{A}} \mathrm{d}^2 x \frac{\mathbf{n}\cdot(\boldsymbol{\nabla}\rho \times \boldsymbol{\nabla} p)}{\rho^2}$	$\rho \frac{\mathrm{D}}{\mathrm{D}t} q_\eta = \frac{1}{\rho^2}(\boldsymbol{\nabla}\rho \times \boldsymbol{\nabla} p) \cdot \boldsymbol{\nabla}\eta$
One-component $\rho = \rho(p, \eta)$	$\frac{\mathrm{D}}{\mathrm{D}t}\hat{\omega} = \hat{\omega} \cdot \boldsymbol{\nabla} u + \frac{1}{\rho^3}\boldsymbol{\nabla}\rho \times \boldsymbol{\nabla} p$	$\frac{\mathrm{d}}{\mathrm{d}t}\Gamma_\eta = 0$	$\frac{\mathrm{D}}{\mathrm{D}t} q_\eta = 0$
Homentropic $\rho = \rho(p)$	$\frac{\mathrm{D}}{\mathrm{D}t}\hat{\omega} = \hat{\omega} \cdot \boldsymbol{\nabla}\mathbf{u}$	$\frac{\mathrm{d}}{\mathrm{d}t}\Gamma = 0$	$\frac{\mathrm{D}}{\mathrm{D}t} q_\eta = -\boldsymbol{\nabla} \cdot (\hat{\omega}\rho^2 \boldsymbol{\nabla} \cdot \boldsymbol{u})$
Incompressible $\rho = \rho(\eta)$	$\frac{\mathrm{D}}{\mathrm{D}t}\hat{\omega} = \hat{\omega} \cdot \boldsymbol{\nabla}\mathbf{u} + \frac{1}{\rho^3}\boldsymbol{\nabla}\rho \times \boldsymbol{\nabla} p$	$\frac{\mathrm{d}}{\mathrm{d}t}\Gamma_\rho = 0$	$\frac{\mathrm{D}}{\mathrm{D}t} q_\rho = 0$
Homogeneous $\rho = \rho_0$	$\frac{\mathrm{D}}{\mathrm{D}t}\hat{\omega} = \hat{\omega} \cdot \boldsymbol{\nabla}\mathbf{u}$	$\frac{\mathrm{d}}{\mathrm{d}t}\Gamma = 0$	

fluids are governed by the equations

$$\nabla \cdot \mathbf{u} = 0 \tag{7.40}$$

$$\frac{\mathrm{D}}{\mathrm{D}t}\mathbf{u} = -\nabla\left(\frac{p}{\rho_0} + \phi\right) \tag{7.41}$$

called *Euler equations*. The potential ϕ can be eliminated from the momentum equation by subtracting out a hydrostatic pressure $-\rho_0\phi$ but will reappear in the dynamic boundary condition $p = 0$ at the surface.

The vorticity equation reduces to

$$\frac{\mathrm{D}}{\mathrm{D}t}\hat{\omega} = \hat{\omega} \cdot \nabla\mathbf{u} \tag{7.42}$$

and the circulation Γ does not change in time, $d\Gamma/dt = 0$.

Irrotational flow There exist solutions for which $\hat{\omega} \equiv 0$. For these

$$\mathbf{u} = \nabla\varphi \tag{7.43}$$

$$\frac{p}{\rho} = -\partial_t\varphi - \phi - \frac{1}{2}\mathbf{u} \cdot \mathbf{u} \tag{7.44}$$

where φ is governed by Laplace equation

$$\Delta\varphi = 0 \tag{7.45}$$

These equations describe inviscid nonlinear surface gravity waves, discussed in Section 21.3.

Table 7.1 summarizes the main theorems.

8

Free wave solutions on a sphere

In this chapter we discuss *free* linear waves on a sphere, again for an ideal fluid. Waves are a basic mechanism by which a fluid adjusts to changes and propagates momentum and energy. Many of the approximations that are applied to the basic equations of oceanic motions and that form the remainder of this book can be characterized by the wave types that they eliminate. Free linear waves, as opposed to forced waves, are solutions of *homogeneous* linear equations. They can be superimposed. Mathematically, these homogeneous linear equations form an eigenvalue problem. The eigenfunctions determine the wave form, the eigenvalues the dispersion relation. Specifically, we consider linear waves on a stably stratified motionless background state. The linearized equations have separable eigensolutions. The vertical eigenvalue is characterized by the speed of sound, gravitation, and stratification. The horizontal eigenvalue problem is characterized by rotation and the spherical geometry of the earth. We then classify the various wave solutions into:

- sound (or acoustic) waves;
- surface and internal gravity waves; and
- barotropic and baroclinic Rossby waves;

and consider the limit of short-wave solutions. This chapter owes much to Kamenkovich (1977) where more details and proofs can be found.

Basic kinematic concepts of waves, such as propagating and standing waves, and the geometric optics approximation are covered in Appendix E.

8.1 Linearized equations of motion

Consider an ideal fluid in a shell of constant depth between $z = 0$ and $z = -H_0$. Assume a background state

$$\tilde{\mathbf{u}} = 0 \tag{8.1}$$

$$\tilde{\theta}(z),\ \tilde{S}(z),\ \tilde{p}(z),\ \tilde{\rho}(z),\ \tilde{c}^2(z) \tag{8.2}$$

that is in hydrostatic balance, $d\tilde{p}/dz = -g_0\,\tilde{\rho}(z)$, and in mechanical equilibrium. The background state is thus a solution of the ideal fluid equations. Infinitesimal deviations $\delta\mathbf{u}$, $\delta\theta$, δS, δp, and $\delta\rho$ from this background state are governed by the linear equations

$$\partial_t\delta u - f\delta v = -\frac{1}{\tilde{\rho}}\frac{1}{r_0\cos\theta}\frac{\partial\delta p}{\partial\varphi} \tag{8.3}$$

$$\partial_t\delta v + f\delta u = -\frac{1}{\tilde{\rho}}\frac{1}{r_0}\frac{\partial\delta p}{\partial\theta} \tag{8.4}$$

$$\partial_t\delta w = -\frac{1}{\tilde{\rho}}\frac{\partial\delta p}{\partial z} - \frac{g_0\delta\rho}{\tilde{\rho}} \tag{8.5}$$

$$\partial_t\delta p - \tilde{\rho}g_0\delta w = -\tilde{\rho}\tilde{c}^2\boldsymbol{\nabla}\cdot\delta\mathbf{u} \tag{8.6}$$

$$\partial_t\delta\theta + \delta w\frac{\mathrm{d}\tilde{\theta}}{\mathrm{d}z} = 0 \tag{8.7}$$

$$\partial_t\delta S + \delta w\frac{\mathrm{d}\tilde{S}}{\mathrm{d}z} = 0 \tag{8.8}$$

$$\delta\rho = \tilde{c}^{-2}\,\delta p - \tilde{\rho}\tilde{\alpha}\,\delta\theta + \tilde{\rho}\tilde{\beta}\,\delta S \tag{8.9}$$

where $f = 2\Omega\sin\theta$ is the vertical component of the planetary vorticity. The meridional component $\tilde{f} = 2\Omega\cos\theta$ is neglected. This constitutes the *traditional approximation*. The linearized boundary conditions are

$$\partial_t\eta = w \quad \text{at} \quad z = 0 \tag{8.10}$$

$$\delta p = \rho_0 g_0\eta \quad \text{or} \quad \partial_t\delta p = \rho_0 g_0\delta w \quad \text{at} \quad z = 0 \tag{8.11}$$

$$\delta w = 0 \quad \text{at} \quad z = -H_0 \tag{8.12}$$

where η is the surface displacement and $\rho_0 = \tilde{\rho}(z = 0)$.

The time derivative of the equation of state (8.9) yields

$$\partial_t\delta\rho = \tilde{c}^{-2}\,\partial_t\delta p + \delta w\frac{\tilde{\rho}}{g_0}N^2 \tag{8.13}$$

where $N^2 := g_0\tilde{\alpha}\,\mathrm{d}\tilde{\theta}/\mathrm{d}z - g_0\tilde{\beta}\,\mathrm{d}\tilde{S}/\mathrm{d}z$ is the square of the buoyancy frequency. The prognostic equations for $(\delta\mathbf{u}, \delta p, \delta\rho)$ form a closed system of equations. One does not need to calculate $\delta\theta$ and δS for its solution.

Assume wave solutions of the form $(\delta\mathbf{u}, \delta p, \delta\rho) = (\hat{\mathbf{u}}, \hat{p}, \hat{\rho})\exp(-\mathrm{i}\omega t)$ with frequency ω. The density is then given by

$$\hat{\rho} = \tilde{c}^{-2}\,\hat{p} - \frac{\hat{w}}{\mathrm{i}\omega}\frac{\tilde{\rho}}{g_0}N^2 \tag{8.14}$$

Substitution into the remaining equations yields

$$-i\omega\hat{u} - f\hat{v} = -\frac{1}{\tilde{\rho}}\frac{1}{r_0\cos\theta}\frac{\partial\hat{p}}{\partial\varphi} \tag{8.15}$$

$$-i\omega\hat{v} + f\hat{u} = -\frac{1}{\tilde{\rho}}\frac{1}{r_0}\frac{\partial\hat{p}}{\partial\theta} \tag{8.16}$$

$$(N^2 - \omega^2)\,\hat{w} = \frac{i\omega}{\tilde{\rho}}\left(\frac{\partial\hat{p}}{\partial z} + \frac{g_0}{\tilde{c}^2}\hat{p}\right) \tag{8.17}$$

$$-i\omega\hat{p} - \hat{w}\tilde{\rho}g_0 = -\tilde{\rho}\tilde{c}^2\boldsymbol{\nabla}\cdot\hat{\mathbf{u}} \tag{8.18}$$

where use has been made of the hydrostatic balance for the background state.

8.2 Separation of variables

The equations (8.15)–(8.18) allow separable solutions of the form

$$\left.\begin{array}{l}\hat{u}(\varphi,\theta,z)\\ \hat{v}(\varphi,\theta,z)\end{array}\right\} = \frac{\phi(z)}{\tilde{\rho}(z)}\left\{\begin{array}{l}U(\varphi,\theta)\\ V(\varphi,\theta)\end{array}\right. \tag{8.19}$$

$$\hat{p}(\varphi,\theta,z) = \phi(z)\,P(\varphi,\theta) \tag{8.20}$$

$$\hat{w}(\varphi,\theta,z) = i\omega\psi(z)\,P(\varphi,\theta) \tag{8.21}$$

This separation is possible because of the traditional approximation. Equations (8.15)–(8.18) do not have separable solutions if $\tilde{f}$ was included.

The horizontal eigenvalue problem is

$$-i\omega U - fV = -\frac{1}{r_0}\frac{1}{\cos\theta}\frac{\partial P}{\partial\varphi} \tag{8.22}$$

$$-i\omega V + fU = -\frac{1}{r_0}\frac{\partial P}{\partial\theta} \tag{8.23}$$

$$-i\omega\epsilon P + \boldsymbol{\nabla}_h\cdot\mathbf{U}_h = 0 \tag{8.24}$$

and the vertical eigenvalue problem is

$$\frac{\tilde{\rho}g_0}{\tilde{c}^2}\psi - \tilde{\rho}\frac{d\psi}{dz} = \left(\epsilon - \frac{1}{\tilde{c}^2}\right)\phi \tag{8.25}$$

$$\tilde{\rho}(N^2 - \omega^2)\,\psi = \frac{d\phi}{dz} + \frac{g_0}{\tilde{c}^2}\phi \tag{8.26}$$

with boundary conditions

$$\phi + \rho_0 g_0\,\psi = 0 \quad \text{at} \quad z = 0 \tag{8.27}$$

$$\psi = 0 \quad \text{at} \quad z = -H_0 \tag{8.28}$$

Here ϵ is the separation constant. It has the dimension of one over a phase speed squared. The horizontal eigenvalue problem contains rotation and the sphericity of the Earth through $f = 2\Omega \sin\theta$. It is the linearized homogeneous version of *Laplace tidal equations* discussed in Chapter 17. The vertical eigenvalue problem contains gravity (g_0), stratification (N^2), and compressibility ($\tilde{c}^2$). If $N^2 \geq 0$ then ω and ϵ are real.

The horizontal and vertical eigenvalue problems have eigenvalues $\epsilon_h(\omega)$ and $\epsilon_v(\omega)$. Solutions to the complete problem are the intersections $\epsilon_h(\omega) = \epsilon_v(\omega)$.

8.3 The vertical eigenvalue problem

By introducing the variables

$$\hat{\phi}(z) := \phi(z) \exp\left\{\int_{-H_0}^{z} dz'\, g_0 \tilde{c}^{-2}\right\} \tag{8.29}$$

$$\hat{\psi}(z) := \psi(z) \exp\left\{-\int_{-H_0}^{z} dz'\, g_0 \tilde{c}^{-2}\right\} \tag{8.30}$$

$$\hat{\rho}(z) := \tilde{\rho}(z) \exp\left\{2\int_{-H_0}^{z} dz'\, g_0 \tilde{c}^{-2}\right\} \tag{8.31}$$

the vertical eigenvalue problem takes the form

$$\frac{d\hat{\psi}}{dz} + \frac{1}{\hat{\rho}}\left(\epsilon - \frac{1}{\tilde{c}^2}\right)\hat{\phi} = 0 \tag{8.32}$$

$$\hat{\rho}(N^2 - \omega^2)\,\hat{\psi} = \frac{d\hat{\phi}}{dz} \tag{8.33}$$

$$\hat{\phi} + \hat{\rho}_0 g_0 \hat{\psi} = 0 \quad \text{at} \quad z = 0 \tag{8.34}$$

$$\hat{\psi} = 0 \quad \text{at} \quad z = -H_0 \tag{8.35}$$

These two coupled first order differential equations for $\hat{\phi}$ and $\hat{\psi}$ can be transformed into one second order differential equation for either $\hat{\phi}$ or $\hat{\psi}$. These equations are

$$\frac{d}{dz}\frac{1}{\hat{\rho}(N^2 - \omega^2)}\frac{d}{dz}\hat{\phi} + \frac{1}{\hat{\rho}}\left(\epsilon - \frac{1}{\tilde{c}^2}\right)\hat{\phi} = 0 \tag{8.36}$$

$$\hat{\phi} + \frac{g_0}{N^2 - \omega^2}\frac{d}{dz}\hat{\phi} = 0 \quad \text{at} \quad z = 0 \tag{8.37}$$

$$\frac{d}{dz}\hat{\phi} = 0 \quad \text{at} \quad z = -H_0 \tag{8.38}$$

and

$$\frac{\mathrm{d}}{\mathrm{d}z}\frac{\hat{\rho}}{\epsilon-\frac{1}{\tilde{c}^2}}\frac{\mathrm{d}}{\mathrm{d}z}\hat{\psi}+(N^2-\omega^2)\,\hat{\rho}\hat{\psi}=0 \tag{8.39}$$

$$\frac{\mathrm{d}}{\mathrm{d}z}\hat{\psi}-\left(\epsilon-\frac{1}{\tilde{c}^2}\right)g_0\hat{\psi}=0 \quad \text{at} \quad z=0 \tag{8.40}$$

$$\hat{\psi}=0 \quad \text{at} \quad z=-H_0 \tag{8.41}$$

One only has to solve one eigenvalue problem. If ψ is known then ϕ can be determined from (8.25). If ϕ is known ψ can be determined from (8.26). Note that only ω^2 enters the vertical eigenvalue problem and that the eigenvalue problem for $\hat{\psi}$ contains the eigenvalue ϵ in the boundary condition. Instead of the separation constant ϵ, one often introduces the equivalent depth $1/g_0\epsilon$.

For orientation consider first the case $\tilde{c}=c_0=$ constant, $N=N_0=$ constant, and $\hat{\rho}=\rho_0=$ constant. The vertical eigenvalue problem for $\hat{\psi}$ then reduces to

$$\frac{\mathrm{d}}{\mathrm{d}z}\frac{\mathrm{d}}{\mathrm{d}z}\hat{\psi}+\Delta\epsilon\left(N_0^2-\omega^2\right)\,\hat{\psi}=0 \tag{8.42}$$

$$\frac{\mathrm{d}}{\mathrm{d}z}\hat{\psi}-\Delta\epsilon g_0\hat{\psi}=0 \quad \text{at} \quad z=0 \tag{8.43}$$

$$\hat{\psi}=0 \quad \text{at} \quad z=-H_0 \tag{8.44}$$

where

$$\Delta\epsilon := \epsilon - c_0^{-2} \tag{8.45}$$

This eigenvalue problem has a denumerable set of eigenfunctions $\hat{\psi}_n(z)$ and eigenvalues $\Delta\epsilon_n(\omega)$ with $n=\ldots,-2,-1,0,1,2,\ldots$ The integer n is called the *mode number*. Four different cases must be distinguished. These are listed in Table 8.1 and sketched in Figure 8.1.

The $n=0$ mode is the *external* or *surface* mode. It needs gravitation for its existence and vanishes if $g_0\to 0$. In the limit $g_0\to\infty$ (rigid lid approximation) the $n=0$ mode has the eigenvalue $\Delta\epsilon_0=0$.

The $n=1,2,3,\ldots$ modes are *internal* modes of the first kind. They need stratification for their existence and vanish if $N_0^2\to 0$.

The $n=-1,-2,-3,\ldots$ modes are internal modes of the second kind. They do not need gravity and stratification. They also exist in a homogeneous ($N^2=0$) ocean. They vanish in the hydrostatic limit where the term $\partial_t\delta w$ is neglected in the vertical momentum balance (8.5).

The external and the internal modes of the first and second kind are well separated in the $(\omega^2,\Delta\epsilon)$-plane as long as $N_0^2H_0^2\ll g_0H_0\ll c_0^2$ or $H_0/\tilde{D}_{\mathrm{d}}$, $H_0/\tilde{D}_{\mathrm{a}}\ll 1$ where $\tilde{D}_{\mathrm{d}}:=g_0/N_0^2$ is the diabatic and $\tilde{D}_{\mathrm{a}}:=c_0^2/g_0$ the adiabatic depth scale of the density stratification (see Section 6.6). This is typically assumed for the ocean.

Table 8.1. *Eigensolutions and eigenvalues of the vertical eigenvalue problem (8.42)–(8.44)*

	Eigensolutions	Eigenvalues
$\Delta\epsilon > 0, \omega > N_0$	$\hat{\psi}(z) = \sinh k(z + H_0)$ $k := \Delta\epsilon^{1/2}\left(\omega^2 - N_0^2\right)^{1/2} \geq 0$	$\Delta\epsilon = \Delta\epsilon_0(\omega)$ solution of $\tanh kH_0 = \frac{kH_0}{\Delta\epsilon g_0 H_0}$ $\lim_{\omega/N_0\to\infty} \Delta\epsilon_0 = \frac{\omega^2}{g_0^2}$
$\Delta\epsilon > 0, \omega \leq N_0$	$\hat{\psi}(z) = \sin k(z + H_0)$ $k := \Delta\epsilon^{1/2}\left(N_0^2 - \omega^2\right)^{1/2} \geq 0$	$\Delta\epsilon = \Delta\epsilon_n(\omega) \quad n = 0, 1, 2, \ldots$ solution of $\tan kH_0 = \frac{kH_0}{\Delta\epsilon g_0 H_0}$ $\lim_{\omega/N_0\to 0} \Delta\epsilon_0 = \frac{1}{g_0 H_0}$ $\lim_{\omega/N_0\to 0} \Delta\epsilon_n = \frac{n^2\pi^2}{(N_0^2-\omega^2)H_0^2}$ $n = 1, 2, 3, \ldots$
$\Delta\epsilon < 0, \omega > N_0$	$\hat{\psi}(z) = \sin k(z + H_0)$ $k^2 = -\Delta\epsilon\left(\omega^2 - N_0^2\right) \geq 0$	$\Delta\epsilon = \Delta\epsilon_n(\omega) \quad n = -1, -2, -3, \ldots$ solution of $\tan kH_0 = \frac{kH_0}{\Delta\epsilon g_0 H_0}$ $\lim_{\Delta\epsilon g_0 H_0\to 0} \Delta\epsilon_n = -\left(\frac{2n+1}{2}\right)^2 \frac{\pi^2}{(\omega^2-N_0^2)H_0^2}$ $n = -1, -2, -3, \ldots$
$\Delta\epsilon < 0, \omega < N_0$	No solutions	No solutions

In the general case the vertical eigenvalue problem has a denumerable set of eigenfunctions $(\phi_n(z), \psi_n(z))$ and eigenvalues $\epsilon_n(\omega)$ with $n = \ldots, -2, -1, 0, 1, 2, \ldots$ as well. The eigenvalues have the following asymptotic behavior in the (ω^2, ϵ)-plane[1] (see Figure 8.2)

$$\begin{array}{rlr} \omega_n^2(\epsilon) \to N_{\min}^2 & \text{for } \epsilon \to -\infty & n = -1, -2, -3, \ldots \\ \epsilon_n(\omega^2) \to 1/c_{\min}^2 & \text{for } \omega^2 \to \infty & n = -1, -2, -3, \ldots \\ \omega_0^2(\epsilon) \to g_0^2\epsilon & \text{for } \epsilon \to \infty & n = 0 \\ \epsilon_0(\omega^2) \to 1/c_{\max}^2 & \text{for } \omega^2 \to -\infty & n = 0 \\ \omega_n^2(\epsilon) \to N_{\max}^2 & \text{for } \epsilon \to \infty & n = 1, 2, 3, \ldots \\ \epsilon_n(\omega^2) \to 1/c_{\max}^2 & \text{for } \omega^2 \to -\infty & n = 1, 2, 3, \ldots \end{array} \tag{8.46}$$

The eigenfunctions have the following properties

- The number of zero crossings of $\phi_n(z)$ in the interval $(-H_0, 0]$ is

$$\begin{array}{llll} n & \text{for } \omega^2 < N_{\min}^2 & \text{and} & n = 0, 1, 2, \ldots \\ |n| & \text{for } \omega^2 > N_{\max}^2 & \text{and} & n = 0, -1, -2, \ldots \end{array} \tag{8.47}$$

[1] For general reference, we present the eigenvalue curves for positive and negative ω^2 although $\omega^2 \geq 0$ for a stably stratified ocean.

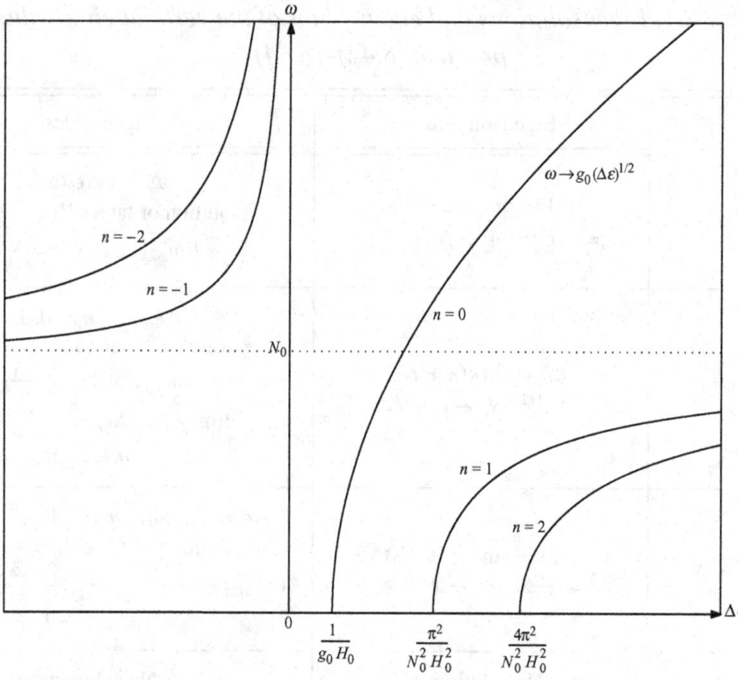

Figure 8.1. Schematic representation of the vertical eigenvalues $\Delta\epsilon_n(\omega)$ for mode numbers $n = -2, \ldots, +2$.

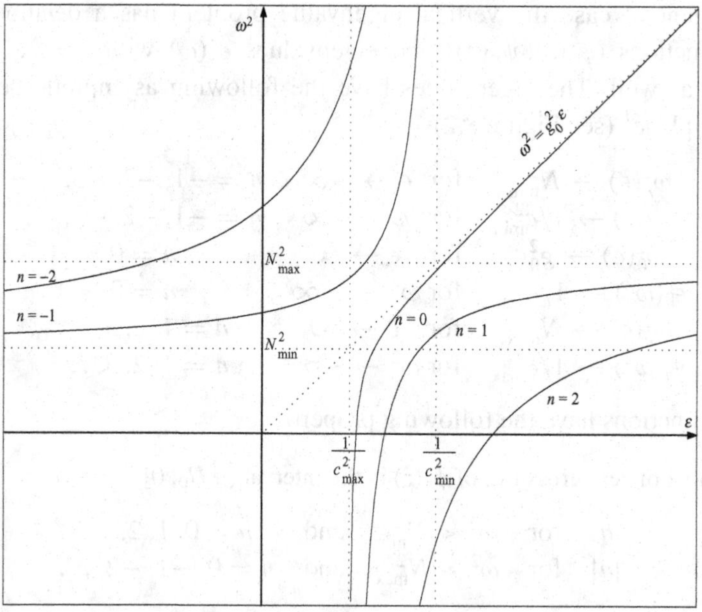

Figure 8.2. Schematic representation of the vertical eigenvalues $\epsilon_n(\omega)$ in the (ω^2, ϵ)-plane for $n = -2, \ldots, +2$. Adapted from Kamenkovich (1977).

- The number of zero crossings of $\psi_n(z)$ in the interval $(-H_0, 0]$ is

$$\begin{array}{llll} n & \text{for} \quad \epsilon > 1/c_{\min}^2 & \text{and} & n = 0, 1, 2, \ldots \\ |n| - 1 & \text{for} \quad \epsilon < 1/c_{\max}^2 & \text{and} & n = -1, -2, \ldots \end{array} \tag{8.48}$$

The vertical eigenvalue problem in the hydrostatic limit will be discussed more explicitly in Section 14.6.

8.4 The horizontal eigenvalue problem

The horizontal eigenvalue problem is most conveniently analyzed by introducing the dimensionless variables

$$(U(\varphi,\theta),\ V(\varphi,\theta),\ P(\varphi,\theta)) := \left(\frac{\hat{U}(\theta)}{\cos\theta}, \frac{\hat{V}(\theta)}{\mathrm{i}\cos\theta}, 2\Omega r_0 \hat{P}(\theta) \right) \exp^{im\varphi} \tag{8.49}$$

where m is the zonal wavenumber. The horizontal eigenvalue problem then transforms into

$$\hat{\omega}\hat{U} - \mu\hat{V} - m\hat{P} = 0 \tag{8.50}$$

$$\hat{\omega}\hat{V} - \mu\hat{U} - D\hat{P} = 0 \tag{8.51}$$

$$-\hat{\omega}\hat{\epsilon}(1-\mu^2)\hat{P} - D\hat{V} + m\hat{U} = 0 \tag{8.52}$$

where

$$\hat{\omega} := \omega/2\Omega \tag{8.53}$$

$$\mu := \sin\theta \tag{8.54}$$

$$D := (1-\mu^2)\frac{\mathrm{d}}{\mathrm{d}\mu} \tag{8.55}$$

$$\hat{\epsilon} := \epsilon\,(2\Omega)^2\, r_0^2 = \text{Lamb parameter} \tag{8.56}$$

The horizontal eigenvalue problem contains the effects of rotation and sphericity. One can proceed in two ways.

Longuet-Higgins approach One eliminates $\hat{U}$ and $\hat{V}$ to arrive at an equation for $\hat{P}$. This is the Longuet-Higgins (1968) approach. From (8.50) and (8.51) one finds

$$\hat{U} = \frac{\mu}{\hat{\omega}}\hat{V} + \frac{m}{\hat{\omega}}\hat{P} \tag{8.57}$$

$$\hat{V} = \frac{\omega}{\hat{\omega}^2 - \mu^2}\left(D + \frac{\mu m}{\hat{\omega}}\right)\hat{P} \tag{8.58}$$

When substituted into (8.52) one arrives at

$$\left[L + \frac{m}{\hat{\omega}} + \frac{2\mu}{\hat{\omega}^2 - \mu^2}\left(D + \frac{\mu m}{\hat{\omega}}\right) + \hat{\epsilon}(\hat{\omega}^2 - \mu^2) \right] \hat{P} = 0 \tag{8.59}$$

where

$$L := \frac{\mathrm{d}}{\mathrm{d}\mu}(1-\mu^2)\frac{\mathrm{d}}{\mathrm{d}\mu} - \frac{m^2}{1-\mu^2} \tag{8.60}$$

Matsuno's approach Alternatively, one obtains from (8.50) and (8.52)

$$\hat{P} = \frac{\hat{\omega}}{m^2 - \hat{\epsilon}\hat{\omega}^2(1-\mu^2)}\left(D - \frac{m\mu}{\hat{\omega}}\right)\hat{V} \tag{8.61}$$

When substituted into (8.51) one arrives at

$$\left[L - \frac{m}{\hat{\omega}} - \frac{2\hat{\epsilon}\hat{\omega}^2\mu}{m^2 - \hat{\epsilon}\hat{\omega}^2(1-\mu^2)}\left(D - \frac{\mu m}{\hat{\omega}}\right) + \hat{\epsilon}(\hat{\omega}^2 - \mu^2)\right]\hat{V} = 0 \tag{8.62}$$

This is Matsuno's representation. It is often rewritten as

$$(1-\chi\mu^2)\left[L + \frac{m}{\hat{\omega}} + \hat{\epsilon}(\hat{\omega}^2 - \mu^2)\right]\hat{V} = -2\left[\chi\mu D - \frac{m}{\hat{\omega}}\right]\hat{V} \tag{8.63}$$

where

$$\chi := \frac{\hat{\omega}^2\hat{\epsilon}}{\hat{\omega}^2\hat{\epsilon} - m^2} \tag{8.64}$$

For $|\chi| \ll 1$ this becomes

$$\left[L - \frac{m}{\hat{\omega}} + \hat{\epsilon}(\hat{\omega}^2 - \mu^2)\right]\hat{V} = 0 \tag{8.65}$$

which is the equation actually studied by Matsuno (1966).

The eigenvalues and eigensolutions of the general horizontal eigenvalue problem were calculated by Longuet-Higgins (1968). He uses $\sqrt{g_0 H}/2\Omega = r_0^2/\hat{\epsilon}$ as the eigenvalue and the symbols σ and s instead of ω and m in his graphs. For general orientation we present first the f-plane limit and then derive asymptotic limits of the eigensolutions and values, separately for the cases $m > 0$ and $m = 0$.

The *f*-plane limit The f-plane limit (see Section 18.1) neglects the sphericity of the Earth and introduces Cartesian coordinates. The eigenvalue problem then reduces to

$$-\mathrm{i}\omega\, U - f_0 V + \frac{\partial P}{\partial x} = 0 \tag{8.66}$$

$$-\mathrm{i}\omega\, V + f_0 U + \frac{\partial P}{\partial y} = 0 \tag{8.67}$$

$$-\mathrm{i}\omega\epsilon\, P + \partial_x U + \partial_y V = 0 \tag{8.68}$$

where $f_0 = 2\Omega \sin\theta_0 =$ constant. Eigensolutions in an unbounded ocean are proportional to $\exp\{\mathrm{i}(kx + ly)\}$ where k and l are the horizontal wavenumber

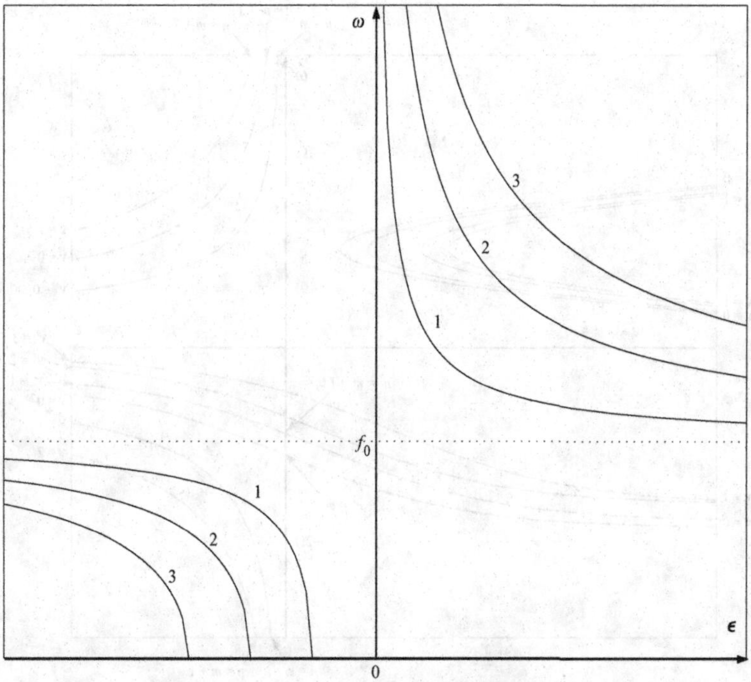

Figure 8.3. Schematic representation of the eigenvalues $\omega(\epsilon)$ in the f-plane limit.

components. The eigenvalues are

$$\omega_s = s\left(f_0^2 + \frac{k^2 + l^2}{\epsilon}\right)^{1/2} \qquad s = 0, +, - \tag{8.69}$$

and are sketched in Figure 8.3. The numbers refer to increasing values of the horizontal wavenumber magnitude $(k^2 + l^2)^{1/2}$. The solutions $\omega_{+,-}$ are called solutions of the first type for $\epsilon > 0$ and solutions of the third type for $\epsilon < 0$. The solutions $\omega_0 = 0$ are called solutions of the second type. We will find the same three types of solutions in the general case.

The general case for $m > 0$ For $m > 0$ the eigenvalue problem depends on the sign of ω. The convention is

$\omega > 0 (\sigma > 0)$ for eastward propagating waves and
$\omega < 0 (\sigma < 0)$ for westward propagating waves.

Figure 8.4 sketches the eigenvalue curves $\hat{\omega}_n(\hat{\epsilon})$ in the $(\hat{\omega}, \hat{\epsilon})$-plane. Explicit solutions can be obtained in the following six limits:

1. $(\hat{\epsilon} \to +0,\ \hat{\omega}^2 \to \infty)$. In this limit the equation for $\hat{P}$ reduces to

$$[L + \hat{\epsilon}\hat{\omega}^2]\,\hat{P} = 0 \tag{8.70}$$

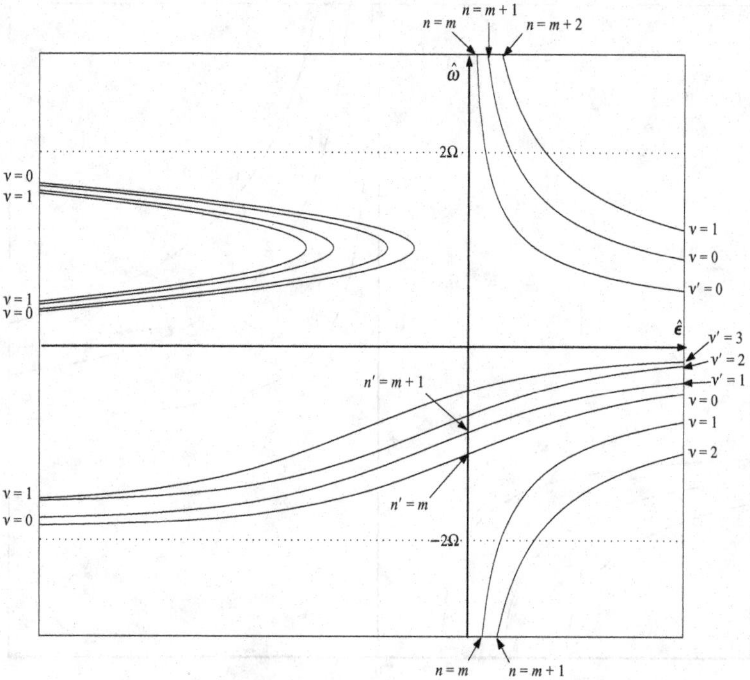

Figure 8.4. Schematic representation of the eigenvalues $\hat{\omega}(\hat{\epsilon})$ for the $m > 0$ case. Adapted from Kamenkovich (1977).

with solutions

$$\hat{P} = P_n^m(\mu) \qquad n = m, m+1, \dots \tag{8.71}$$

$$\hat{\omega}^2 = \frac{n(n+1)}{\hat{\epsilon}} \tag{8.72}$$

where $P_n^m(\mu)$ are the associated Legendre functions of the first kind with $n = m, m+1, \dots$ The number of zero crossings of $P_n^m(\mu)$ in the interval (0, 1) is $n - m$. This is the non-rotating limit, $2\Omega = 0$.

2. $\hat{\epsilon} \to 0$, $\hat{\omega}$ bounded. In this limit the equation for $\hat{V}$ reduces to

$$\left[L - \frac{m}{\hat{\omega}}\right] \hat{V} = 0 \tag{8.73}$$

with solution

$$\hat{P} = P_{n'}^m(\mu) \qquad n' = m, m+1, m+2, \dots \tag{8.74}$$

$$\hat{\omega} = -\frac{m}{n'(n'+1)} \tag{8.75}$$

This limit describes horizontally non-divergent flows $\nabla_h \cdot \mathbf{u}_h = 0$.

3. $\hat{\epsilon} \to +\infty$, $\hat{\omega}^2\hat{\epsilon}$ bounded, $\mu^2 \ll 1$. In this limit the equation for $\hat{V}$ reduces to

$$\left[\frac{d^2}{d\mu^2} - m^2 - \frac{m}{\hat{\omega}} + \hat{\epsilon}(\hat{\omega}^2 - \mu^2)\right] \hat{V} = 0 \tag{8.76}$$

This is the equatorial beta-plane (see Section 18.1). Solutions are

$$\hat{V} = \exp\left\{-\tfrac{1}{2}\hat{\mu}^2\right\} H_\nu(\hat{\mu}) \qquad \nu = 0, 1, 2, \ldots \tag{8.77}$$

$$\hat{\epsilon}\hat{\omega}^2 - \frac{m}{\hat{\omega}} - m^2 = (2\nu + 1)\hat{\epsilon}^{1/2} \tag{8.78}$$

where

$$\hat{\mu} := \hat{\epsilon}^{1/4}\mu \tag{8.79}$$

and H_ν are the Hermite polynomials. The argument can also be written as $\hat{\mu} = y/a$ where $y = r_0 \sin\theta$ and $a := r_0/\hat{\epsilon}^{1/4}$ is the equatorial Rossby radius of deformation. The critical latitude μ_c is defined as the latitude where the meridional structure of $\hat{V}$ changes from oscillatory to exponential. This latitude is given by $d^2\hat{V}/d\mu^2 = 0$. Equation (8.76) then implies

$$\hat{\epsilon}\mu_c^2 = \hat{\epsilon}\hat{\omega}^2 - \frac{m}{\hat{\omega}} - m^2 \tag{8.80}$$

or

$$\mu_c^2 = (2\nu + 1)\hat{\epsilon}^{-1/2} \tag{8.81}$$

if the dispersion relation is substituted. The critical latitude depends only on the mode index ν.

Asymptotic solutions of the eigenvalue equation (8.78) are

$$\hat{\omega} = \pm\frac{(2\nu + 1)^{1/2}}{\hat{\epsilon}^{1/4}} + \frac{m}{4\nu + 2}\frac{1}{\hat{\epsilon}^{1/2}} + \ldots \qquad \nu = 0, 1, 2, \ldots \tag{8.82}$$

$$\hat{\omega} = -\frac{m}{2\nu' + 1}\frac{1}{\hat{\epsilon}^{1/2}} + \ldots \qquad \nu' = 1, 2, 3, \ldots \tag{8.83}$$

The solution $\nu' = 0$ must be excluded since it implies $m^2 - \hat{\epsilon}\hat{\omega}^2 = 0$ and hence a singularity in (8.62) (or (8.63)) for $\hat{V}$ when expanded to first order in μ. A solution for which $\hat{V} \equiv 0$ to first order in μ and $\hat{\omega} = \pm m/\hat{\epsilon}^{1/2}$ can be obtained from the first order of (8.58), which takes the form

$$\left(\frac{d}{d\mu} \pm \mu\hat{\epsilon}^{1/2}\right) \hat{P} = 0 \tag{8.84}$$

It has the solution

$$\hat{P}(\hat{\mu}) = \exp^{\mp\frac{1}{2}\hat{\mu}^2} \tag{8.85}$$

Only the solution corresponding to

$$\hat{\omega} = +m/\hat{\epsilon}^{1/2} \tag{8.86}$$

is bounded and denoted by $\nu' = 0$.

The (ν, ν') solutions of this limit connect to (n, n') solutions of the previous limits (1) and (2) for eastward propagating waves as

$$\begin{aligned} \nu' = 0 &\rightarrow n = m \\ \nu &\rightarrow n = m + \nu + 1 \quad \text{for} \quad \nu \geq 0 \end{aligned} \tag{8.87}$$

and for westward propagating waves as

$$\begin{aligned} \nu = 0 &\rightarrow n' = m \\ \nu &\rightarrow n \ = m + \nu - 1 \quad \text{for} \quad \nu \geq 1 \\ \nu' &\rightarrow n' = m + \nu' \quad\ \ \text{for} \quad \nu' \geq 1 \end{aligned} \tag{8.88}$$

4. $\hat{\epsilon} \rightarrow -\infty$, $\hat{\omega} \rightarrow -1$. This limit requires $(1 - \mu^2) \ll 1$. The solutions are localized about the pole. Let

$$\xi := 2(-\hat{\epsilon})^{1/2}(1 - \mu) \tag{8.89}$$

and

$$\hat{\omega} = -1 + \frac{q}{(-\hat{\epsilon})^{1/2}} + O\left(\frac{1}{\hat{\epsilon}}\right) \tag{8.90}$$

with q to be determined. Then the equation for $\hat{V}$ reduces to

$$\left[\frac{\mathrm{d}^2}{\mathrm{d}\xi^2} - \frac{1}{4} + \frac{q}{2\xi} - \frac{m^2 - 2m}{4\xi^2}\right] \hat{V} = 0 \tag{8.91}$$

or

$$\frac{\mathrm{d}}{\mathrm{d}\xi}\left(\xi \frac{\mathrm{d}}{\mathrm{d}\xi}\phi\right) + \left(\frac{q}{2} - \frac{\xi}{4} - \frac{(m-1)^2}{4\xi}\right)\phi = 0 \tag{8.92}$$

for $\phi := \xi^{-1/2}\hat{V}$. If $q = m + 2\nu$ these equations have solutions

$$\hat{V} = \exp(-\xi/2)\ \xi^{m/2} \frac{\mathrm{d}^{m-1}}{\mathrm{d}\xi^{m-1}} L_{m-1+\nu}(\xi) \quad \nu = 0, 1, 2, \ldots \tag{8.93}$$

where L_n are Laguerre polynomials of order n. The eigenvalue equation becomes

$$\hat{\omega} = -1 + \frac{m + 2\nu}{(-\hat{\epsilon})^{1/2}} \tag{8.94}$$

There are corresponding solutions about the other pole with the same eigenvalue.

The original problem for $\hat{V}$ has solutions that are either even or odd functions of μ. To connect to the $\hat{\epsilon} \rightarrow 0$ solutions one has to combine the solutions at the two poles to form either an even or odd function. Therefore two n' solutions in the $\hat{\epsilon} \to 0$ limit connect to one ν solution in the $\hat{\epsilon} \to -\infty$ limit

$$\left.\begin{array}{r} n' = m \\ n' = m+1 \end{array}\right\} \to \nu = 0 \tag{8.95}$$

$$\left.\begin{array}{r} n' = m+2 \\ n' = m+3 \end{array}\right\} \to \nu = 1 \tag{8.96}$$

etc.

5. $\hat{\epsilon} \to -\infty$, $\hat{\omega} \to +1$. An analysis similar to the one of case 4 leads to the solutions

$$\hat{V} = \exp(-\xi/2)\ \xi^{(m+2)/2} \frac{\mathrm{d}^{m+1}}{\mathrm{d}\xi^{m+1}} L_{m+1+\nu}(\xi) \qquad \nu = 0, 1, 2, \ldots \tag{8.97}$$

$$\hat{\omega} = 1 - \frac{m + 2\nu + 2}{(-\hat{\epsilon})^{1/2}} \tag{8.98}$$

6. $\hat{\epsilon} \to -\infty$, $\hat{\omega} \to 0$. An analysis similar to the one of case 4 leads to the solutions

$$\hat{V} = \exp(-\xi/2)\ \xi^{m/2} \frac{\mathrm{d}^{m}}{\mathrm{d}\xi^{m}} L_{m+\nu}(\xi) \qquad \nu = 0, 1, 2, \ldots \tag{8.99}$$

$$\hat{\omega} = \frac{m}{-\hat{\epsilon}} + \frac{2m(m + 2\nu + 1)}{(-\hat{\epsilon})^{3/2}} \tag{8.100}$$

The solutions of case 5 and case 6 connect in a one-to-one fashion, $\nu \to \nu$. The complete eigenvalue curves are sketched in Figure 8.4.

The general case for $m = 0$ The zonally uniform solutions $m = 0$ depend on $\hat{\omega}^2$. Graphs for $\hat{\omega} > 0$ are thus sufficient. The zonally uniform solutions have asymptotic solutions:

1. $\hat{\epsilon} \to +0$. In this limit, the equation for $\hat{P}$ has solutions

$$\hat{P} = P_n^0(\mu) \tag{8.101}$$

with

$$\hat{\omega} = \pm \left(\frac{n(n+1)}{\hat{\epsilon}} \right)^{1/2} \qquad n = 1, 2, 3, \ldots \tag{8.102}$$

where $P_n^0(\mu)$ are associated Legendre functions of the first kind.

2. $\hat{\epsilon} \to +\infty$. In this limit, the equation for $\hat{V}$ has solutions

$$\hat{V} = \exp\left\{ -\frac{1}{2}\hat{\mu}^2 \right\} H_\nu(\mu) \tag{8.103}$$

with

$$\hat{\omega} = \pm \left(\frac{2\nu + 1}{\hat{\epsilon}^{1/2}} \right)^{1/2} \qquad \nu = 0, 1, 2, \ldots \tag{8.104}$$

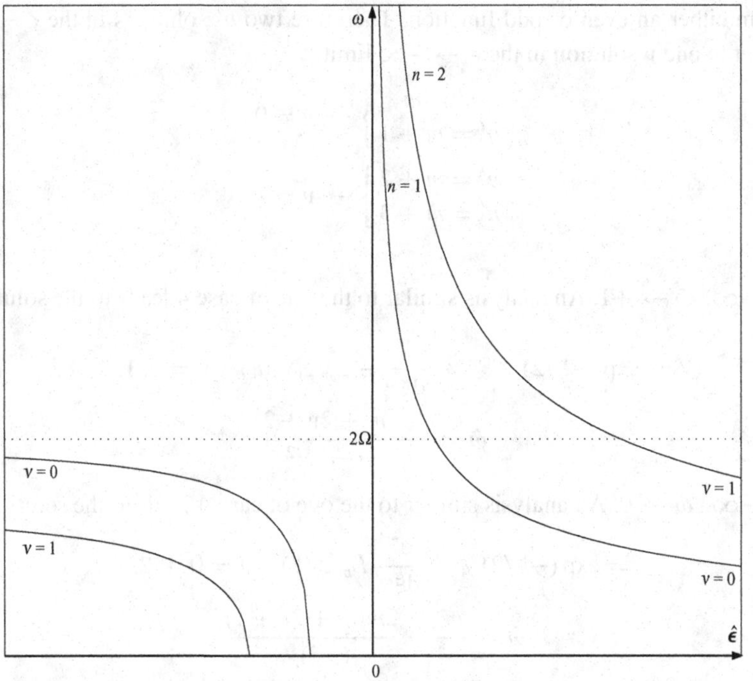

Figure 8.5. Schematic representation of the eigenvalues $\hat{\omega}(\hat{\epsilon})$ for the $m = 0$ case.

where $H_\nu(\mu)$ are Hermite polynomials. These solutions connect to the $\hat{\epsilon} \to +0$ solutions by $\nu \to n = \nu + 1$.

3. $\hat{\epsilon} \to -\infty$. In this limit, the equation for $\hat{V}$ has solutions

$$\hat{V} = \exp\left\{-\frac{\xi}{2}\right\} \xi \frac{\mathrm{d}}{\mathrm{d}\xi} L_{\nu+1}(\xi) \tag{8.105}$$

with

$$\hat{\omega} = \pm 1 \mp \frac{2 + 2\nu}{(-\hat{\epsilon})^{1/2}} \qquad \nu = 0, 1, 2, \ldots \tag{8.106}$$

where $L_\nu(\xi)$ are Laguerre polynomials.

The complete eigenvalue curves in the $(\hat{\omega}, \hat{\epsilon})$-plane are sketched in Figure 8.5.

8.5 Short-wave solutions

Here we construct another set of asymptotic solutions to the horizontal eigenvalue problem, namely short-wave solutions. We do so for three cases.

Type I solutions Let $\hat{\omega}$ be constant. Then the index ν of the eigenfunction and hence the number of zero-crossings in the interval $(-1, +1)$ increases as $\hat{\epsilon}$ increases.

Short-wave solutions are thus obtained in the limit $\hat{\epsilon} \to \infty$. Such solutions are sought by an asymptotic expansion of the form

$$\hat{P}(\mu) = \left[\hat{P}^{(0)}(\mu) + \hat{\epsilon}^{-1/2}\hat{P}^{(1)}(\mu) + \ldots\right] \exp\left\{ i\hat{\epsilon}^{1/2} \int_0^\mu d\mu' \, \nu_*(\mu') \right\} \quad (8.107)$$

where $\nu = \hat{\epsilon}^{1/2}\nu_*$ with $\nu_* = O(1)$. When this expansion is substituted into (8.59) for $\hat{P}$ and different orders of $\hat{\epsilon}$ are equated then one obtains at lowest order

$$\left\{ -(1-\mu^2)\nu_*^2 - \frac{m_*^2}{(1-\mu^2)} + (\hat{\omega}^2 - \mu^2) \right\} \hat{P}^{(0)} = 0 \quad (8.108)$$

or

$$\nu_*^2 = \frac{\mu^4 - \mu^2(1+\hat{\omega}^2) + \hat{\omega}^2 - m_*^2}{(1-\mu^2)^2} \quad (8.109)$$

where

$$m = \hat{\epsilon}^{1/2}\, m_* \quad (8.110)$$

with $m_* = O(1)$. For $\hat{\omega}^2 < m_*^2$ the numerator is negative in the interval $(-1, 1)$. There exist no oscillatory waves. For $\hat{\omega}^2 > m_*^2$ the behavior is oscillatory in the interval $(-\mu_1, \mu_1)$ and exponential in the intervals $(-1, -\mu_1)$ and $(\mu_1, 1)$ where

$$\mu_{1,2}^2 = \frac{1+\hat{\omega}^2}{2} \mp \left(\frac{(1-\hat{\omega}^2)^2}{4} + m_*^2 \right)^{1/2} \quad (8.111)$$

are the two roots of the numerator and μ_1 the root for which $\mu_1^2 < 1$.

Local coordinates are introduced by

$$x = (\varphi - \varphi_0) r_0 \cos\theta_0 \quad (8.112)$$

$$y = r_0(\theta - \theta_0) \quad (8.113)$$

and local wavenumbers by

$$k = \frac{m_* \hat{\epsilon}^{1/2}}{r_0 \cos\theta_0} \quad (8.114)$$

$$l = \frac{\nu_* \hat{\epsilon}^{1/2}}{r_0} \cos\theta_0 \quad (8.115)$$

where θ_0 is a reference latitude. Then the solution varies proportional to $\exp\{i(kx + ly)\}$ and the eigenvalue equation is given by

$$\omega^2 = f_0^2 + \frac{k^2 + l^2}{\epsilon} \quad (8.116)$$

where $f_0 = 2\Omega \sin\theta_0$ is the local value of the Coriolis parameter. This result is identical to the f-plane approximation discussed in the previous section.

Type II and III solutions for $\hat{\epsilon} < 0$ In this case the short-wave solutions are obtained in the limit $\hat{\epsilon} \to -\infty$. The asymptotic expansion (8.107) with $\hat{\epsilon}$ replaced by $-\hat{\epsilon}$ then leads to

$$\nu_*^2 = -\frac{\mu^4 - \mu^2(1+\hat{\omega}^2) + \hat{\omega}^2 + m_*^2}{(1-\mu^2)^2} \tag{8.117}$$

with an overall minus sign in front and a plus sign in front of m_*^2 instead of a minus sign. If $\hat{\omega}^2 > 1 - 2m_*$ there are no oscillatory solutions. For $\hat{\omega}^2 < 1 - 2m_*$ the roots of the numerator are

$$\mu_{1,2}^2 = \frac{1+\hat{\omega}^2}{2} \mp \left(\frac{(1-\hat{\omega}^2)^2}{4} - m_*^2\right)^{1/2} \tag{8.118}$$

and there exist solutions that are oscillatory in the intervals $(-\mu_2, -\mu_1)$ and (μ_1, μ_2) and exponential in the intervals $(-1, -\mu_2)$, $(-\mu_1, \mu_1)$, and $(\mu_2, 1)$. The oscillatory part again varies locally as $\exp\{\mathrm{i}(kx + ly)\}$ with $\omega(\epsilon)$ given by (8.116). This relation thus describes short type II and III waves for negative ϵ and short type I waves for positive ϵ.

Type II solutions for $\epsilon > 0$ In this case one has to seek solutions for $\hat{\omega}\hat{\epsilon}^{1/2}$ being fixed. An asymptotic expansion

$$\hat{V}(\mu) = \left[\hat{V}^{(0)}(\mu) + \hat{\epsilon}^{-1/2}\hat{V}^{(1)}(\mu) + \ldots\right] \exp\left\{\mathrm{i}\hat{\epsilon}^{1/2} \int_0^{\mu} \mathrm{d}\mu'\, \nu_*(\mu')\right\} \tag{8.119}$$

then leads to

$$\nu_*^2 = \frac{\mu^4 - \mu^2\left(1 + \frac{m_*}{\hat{\omega}}\right) + \hat{\omega}^2 - m_*^2}{(1-\mu^2)^2} \tag{8.120}$$

For $m_*\hat{\omega} > 1$ no oscillatory solutions exist. For $m_*\hat{\omega} < 1$ the solutions are oscillatory in the interval $(-\mu_1, \mu_1)$ and exponential in the intervals $(-1, -\mu_1)$ and $(\mu_1, 1)$. The oscillatory part varies locally as $\exp\{\mathrm{i}(kx + ly)\}$ with

$$\omega = -\frac{\beta_0 k}{k^2 + l^2 + f_0^2 \epsilon} \tag{8.121}$$

where

$$\beta_0 = \frac{1}{r_0} 2\Omega \cos\theta_0 \tag{8.122}$$

is the local value of the beta-parameter. This limit constitutes the midlatitude beta-plane approximation discussed in Section 18.1.

Table 8.2. *Classification of waves based on the intersection of vertical and horizontal eigenvalue curves*

Horizontal/vertical eigenvalue curves	*Type I*	*Type II*	*Type III*
n = −1, −2, −3, ...	Acoustic waves if $c_{\text{min}}^{-2} > 0$	Gyroscopic waves if $N_{\text{min}} < 2\Omega$	Gyroscopic waves if $N_{\text{min}} < 2\Omega$
n = 0	Surface gravity waves if $g_0 > 0$	Barotropic Rossby waves if $g_0 > 0$	
n = 1, 2, 3, ...	Internal gravity waves if $N_{\text{max}}^2 > 0$	Baroclinic Rossby waves if $N_{\text{max}}^2 > 0$	

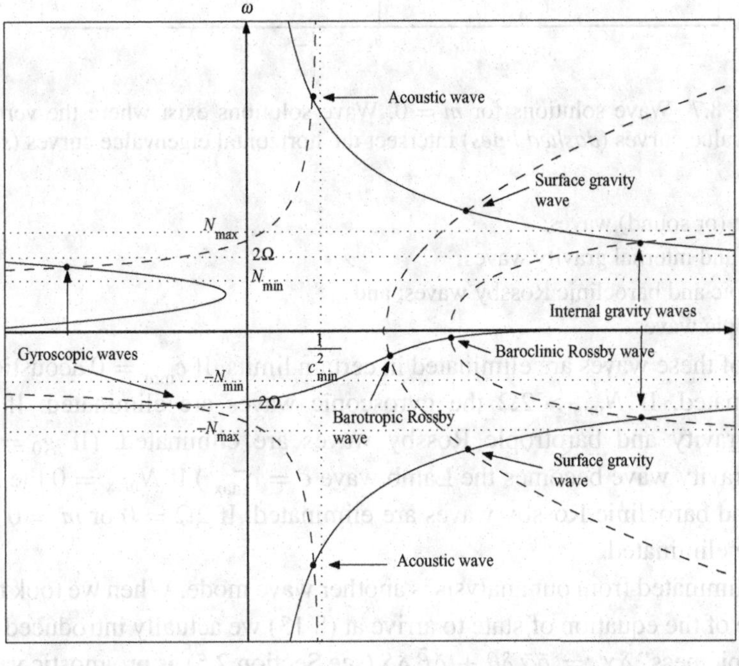

Figure 8.6. Wave solutions for $m > 0$. Wave solutions exist where the vertical eigenvalue curves (*dashed lines*) intersect the horizontal eigenvalue curves (*solid lines*). Adapted from Kamenkovich (1977).

8.6 Classification of waves

Solutions to the wave equation exist where the eigenvalues of the horizontal and vertical problems intersect. These intersections are given in Table 8.2 and shown in Figures 8.6 and 8.7 for $m > 0$ and $m = 0$. One finds:

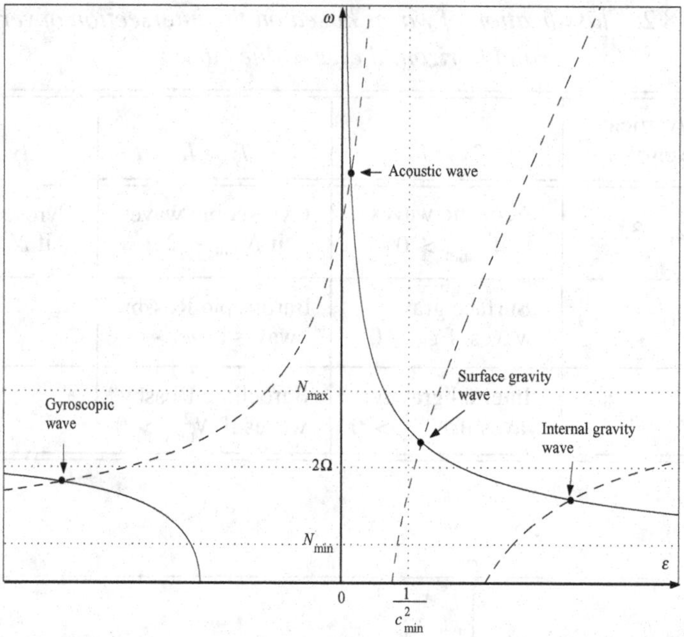

Figure 8.7. Wave solutions for $m = 0$. Wave solutions exist where the vertical eigenvalue curves (*dashed lines*) intersect the horizontal eigenvalue curves (*solid lines*).

- acoustic (or sound) waves;
- surface and internal gravity waves;
- barotropic and baroclinic Rossby waves; and
- gyroscopic waves.

Some of these waves are eliminated in certain limits. If $c_{\min}^{-2} = 0$ acoustic waves are eliminated. If $N_{\min} > 2\Omega$ the gyroscopic waves are eliminated. If $g_0 = 0$ surface gravity and barotropic Rossby waves are eliminated. (If $g_0 = \infty$ the surface gravity wave becomes the Lamb wave $\epsilon = c_{\max}^{-2}$.) If $N_{\max} = 0$ the internal gravity and baroclinic Rossby waves are eliminated. If $2\Omega = 0$ or $m = 0$ Rossby waves are eliminated.

Also eliminated from our analysis is another wave mode. When we took the time derivative of the equation of state to arrive at (8.13) we actually introduced density $\delta\rho$ and "spiciness" $\delta\chi := \tilde{\rho}\tilde{\alpha}\,\delta\theta + \tilde{\rho}\tilde{\beta}\,\delta S$ (see Section 2.5) as prognostic variables, instead of $\delta\theta$ and δS. This change of representation leads to the equations

$$\partial_t \delta\rho = \tilde{c}^{-2}\,\partial_t \delta p + \delta w \frac{\tilde{\rho}}{g_0} N^2 \tag{8.123}$$

$$\partial_t \delta\chi = \delta w \frac{\tilde{\rho}}{g_0} M^2 \tag{8.124}$$

where $M^2 := -g_0\tilde{\alpha}\,\mathrm{d}\tilde{\theta}/\mathrm{d}z - g_0\tilde{\beta}\,d\tilde{S}/dz$ is a "spiciness" frequency, similar to the buoyancy frequency. However, under the assumptions of this chapter the variable

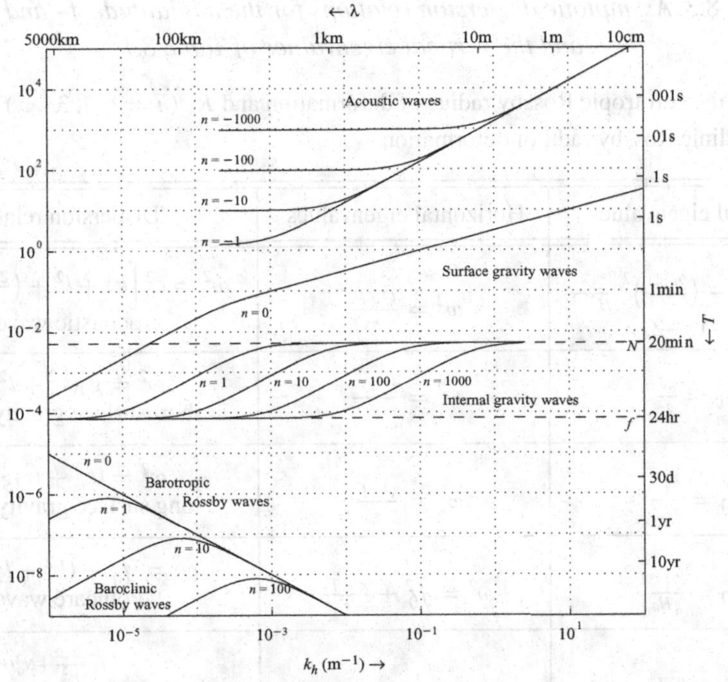

Figure 8.8. Asymptotic dispersion relations for the midlatitude f- and beta-plane and the vertical eigenvalues of Table 8.1. The background parameters are $c_0 = 1500\,\mathrm{m\,s^{-1}}$, $H_0 = 5\,\mathrm{km}$, $f_0 = 7 \cdot 10^{-5}$, and $N_0 = 5 \cdot 10^{-3}$.

$\delta\chi$ does not affect the dynamics. The prognostic equations for ($\delta\mathbf{u}$, δp, $\delta\rho$) form a closed system of equations. One does not need to calculate $\delta\chi$ for their solution. Spiciness is a passive tracer. Whenever there is a vertical velocity δw the value of $\delta\chi$ changes according to (8.124), but the value of $\delta\chi$ does not affect δw. This passive behavior of $\delta\chi$ is not very interesting. One may consider other passive tracers as well. In Section 21.2 we discuss circumstances under which $\delta\chi$ becomes an active tracer. In this case both density and spiciness or both temperature and salinity are needed to determine the dynamic evolution. An additional wave type, the temperature–salinity mode, becomes active. This mode is suppressed here.

Explicit dispersion relations using the midlatitude f- and beta-plane horizontal eigenvalues and the vertical eigenvalues of Table 8.1 are given in Table 8.3 and are shown in Figure 8.8 for $N^2 \gg f_0^2$. The dispersion relations contain various background parameters:

- The phase speeds c_0, $\sqrt{g_0 H_0}$, and $N_0 H_0$ that characterize the phase speeds of acoustic, long surface gravity, and long internal gravity waves. In oceanography one generally assumes $N_0^2 H_0^2 \ll g_0 H_0 \ll c_0^2$, which is equivalent to $H_0/\tilde{D}_a \ll 1$ and $H_0/\tilde{D}_d \ll 1$ where $\tilde{D}_a := c_0^2/g_0$ and $\tilde{D}_d := g_0/N_0^2$ are the adiabatic and diabatic depth scales of the background density field.

Table 8.3. *Asymptotic dispersion relations for the midlatitude f- and β-plane and the vertical eigenvalues of Table 8.1.*

R_0 is the barotropic Rossby radius of deformation and R_n $(n = 1, 2, 3, \ldots)$ are the baroclinic Rossby radii of deformation

Vertical eigenvalues	Horizontal eigenvalues	Dispersion relation
$\epsilon_n = \frac{1}{c_0^2} - \left(\frac{2n+1}{2}\right)^2 \frac{\pi^2}{H_0^2 \omega^2}$ $n = -1, -2, \ldots$	$\omega^2 = \frac{k^2+l^2}{\epsilon}$	$\omega^2 = c_0^2 \left[k^2 + l^2 + \left(\frac{2n+1}{2}\right)^2 \frac{\pi^2}{H_0^2}\right]$ acoustic waves
$\epsilon_0 = \frac{\omega^2}{g^2}$	$\omega^2 = \frac{k^2+l^2}{\epsilon}$	$\omega^2 = g_0(k^2 + l^2)^{1/2}$ short surface gravity waves
$\epsilon_0 = \frac{1}{g H_0}$	$\omega^2 = \frac{k^2+l^2}{\epsilon}$	$\omega^2 = (k^2 + l^2) g_0 H_0$ long surface gravity waves
$\epsilon_0 = \frac{1}{g H_0}$	$\omega^2 = f_0^2 + \frac{k^2+l^2}{\epsilon}$	$\omega^2 = f_0^2 + (k^2 + l^2) g_0 H_0$ Poincaré waves
$\epsilon_n = \frac{n^2\pi^2}{\left(N_0^2-\omega^2\right) H_0^2}$ $n = 1, 2, \ldots$	$\omega^2 = f_0^2 + \frac{k^2+l^2}{\epsilon}$	$\omega^2 = \frac{f_0^2 \frac{n^2\pi^2}{H_0^2} + N_0^2(k^2+l^2)}{k^2+l^2+\frac{n^2\pi^2}{H_0^2}}$ internal gravity waves $\left(N_0^2 > f_0^2\right)$
$\epsilon_0 = \frac{1}{g H_0}$	$\omega = -\frac{\beta k}{k^2+l^2+f_0^2\epsilon}$	$\omega = -\frac{\beta k}{k^2+l^2+R_0^{-2}}$ $R_0^2 := \frac{g_0 H_0}{f_0^2}$ barotropic Rossby waves
$\epsilon_n = \frac{n^2\pi^2}{N_0^2 H_0^2}$ $n = 1, 2, \ldots$	$\omega = -\frac{\beta k}{k^2+l^2+f_0^2\epsilon}$	$\omega = -\frac{\beta k}{k^2+l^2+R_n^{-2}}$ $R_n^2 := \frac{N_0^2 H_0^2}{f_0^2 n^2 \pi^2}$ baroclinic Rossby waves
$\epsilon_n = -\frac{n^2\pi^2}{\left(\omega^2-N_0^2\right) H_0^2}$ $n = -1, -2, \ldots$	$\omega^2 = f_0^2 + \frac{k^2+l^2}{\epsilon}$	$\omega^2 = \frac{f_0^2 \frac{n^2\pi^2}{H_0^2} + N_0^2(k^2+l^2)}{k^2+l^2+\frac{n^2\pi^2}{H_0^2}}$ gyroscopic waves $\left(N_0^2 < f_0^2\right)$

- The mid latitude Rossby radii of deformation $R_0^2 := g_0 H_0 / f_0^2$ and $R_n^2 := N_0^2 H_0^2 / n^2 \pi^2 f_0^2$. In oceanography one generally assumes $r_0 \geq R_0 \gg R_1$. This condition puts an upper limit on stratification, $N_0^2 / f_0^2 \ll \pi^2 r_0^2 / H_0^2$.
- The buoyancy and Coriolis frequency N_0 and f_0. If one assumes $N_0^2 \geq f_0^2$ one puts a lower limit on stratification and eliminates gyroscopic waves.

9

Asymptotic expansions

The basic equations of oceanic motions describe *all* fluid motions on *all* space and time scales. All these motions and scales cannot be resolved simultaneously, neither observationally nor computationally. In the remainder of this book we therefore consider approximations to these equations. These approximations reduce the physics. They suppress certain processes, modes of motions, and scales. All these approximations can be obtained in a systematic manner by asymptotic expansions. These expansions clearly identify the circumstances under which a certain approximation holds and the errors involved in making this approximation. These approximations should thus be distinguished from idealizations. These merely assign specific values to parameters that describe the fluid and its environment. Typical idealizations are ideal fluids ($\lambda = D = D' = \nu = \nu' = 0$), non-rotating systems ($\boldsymbol{\Omega} = 0$), incompressible fluids ($c = \infty$), and two-dimensional flows ($w = \partial/\partial z = 0$). Such idealized systems might be much more amenable to analysis. However, it is unclear how they relate to the real ocean. Real fluids are not ideal. They diffuse salt, momentum, and heat. Ideal fluids do not. Real fluid flows are irreversible. Ideal fluid flows are time-reversible. Similar remarks apply to the other idealizations. There are, of course, oceanic motions that behave approximately as if the fluid were ideal or non-rotating or incompressible or two-dimensional. One would like to understand the circumstances under which this is the case and the errors involved in any such approximation. Asymptotic expansions address this question in a systematic manner.

We first outline the general method and then apply it to an important problem, the elimination of fast variables. Fast variables are usually eliminated by arguing that they adjust so rapidly to a quasi-steady equilibrium value that their time derivatives can be neglected. A systematic asymptotic expansion reveals the conditions when this heuristic argument is valid and when it must be abandoned. It also motivates the concept of stochastic forcing. Fast motions can also be eliminated by Reynolds averaging, considered in the next chapter.

9.1 General method

Asymptotic expansions are Taylor expansions of the equations of oceanic motions with respect to a small dimensionless parameter. An asymptotic expansion involves the following steps:

1. *Scaling.* First one scales all independent and dependent variables

$$t = T\hat{t}, \cdots \tag{9.1}$$

$$u = U\hat{u}, \cdots \tag{9.2}$$

   Here T is the typical time scale over which a dependent variable, for example u, changes significantly and U is the magnitude of this change. The variables with a hat are non-dimensional and assumed to be $O(1)$. Derivatives then become $\partial u/\partial t = U/T\ \partial\hat{u}/\partial\hat{t}, \ldots$ External fields like the buoyancy frequency $N(z)$ have to be scaled as well. Thus $N(z) = N_*\hat{N}(z)$ where N_* is the magnitude of $N(z)$ and $\hat{N}(z)$ is $O(1)$.
2. *Non-dimensionalization.* Next, one substitutes the scaling into the equations of motion. This leads to a set of equations for the dimensionless variables with dimensionless coefficients $(\epsilon_1, \cdots, \epsilon_N)$. The result is not unique. Different dimensionless parameters can be introduced, although the number N of independent dimensionless parameters is unique and determined by the number of scales and the number of physical dimensions according to Buckingham's Π-theorem.The solution $\hat{\psi}(\hat{\mathbf{x}}, \hat{t})$ of the non-dimensional equations is a function of these non-dimensional parameters, $\hat{\psi}(\hat{\mathbf{x}}, \hat{t}) = \hat{\psi}(\hat{\mathbf{x}}, \hat{t}; \epsilon_1, \cdots, \epsilon_N)$.
3. *Expansion.* If any of the dimensionless parameters are assumed to be small, say $\epsilon_1 \ll 1$, then one can expand the solution with respect to ϵ_1

$$\hat{\psi}(\hat{\mathbf{x}}, \hat{t}; \epsilon_1, \cdots, \epsilon_N) = \sum_{n=0}^{\infty} \epsilon_1^n \hat{\psi}^{(n)}(\hat{\mathbf{x}}, \epsilon_1\hat{\mathbf{x}}, \epsilon_1^2\hat{\mathbf{x}}, \cdots, \hat{t}, \epsilon_1\hat{t}, \epsilon_1^2\hat{t}, \cdots; \epsilon_2, \cdots, \epsilon_N) \tag{9.3}$$

   This expansion includes an expansion of the independent variables into multiple scales. Derivatives then take the form

$$\frac{\partial\hat{\psi}}{\partial\hat{t}} = \sum_{n=0}^{\infty} \epsilon_1^n \frac{\partial\hat{\psi}}{\partial\hat{t}^{(n)}} \tag{9.4}$$

   where $\hat{t}^{(n)} = \epsilon_1^n\hat{t}$ and $\partial\hat{\psi}/\partial\hat{t}^{(n)} = O(1)$. Substituting this expansion into the dimensionless equations of motion and equating terms of like order in the expansion parameter ϵ_1 then leads to a hierarchy of equations that needs to be analyzed further.

Most of the approximations given in this book will not be derived by carrying out such a systematic expansion. Instead, they will be derived by identifying small dimensionless coefficients and neglecting the associated terms. This procedure can of course be viewed as the lowest order of a (trivial) asymptotic expansion. The spherical approximation in Section 5.3 consisted of such trivial "asymptotic expansions," simply neglecting terms of $O(d_0^2/r_0^2)$ and $O(H_0/r_0)$ in the scale factors. A proper asymptotic expansion is carried out in Chapter 19 when we derive the quasi-geostrophic potential vorticity equation and in the next section when we eliminate fast variables by a two-time scale expansion.

9.2 Adiabatic elimination of fast variables

Consider a system with state vector $\boldsymbol{\psi}$ and evolution equation[1]

$$\frac{\mathrm{d}}{\mathrm{d}t}\boldsymbol{\psi}(t) = \mathbf{F}\left[\boldsymbol{\psi}(t)\right] \tag{9.5}$$

Assume that the complete system consists of two subsystems, $\boldsymbol{\psi} = (\boldsymbol{\psi}_{\mathrm{s}}, \boldsymbol{\psi}_{\mathrm{f}})$, where $\boldsymbol{\psi}_{\mathrm{s}}$ is a slow responding system with a long response time T_{s} and $\boldsymbol{\psi}_{\mathrm{f}}$ a fast responding system with a short response time T_{f}. After suitable scaling, the evolution of these subsystems is governed by

$$\frac{\mathrm{d}}{\mathrm{d}\hat{t}}\hat{\boldsymbol{\psi}}_{\mathrm{s}}(t) = \hat{\mathbf{f}}(\hat{\boldsymbol{\psi}}_{\mathrm{s}}, \hat{\boldsymbol{\psi}}_{\mathrm{f}}) \tag{9.6}$$

$$\frac{\mathrm{d}}{\mathrm{d}\hat{t}}\hat{\boldsymbol{\psi}}_{\mathrm{f}}(t) = \frac{1}{\epsilon}\hat{\mathbf{g}}(\hat{\boldsymbol{\psi}}_{\mathrm{s}}, \hat{\boldsymbol{\psi}}_{\mathrm{f}}) \tag{9.7}$$

where all quantities with a hat are $O(1)$. The factor $1/\epsilon$ with $\epsilon := T_{\mathrm{f}}/T_{\mathrm{s}} \ll 1$ in front of $\hat{\mathbf{g}}$ indicates that $\hat{\boldsymbol{\psi}}_{\mathrm{f}}$ is the fast responding subsystem. The functions $\hat{\mathbf{f}}$ and $\hat{\mathbf{g}}$ are determined uniquely by $\mathbf{F}$. No approximation has been made so far. We envision a situation where the fast variable $\hat{\boldsymbol{\psi}}_{\mathrm{f}}$ cannot be resolved and seek to obtain an evolution equation for $\hat{\boldsymbol{\psi}}_{\mathrm{s}}$ alone. In order to achieve this one takes advantage of the smallness of ϵ. Heuristically, one argues that the fast variables adjust rapidly to quasi-steady equilibrium values for which $\mathrm{d}\hat{\boldsymbol{\psi}}_{\mathrm{f}}/\mathrm{d}t = 0$. During this rapid adjustment the slow variable $\hat{\boldsymbol{\psi}}_{\mathrm{s}}$ can be regarded as constant such that the final adjustment state can be calculated from $\hat{\mathbf{g}}(\hat{\boldsymbol{\psi}}_{\mathrm{s}}, \hat{\boldsymbol{\psi}}_{\mathrm{f}}) = 0$ for fixed $\hat{\boldsymbol{\psi}}_{\mathrm{s}}$. Denote this solution by $\hat{\boldsymbol{\psi}}_{\mathrm{f}}^{\infty}(\hat{\boldsymbol{\psi}}_{\mathrm{s}})$ and substitute it into the equation for the slow variable

$$\frac{\mathrm{d}}{\mathrm{d}\hat{t}}\hat{\boldsymbol{\psi}}_{\mathrm{s}}(t) = \hat{\mathbf{f}}(\hat{\boldsymbol{\psi}}_{\mathrm{s}}, \hat{\boldsymbol{\psi}}_{\mathrm{f}}^{\infty}(\hat{\boldsymbol{\psi}}_{\mathrm{s}})) \tag{9.8}$$

In this way one obtains a closed equation for the slow variable $\hat{\boldsymbol{\psi}}_{\mathrm{s}}$. The fast variables have been eliminated. They are given by $\hat{\boldsymbol{\psi}}_{\mathrm{f}} = \hat{\boldsymbol{\psi}}_{\mathrm{f}}^{\infty}(\hat{\boldsymbol{\psi}}_{\mathrm{s}})$. They are *slaved* by the slow variables. This elimination is achieved by neglecting the time derivative of the fast variables. We will follow this heuristic approach in this book but discuss next its intricacies and limitations.

The elimination of the fast variable can be achieved formally by a two-time scale expansion. For simplicity of notation assume that $\hat{\boldsymbol{\psi}}_{\mathrm{s}}$ and $\hat{\boldsymbol{\psi}}_{\mathrm{f}}$ are scalars, denote the

[1] The text for this section was to a large extent provided by Rupert Klein, Berlin.

slow variable by x and the fast variable by y, and delete the hats on the dimensionless variables. Introduce a fast time scale

$$\tau := t/\epsilon \tag{9.9}$$

and assume an expansion

$$\begin{aligned} x(t;\epsilon) &= x^{(0)}(\tau, t) + \epsilon\, x^{(1)}(\tau, t) + \epsilon^2\, x^{(2)}(\tau, t) + \ldots \\ y(t;\epsilon) &= y^{(0)}(\tau, t) + \epsilon\, y^{(1)}(\tau, t) + \epsilon^2\, y^{(2)}(\tau, t) + \ldots \end{aligned} \tag{9.10}$$

where each $x^{(i)}$ and $y^{(i)}$ depends on both the slow time scale t and the fast time scale τ. Expansion of the time derivatives then yields

$$\begin{aligned} \tfrac{\mathrm{d}}{\mathrm{d}t} x &= \tfrac{1}{\epsilon n}\, x^{(0)}_\tau + (x^{(1)}_\tau + x^{(0)}_t) + \epsilon(x^{(2)}_\tau + x^{(1)}_t) + \ldots \\ \tfrac{\mathrm{d}}{\mathrm{d}t} y &= \tfrac{1}{\epsilon}\, y^{(0)}_\tau + (y^{(1)}_\tau + y^{(0)}_t) + \epsilon(y^{(2)}_\tau + y^{(1)}_t) + \ldots \end{aligned} \tag{9.11}$$

where subscripts denote differentiation. Insert this expansion into the original equations and collect equal powers of ϵ. At order $O(1/\epsilon)$ one finds

$$\begin{aligned} x^{(0)}_\tau &= 0 \\ y^{(0)}_\tau &= g\big(x^{(0)}(\tau, t), y^{(0)}(\tau, t)\big) \end{aligned} \tag{9.12}$$

From the first equation one concludes that $x^{(0)}$ does not depend on the fast time coordinate τ, so that

$$x^{(0)} = x^{(0)}(t) \tag{9.13}$$

Then the second equation reads

$$y^{(0)}_\tau = g\big(x^{(0)}(t), y^{(0)}(\tau, t)\big) \tag{9.14}$$

The variables (τ, t) are *independent* arguments of the expansion functions $x^{(i)}(\tau, t)$ and $y^{(i)}(\tau, t)$. As a consequence, (9.14) is a closed first order ordinary differential equation for the dependence of $y^{(0)}$ on τ *at fixed* t. Given the function $g(\cdot, \cdot)$ this ordinary differential equation must now be solved. This is in general an easier problem than the original one.

The result is, if this solution has been obtained successfully, the functional dependence of $y^{(0)}(\tau, t)$ on the fast variable τ, with t as a parameter. Denote this partial solution by $Y(\tau; t)$ and consider it given.

At order $O(1)$ we then find

$$x^{(1)}_\tau(\tau, t) = -x^{(0)}_t(t) + f\big(x^{(0)}(t), Y(\tau; t)\big) \tag{9.15}$$

$$y^{(1)}_\tau(\tau, t) = -Y_t(\tau; t) + x^{(1)} \frac{\partial g}{\partial x}\big(x^{(0)}(t), Y(\tau; t)\big) + y^{(1)} \frac{\partial g}{\partial y}\big(x^{(0)}(t), Y(\tau; t)\big) \tag{9.16}$$

The evolution equation for $x^{(0)}$ is now obtained by the so-called "sublinear growth" condition: taking again into account that t and τ are *independent* arguments of the expansion functions, we integrate the first equation with respect to τ from, for example, τ_0 to τ_1 at fixed t. This yields

$$x^{(1)}(\tau_1, t) - x^{(1)}(\tau_0, t) = -(\tau_1 - \tau_0)\, x_t^{(0)} + \int_{\tau_0}^{\tau_1} f\big(x^{(0)}(t), Y(\tau; t)\big)\, \mathrm{d}\tau \tag{9.17}$$

For the asymptotic expansion to remain meaningful even for ranges of the fast time of the order $|\tau_1 - \tau_0| = O(1/\epsilon)$, we must require that $|x^{(1)}(\tau_1, t) - x^{(1)}(\tau_0, t)|$ grows slower than linear with $|\tau_1 - \tau_0|$. Otherwise, the perturbation assumption $|\epsilon x^{(1)}| \ll x^{(0)}$ would be violated and the whole approach of collecting "terms of equal order of magnitude" by grouping expressions multiplied by the same power of ϵ would be flawed. This statement is equivalent to requiring that

$$\lim_{|\tau_1 - \tau_0| \to \infty} \left(\frac{x^{(1)}(\tau_1, t) - x^{(1)}(\tau_0, t)}{\tau_1 - \tau_0} \right) = 0 \tag{9.18}$$

which is achieved in (9.17) by setting

$$x_t^{(0)} = \lim_{|\tau_1 - \tau_0| \to \infty} \frac{1}{\tau_1 - \tau_0} \int_{\tau_0}^{\tau_1} f\big(x^{(0)}(t), Y(\tau; t)\big)\, \mathrm{d}\tau \tag{9.19}$$

provided the limit does exist.

This is the classical result of the "method of averaging," which states that one should:

- solve the "fast problem" for fixed slow coordinates (here the result is $Y(\tau; t)$, which solves (9.14) for fixed t);
- substitute the result into the right-hand side of the next order equation (here this means substitution into the right-hand side of (9.16)); and
- average the resulting equation over the fast time scale, which results in the slow time derivative of the leading order function via the sub-linear growth condition (here this result is (9.19)).

Notice that, in general, the result is *not* the heuristic one, which amounts to setting $g(x, y) = 0$, thus determining y as a function of x, and inserting the result into the original equation for x. This heuristic solution is valid only under special conditions regarding the behavior of the fast solution $Y(\tau; t)$ as a function of τ. If, in particular, the fast process is a rapid relaxation to a quasi-steady equilibrium, satisfying $Y_\tau \to 0$ as $\tau \to \infty$, and $Y \to Y_\infty(t) = \lim_{\tau\to\infty} Y(\tau, t)$, then the heuristic approach matches the above more general theory. In that case the τ-dependence of Y vanishes rapidly, the second equation in (9.12) becomes the steady state constraint

$g(x(t), Y_\infty(t)) = 0$ for sufficiently large τ, and the average in (9.19) simply becomes $f(x^{(0)}(t), Y_\infty(t))$.

The multiple scales approach can, however, also deal with more complex cases, in which the fast component exhibits long-time oscillations in τ and the average in (9.19) will generally be very different from the heuristic quasi-equilibrium value $f(x^{(0)}(t), Y_\infty(t))$. In particular, the multi-scale approach provides a formalism for studying nonlinear effects of these fast oscillations on the slow variables. Note that the method relies on asymptotic scale separation and cannot address problems in which a continuous spectrum of scales plays a crucial role.

9.3 Stochastic forcing

Consider the case that the fast components of the system are sufficiently "chaotic." In this case it is not meaningful to calculate the exact evolution of the fast variables, i.e., $y^{(0)}(\tau, t)$ in the previous section, because it depends sensitively on the initial conditions. In this case it is more appropriate to treat the fast variable as a random variable ξ. Its probability distribution may depend on the slow time scale t or the slow variable $x^{(0)}$ (Since there is no explicit t-dependence in $g(x, y)$, the dependence of $y^{(0)}$ on t is only through $x^{(0)}(t)$.) We thus write $\xi(x^{(0)})$. If "ergodicity" is assumed then the time integral in (9.19) can be replaced by an ensemble average

$$x_t^{(0)} = \lim_{|\tau_1-\tau_0|\to\infty} \frac{1}{\tau_1 - \tau_0} \int_{\tau_0}^{\tau_1} f\left(x^{(0)}(t), Y(\tau; t)\right) \mathrm{d}\tau \tag{9.20}$$

$$= \int f\left(x^{(0)}, \xi\right) \rho(\xi | x^{(0)}) \mathrm{d}\xi =: F\left(x^{(0)}\right) \tag{9.21}$$

where $\rho(\xi | x^{(0)})$ is the probability density function of the random variable ξ, conditioned on $x^{(0)}$. One then proceeds as follows:

1. Denote by $\tilde{x}(t)$ the solution of the averaged equation

$$\frac{\mathrm{d}}{\mathrm{d}t} \tilde{x} = F[\tilde{x}] \tag{9.22}$$

This is basically the result given by (9.19), with $\tilde{x}(t)$ corresponding to $x^{(0)}(t)$.

2. The deviation $\Delta x = x - \tilde{x}$ now becomes a random function governed by

$$\frac{\mathrm{d}}{\mathrm{d}t} \Delta x = f(\tilde{x} + \Delta x, \xi) - F(\tilde{x}) \tag{9.23}$$

Taylor expansion in Δx then yields to lowest order

$$\frac{\mathrm{d}}{\mathrm{d}t} \Delta x = \lambda \Delta x + \eta \tag{9.24}$$

where $\lambda := \mathrm{d}f/\mathrm{d}x$ is the derivative resulting from the Taylor expansion and

$$\eta := f(\tilde{x}, \xi) - F(\tilde{x}) \tag{9.25}$$

a rapidly varying stochastic process with zero mean by construction.

The deviation Δx from the averaged solution $\tilde{x}$ is a random function governed by the stochastic differential equation (9.24). This equation is called the *Langevin equation* if η is white noise. The Langevin equation is used to describe heuristically phenomena such as Brownian motion. In this case, the variable Δx describes the position of an ink particle, the stochastic forcing η the random impacts of fluid molecules on the ink particle, and the term $\lambda \Delta x$ with $\lambda < 0$ the viscous drag on the ink particle.

10

Reynolds decomposition

In this chapter we consider a second method to eliminate fast motions, Reynolds averaging. It is much more straightforward. It is the basis for our distinction between large-, medium-, and small-scale motions.

Reynolds averaging does not require any assumptions. It simply decomposes the flow field into a mean and a fluctuating component, and derives the equations for these two components, the Reynolds equations. The equations are coupled. Decoupling requires a closure. Eddy or subgridscale fluxes need to be parametrized in the equations for the mean component. Mean fields need to be regarded as given in the equations for the fluctuating component. The Reynolds equations clearly expose the fact that all scales of motion are coupled and cannot be considered separately, unless closure hypotheses are applied. Other methods that purport to achieve such a decoupling also apply closures, however indirectly. The particulars of the Reynolds equations depend on the choice of independent and dependent variables, and on the form of the governing equations.

10.1 Reynolds decomposition

Any state variable $\psi(\mathbf{x}, t)$ can be decomposed into a mean part $\overline{\psi}(\mathbf{x}, t)$ and a fluctuating part $\psi'(\mathbf{x}, t)$

$$\psi(\mathbf{x}, t) = \overline{\psi}(\mathbf{x}, t) + \psi'(\mathbf{x}, t) \tag{10.1}$$

This decomposition is called Reynolds decomposition. It is *not an approximation.* It is simply a decomposition. There are two ways to define the mean, either as an ensemble average or as a space-time average.

Ensemble average When applying the ensemble mean one assumes that $\psi(\mathbf{x}, t)$ is a random field and defines the mean field as the *expectation value*

$$\overline{\psi}(\mathbf{x}, t) := \langle \psi(\mathbf{x}, t) \rangle \tag{10.2}$$

denoted by cornered brackets, $\langle\psi(\mathbf{x},t)\rangle := E[\psi(\mathbf{x},t)]$. The ensemble mean is a linear operation

$$\langle\psi_1 + \psi_2\rangle = \langle\psi_1\rangle + \langle\psi_2\rangle \tag{10.3}$$
$$\langle\lambda\psi\rangle = \lambda\langle\psi\rangle \qquad \mathit{for} \;\; \lambda \in C \tag{10.4}$$

that commutes with differentiation

$$\langle\partial_t\psi\rangle = \partial_t\langle\psi\rangle \tag{10.5}$$
$$\langle\boldsymbol{\nabla}\psi\rangle = \boldsymbol{\nabla}\langle\psi\rangle \tag{10.6}$$

and has the property

$$\langle\langle\psi_1\rangle\psi_2\rangle = \langle\psi_1\rangle\langle\psi_2\rangle \tag{10.7}$$

These properties imply that the ensemble average is idempotent

$$\langle\langle\psi\rangle\rangle = \langle\psi\rangle \tag{10.8}$$

and

$$\langle\psi'\rangle = 0 \tag{10.9}$$
$$\langle\psi_1\psi_2\rangle = \langle\psi_1\rangle\langle\psi_2\rangle + \langle\psi_1'\psi_2'\rangle \tag{10.10}$$

Thus $\langle f(\psi)\rangle \neq f(\langle\psi\rangle)$ unless f is linear in ψ. If an averaging operation satisfies these properties it is said to satisfy the rules of Reynolds averaging. The ensemble average can be mimicked in the laboratory by repeating experiments.

Space-time average In oceanography the mean is usually defined by an Eulerian space-time mean

$$\overline{\psi}(\mathbf{x},t) := \int \mathrm{d}t' \iiint \mathrm{d}^3x' \;\; \psi(\mathbf{x}',t')\, H(\mathbf{x}-\mathbf{x}',t-t') \tag{10.11}$$

that is the convolution between the field and an appropriate filter function $H(\mathbf{x},t)$. The space-time average is also a linear operation and commutes with differentiation since

$$\begin{aligned}
\partial_t\overline{\psi} &= \int \mathrm{d}t' \iiint \mathrm{d}^3x' \;\; \psi(\mathbf{x}',t')\partial_t H(\mathbf{x}-\mathbf{x}',t-t') \\
&= -\int \mathrm{d}t' \iiint \mathrm{d}^3x' \;\; \psi(\mathbf{x}',t')\partial_{t'} H(\mathbf{x}-\mathbf{x}',t-t') \\
&= \int \mathrm{d}t' \iiint \mathrm{d}^3x' \;\; \partial_{t'}\psi(\mathbf{x}',t')\, H(\mathbf{x}-\mathbf{x}',t-t') = \overline{\partial_t\psi}
\end{aligned} \tag{10.12}$$

if the filter function vanishes at $t = \pm\infty$. The space-time average is, however, not idempotent. Hence

$$\overline{\overline{\psi}} \neq \overline{\psi} \tag{10.13}$$

and

$$\overline{\psi_1 \psi_2} \neq \overline{\psi_1}\,\overline{\psi_2} + \overline{\psi_1' \psi_2'} \tag{10.14}$$

in general. The differences become small if mean and fluctuations are separated by a "spectral gap." In this case the rules of Reynolds averaging are also applied to the space-time average.

The effective width of the filter function H determines which scales contribute to the mean part and which scales contribute to the fluctuating part. Scales that are part of the fluctuating field become part of the mean field when the averaging width becomes smaller.

Moments Under certain fairly general conditions the random field $\psi(\mathbf{x}, t)$ can be described by its moments

$$M_n := \langle \psi^n \rangle \qquad n = 1, 2, 3, \ldots \tag{10.15}$$

or by M_1 and its central moments

$$\hat{M}_n := \langle (\psi - \langle \psi \rangle)^n \rangle \qquad n = 2, 3, \ldots \tag{10.16}$$

10.2 Reynolds equations

To understand the effect of Reynolds decomposition and Reynolds averaging on the equations of oceanic motions consider the simple evolution equation

$$\partial_t \psi + a\,\psi + b\,\psi^2 = 0 \tag{10.17}$$

with a quadratic nonlinearity. Substituting the Reynolds decomposition, averaging, and applying the rules of Reynolds averaging results in

$$\partial_t \overline{\psi} + a\,\overline{\psi} + b\,\overline{\psi}\,\overline{\psi} + b\,\overline{\psi' \psi'} = 0 \tag{10.18}$$

Owing to the quadratic nonlinearity the equation for $\overline{\psi}$ is not closed. It depends on the fluctuating field ψ' through the term $\overline{\psi' \psi'}$. Subtracting the equation for the mean from the original equation gives the equation for ψ'

$$\partial_t \psi' + a\,\psi' + 2b\,\overline{\psi}\,\psi' + b\,\psi'\psi' - b\,\overline{\psi' \psi'} = 0 \tag{10.19}$$

The equation for ψ' is also not closed. Its evolution depends on $\overline{\psi}$. The equations for $\overline{\psi}$ and ψ' are coupled through the nonlinear term. Nonlinearities couple different scales of motion. Note that the Reynolds decomposition does not cause any loss of information. It is a simple decomposition. The original field can be obtained by adding $\overline{\psi}$ and ψ'. The original equation for ψ can be re-established by adding the equations for $\overline{\psi}$ and ψ'.

If one is only interested in $\overline{\psi}$ then one does not need ψ' but $\overline{\psi'\psi'}$. Its evolution equation is

$$\partial_t \overline{\psi'\psi'} + 2a\, \overline{\psi'\psi'} + 4b\, \overline{\psi}\ \overline{\psi'\psi'} + 2b\, \overline{\psi'\psi'\psi'} = 0 \qquad (10.20)$$

and contains the triple product $\overline{\psi'\psi'\psi'}$. The time evolution of this triple product is governed by an equation that contains fourth order products, etc. One obtains an infinite hierarchy of coupled equations for the moments. In terms of the moments $M_n := \langle \psi^n \rangle$, this hierarchy is given by

$$\partial_t M_n + na\, M_n + nb\, M_{n+1} = 0 \qquad n = 1, 2, 3, \ldots \qquad (10.21)$$

and in terms of the central moments $\hat{M}_n := \langle (\psi - \langle \psi \rangle)^n \rangle$ by

$$\partial_t \hat{M}_n + na\, \hat{M}_n + 2nb\, \langle \psi \rangle \hat{M}_n + nb\, \hat{M}_{n+1} - nb\, \hat{M}_{n-1} \hat{M}_2 = 0 \qquad n = 2, 3, \ldots \qquad (10.22)$$

To obtain a finite closed system of equations one has to truncate this hierarchy by a *closure hypothesis*. Closure at the nth order means that all moments of order $n+1$ and higher are expressed in terms of moments of order n or smaller. A first order closure thus expresses $\overline{\psi'\psi'}$ in terms of $\overline{\psi}$.

10.3 Eddy fluxes

The equations of oceanic motions are not of the simple form (10.17) but are generally a much more complicated set of partial differential equations. When Reynolds averaging is applied to these equations, moments of the fluctuating fields appear wherever there is a nonlinearity in the equations. The most important nonlinearity is the advective nonlinearity. If the advective flux is given by $\psi \mathbf{u}$ then the averaged advective flux is given by

$$\overline{\psi \mathbf{u}} = \overline{\psi}\ \overline{\mathbf{u}} + \overline{\psi' \mathbf{u}'} \qquad (10.23)$$

In addition to the flux $\overline{\psi}\ \overline{\mathbf{u}}$ there occurs an eddy or turbulent flux

$$\mathbf{I}_\psi^{\text{eddy}} = \overline{\psi' \mathbf{u}'} \qquad (10.24)$$

The qualifier "eddy" or "turbulent" is used because the fluctuating fields are generally regarded as turbulent eddies. The eddy fluxes appear in the averaged balance equation for the same reason that molecular fluxes appear in the basic equations. There a property was transported by a velocity that differed from the fluid velocity (see equations (3.13) and (3.14)). The difference had to be accounted for by a molecular flux. Here, a property is transported by a velocity $\mathbf{u}$ that differs from the mean velocity $\overline{\mathbf{u}}$. The difference has to be accounted for by an eddy flux. Note that an eddy flux $\overline{\rho' \mathbf{u}'}$ will appear in the Reynolds averaged continuity equation (3.18) although there is no molecular flux in the original equation. One expects the eddy fluxes, which are caused by macroscopic fluid motions, to be larger than the molecular fluxes, which are caused by microscopic molecular motions. One therefore generally neglects the molecular fluxes in Reynolds averaged equations and only retains the eddy fluxes.

The basic equations of oceanic motion contain nonlinearities in other places as well. The pressure force $-\rho^{-1}\nabla p$ is the product of ρ^{-1} and ∇p. The thermodynamic coefficients and the phenomenological coefficients are generally nonlinear functions of (p, T, S).

Reynolds decomposition must be applied also to tracer equations, theorems, and other quantities of interest. The mean of any nonlinear quantity then contains eddy contributions. One example is the specific kinetic energy $e_{\text{kin}} = \mathbf{u} \cdot \mathbf{u}/2$. Its mean is given by

$$\overline{e_{\text{kin}}} = \frac{1}{2}\overline{\mathbf{u}} \cdot \overline{\mathbf{u}} + \frac{1}{2}\overline{\mathbf{u}' \cdot \mathbf{u}'} \tag{10.25}$$

and consists of two terms, one describing the mean kinetic energy of the mean flow and the other describing the mean kinetic energy of the fluctuating flow.

The specific form of the Reynolds equations and eddy terms depends on the specific set of dependent variables that one chooses (e.g., height versus isopycnal coordinates), on the specific set of dependent variables that one decomposes (e.g., quantities per unit mass versus per unit volume), and on the specific form of equations that one averages (e.g., flux versus advective form). We will give explicit results for the equations in the Boussinesq approximation in Section 12.1.

To close the Reynolds equations for the mean flow one needs to express the eddy fluxes and moments in terms of mean flow quantities. Such a specification is called a *parametrization*. We will discuss the standard diffusion parametrization in Section 12.2. It is emphasized that there exists no general theory to *derive* such parametrizations. This is one of the major unsolved problems in oceanography and an area of active research.

To close the equations for the fluctuating component one needs to prescribe the mean component or regard it as given. This is a closure as well since in actuality the mean component depends on and evolves with the fluctuating component.

10.4 Background and reference state

Another decomposition that is used in this book is the decomposition into a background state $\tilde{\psi}$ and perturbations $\delta\psi$ about it. This decomposition is substituted into the governing equations. A dynamic balance for the background state is then *assumed* and subtracted out. In the resulting equation, the background state is assumed to be given and to be a solution of the background dynamic balance. The key step is the specification of the dynamic balance of the background state. Different choices present themselves, as discussed in Section 18.2. An example of such a decomposition was our analysis of linear waves on a motionless stratified background state in Chapter 8. The background stratification was assumed to be a mechanical equilibrium solution of the ideal fluid equations.

Often the decomposition reduces to a mere decomposition into a specified reference state $\psi_{\rm r}$ and the deviation $\psi_{\rm d}$ therefrom. One typically subtracts out such a reference state when equations contain terms of vastly different magnitude. We will do so in Chapter 11 when we derive the Boussinesq approximation. We subtract out a hydrostatically balanced reference state from the vertical momentum balance in order to eliminate the two largest terms in the balance, without ascribing any physical significance to the reference state.

The decomposition into background state and perturbation differs from the Reynolds decomposition in that it does not require any closure. Indeed, nothing is really gained. Instead of having to calculate the evolution of ψ one has to calculate the evolution of the deviation $\delta\psi$ from a given state $\tilde{\psi}$. The decomposition does require the specification of the background dynamics and one generally strives for a physically meaningful balance, as discussed in section 18.2. Nevertheless, the background state usually does not represent the mean state. The perturbations usually develop mean components as well. If the background state is regarded as the mean state then the specification of its dynamic balance represents a closure.

10.5 Boundary layers

Reynolds averaging must be applied also to boundaries. In general, these boundaries also consist of a mean and fluctuating component. Reynolds decomposition and averaging then requires expanding the boundary condition at the actual surface around the mean surface. This introduces nonlinearities of infinite order. We demonstrate this behavior for the boundary equation $(\partial_t + \mathbf{u} \cdot \boldsymbol{\nabla})G = 0$ (see Tables 4.1 and 4.2). If we write the surface as $G = \eta(\mathbf{x}_h, t) - z = 0$ where η is the surface elevation then this boundary equation takes the form

$$\partial_t \eta + \mathbf{u}_h \cdot \boldsymbol{\nabla}_h \eta - w = 0 \quad \text{at} \quad z = \eta \tag{10.26}$$

or

$$\partial_t \eta(\mathbf{x}_h, t) + \mathbf{u}_h(\mathbf{x}_h, z = \eta, t) \cdot \boldsymbol{\nabla}_h \eta(\mathbf{x}_h, t) - w(\mathbf{x}_h, z = \eta, t) = 0 \tag{10.27}$$

Reynolds decomposition $\eta = \overline{\eta} + \eta'$, $\mathbf{u} = \overline{\mathbf{u}}_h + \mathbf{u}'$, and $w = \overline{w} + w'$ leads to an expression where the fluctuating component η' of the surface elevation appears as an argument of $\mathbf{u}_h$ and w. This expression can hence not be Reynolds-averaged in any simple way. When it is Taylor-expanded about the mean surface elevation $z = \overline{\eta}$, the Reynolds equations take the form

$$\partial_t \overline{\eta} = \overline{w} + \overline{\eta' \partial_z w'} + \ldots \quad \text{at } z = \overline{\eta} \tag{10.28}$$

$$\partial_t \eta' = w' + \eta' \partial_z \overline{w} + \eta' \partial_z w' - \overline{\eta' \partial_z w'} + \ldots \quad at\ z = \overline{\eta} \tag{10.29}$$

They are infinite series and contain moments of any order (for ease of notation, we neglected the advective term $\mathbf{u}_h \cdot \nabla \eta$ in the equations). The mean and fluctuating components are coupled through these boundary conditions. Decoupling again requires closure hypotheses and specifications. If $\eta'/H \ll 1$ or $\eta'/D \ll 1$ where H is the ocean depth and D the vertical scale of the flow then asymptotic expansion methods can be applied.

Such Taylor expansions are generally not applied to the flux boundary conditions. Instead, one introduces the concept of a boundary layer. Water, momentum, and energy are transported across interfaces by molecular processes. They are also transported by molecular diffusion in thin molecular boundary layers on both sides of the interface. Further away from the interface, within a turbulent boundary layer, these quantities are transported by turbulent eddies. If one assumes that the fluxes through these layers are constant and equal to the flux across the interface one arrives at the concept of the *constant flux layer*. The general conditions for the validity of this concept can be derived by integrating the balance equations over the constant flux layers surrounding the interface (see Section 3.9 and Figure 3.1)

$$\frac{\mathrm{d}}{\mathrm{d}t} \int\int\int_{\Delta V} \mathrm{d}^3 x \rho \psi = \int\int\int_{\Delta V} \mathrm{d}^3 x S_\psi - \int\int_{\delta \Delta V} \mathrm{d}^2 x \mathbf{n} \cdot \tilde{\mathbf{I}}_\psi \tag{10.30}$$

with $\tilde{\mathbf{I}}_\psi := \mathbf{I}_\psi - \rho \mathbf{v}$ where $\mathbf{n}$ and $\mathbf{v}$ are the normal vector and the velocity of the surface $\delta \Delta V$ enclosing the volume ΔV. These conditions are:

- Stationarity. No accumulation of the transported quantity ψ within the constant flux layer. The term on the left-hand side then does not contribute.
- No sources and sinks. The first term on the right-hand side then does not contribute.
- Horizontal homogeneity. No fluxes through the side boundaries of ΔV. Then only fluxes through the top and bottom boundary of ΔV contribute to the last term on the right-hand side (see Figure 3.1) and one must require

$$\mathbf{n}^1 \cdot \tilde{\mathbf{I}}^1_\psi + \mathbf{n}^2 \cdot \tilde{\mathbf{I}}^2_\psi = 0 \tag{10.31}$$

The boundary condition for Reynolds-averaged equations that do not resolve the constant flux layer is the continuity of the flux $\overline{\mathbf{n}} \cdot \overline{\tilde{\mathbf{I}}}_\psi$ across the interface.

11

Boussinesq approximation

Here we present the Boussinesq approximation. It consists of two steps: the anelastic approximation and a set of additional approximations exploiting the characteristics of the oceanic density field.

In many oceanographic problems one is only interested in time scales slower than those associated with acoustic phenomena. The simplest and most straightforward way to eliminate sound waves is to regard sea water as incompressible, i.e., to consider the limit $c^2 \to \infty$ in the basic equations of oceanic motions. While this limit is adequate for many purposes compressibility (though not sound waves) plays a crucial role in the deep thermohaline circulation. For this reason one introduces the anelastic approximation, which assumes that the pressure field adjusts instantaneously. The anelastic approximation eliminates sound waves but does not remove compressibility effects. It is the lowest order in an expansion with respect to the small parameter $\epsilon_a := T_a/T$ where T_a is the fast time scale of acoustic waves and T the slow time scale of the motions under consideration. The main structural change in the equations brought about by the anelastic approximation is that the pressure and vertical velocity become diagnostic variables.

The density field of the ocean is characterized by the facts that:

- the density of the ocean at a point does not differ very much from a reference density; and
- the reference density does not change very much from surface to bottom.

These two facts are exploited in an additional set of approximations. Together with the anelastic approximation they constitute the *Boussinesq approximation*. Its main results are:

- the velocity field can be assumed to be non-divergent; mass conservation is replaced by volume conservation; and
- the density in the momentum equation can be replaced everywhere by a constant reference value ρ_0, except when multiplied by g_0.

Note that the anelastic approximation may also be applied to the atmosphere but not the approximations to the density field.

11.1 Anelastic approximation

The starting point is the basic equations for oceanic motions in the $\{p, \theta, S\}$-representation. First, one introduces a motionless hydrostatically balanced reference state, denoted by a subscript "r," that depends only on the vertical coordinate z

$$\mathbf{u}_\mathrm{r} = 0 \tag{11.1}$$

$$\theta_\mathrm{r} = \mathit{constant} = (0°\mathrm{C}) \tag{11.2}$$

$$S_\mathrm{r} = \mathit{constant} = (35\ \mathrm{psu}) \tag{11.3}$$

$$p_\mathrm{r} = p_\mathrm{r}(z) \tag{11.4}$$

$$\rho_\mathrm{r} = \rho(p_\mathrm{r}, \theta_\mathrm{r}, S_\mathrm{r}) \tag{11.5}$$

Though the reference state is hydrostatically balanced, $\mathrm{d}p_\mathrm{r}/\mathrm{d}z = -g_0\rho_\mathrm{r}$, it is not a solution of the basic equations of oceanic motion. It has a constant potential density $\rho^\mathrm{r}_\mathrm{pot} = \rho(p_0, \theta_\mathrm{r}, S_\mathrm{r})$ rather than constant temperature T. Deviations from this reference state are denoted by a subscript "d." The pressure equation then becomes

$$\frac{\mathrm{D}}{\mathrm{D}t} p_\mathrm{d} + \mathbf{u}_\mathrm{d} \cdot \nabla p_\mathrm{r} = -\rho c^2 \nabla \cdot \mathbf{u}_\mathrm{d} + D^\mathrm{mol}_\mathrm{p} \tag{11.6}$$

The anelastic approximation consists of taking the steady state or diagnostic limit of this equation. It neglects the term $\mathrm{D}p_\mathrm{d}/\mathrm{D}t$. The deviation pressure adjusts instantaneously. This is the heuristic way to eliminate fast variables (see Section 9.2). In this case "fast" sound waves are eliminated. This can be seen explicitly by neglecting the term $\partial_t \delta p$ in (8.6). The vertical eigenvalue problem (8.25) and (8.26) then reduces to

$$\frac{\tilde{\rho} g_0}{\tilde{c}^2} \psi - \tilde{\rho} \frac{\mathrm{d}\psi}{\mathrm{d}z} = \epsilon \phi \tag{11.7}$$

$$\tilde{\rho}(N^2 - \omega^2)\, \psi = \frac{\mathrm{d}\phi}{\mathrm{d}z} + \frac{g_0}{\tilde{c}^2} \phi \tag{11.8}$$

which transforms into

$$-\tilde{\rho} \frac{\mathrm{d}\hat{\psi}}{\mathrm{d}z} = \epsilon \hat{\phi} \tag{11.9}$$

$$\tilde{\rho}(N^2 - \omega^2)\, \hat{\psi} = \frac{\mathrm{d}\hat{\phi}}{\mathrm{d}z} \tag{11.10}$$

Its eigenvalues $\epsilon_n(\omega)$ are identical to the ones for an incompressible fluid ($c^2 = \infty$).

The anelastic approximation results in the diagnostic relation

$$\nabla \cdot \mathbf{u} = -\frac{\mathbf{u} \cdot \nabla p_r}{\rho c^2} + \frac{d_p^{\text{mol}}}{\rho c^2} \tag{11.11}$$

There is no prognostic equation for the pressure anymore. The pressure must be determined from (11.11), which is a constraint on the velocity field. To maintain mass conservation one has to neglect the deviation pressure in the equation of state, which then takes the form

$$\rho = \rho(p_r, \theta_r + \theta_d, S_r + S_d) \tag{11.12}$$

The deviation pressure does not affect the density.

The basic result of the anelastic approximation is the elimination of sound waves. The number of prognostic equations is reduced by two. There is no prognostic pressure equation and only two of the momentum equations are independent because of the constraint (11.11).

11.2 Additional approximations

Next we introduce approximations that arise from the fact that the oceanic density field does not differ very much from the reference density and that the reference density does not change very much from the surface to the bottom. The latter fact implies that the depth scale D of the oceanic flow and the depth H of the ocean are much smaller than the adiabatic depth scale $\tilde{D}_a = c_r^2/g_0 = O(200\,\text{km})$ of the density field.

First consider the terms on the right-hand side of the divergence equation (11.11). The first term scales as

$$\frac{|\mathbf{u} \cdot \nabla p_r|}{\rho c^2} \sim \frac{W}{\tilde{D}_a} \tag{11.13}$$

where W is the scale of the vertical velocity component. It is hence $D/\tilde{D}_a$ times smaller than the corresponding term $\partial_z w \sim W/D$ of the velocity divergence and therefore neglected as part of the Boussinesq approximation.

The second term on the right-hand side of (11.11) is given by

$$\frac{\left|D_p^{\text{mol}}\right|}{\rho c^2} = \tilde{\alpha} \frac{\mathrm{D}}{\mathrm{D}t} \theta_d - \tilde{\beta} \frac{\mathrm{D}}{\mathrm{D}t} S_d \tag{11.14}$$

and therefore scales as

$$\frac{\left|D_p^{\text{mol}}\right|}{\rho c^2} \sim \frac{1}{\rho_r} \left| \frac{\mathrm{D}}{\mathrm{D}t} \rho_d \right| \sim \frac{W}{D} \frac{\rho_d}{\rho_r} \tag{11.15}$$

Since the oceanic density field deviates little from the reference density, $\rho_{\mathrm{d}}/\rho_{\mathrm{r}} = O(10^{-3})$, this term is $\rho_{\mathrm{d}}/\rho_{\mathrm{r}}$ times smaller than the corresponding term $\partial_z w \sim W/D$ of the velocity divergence. It is therefore also neglected as part of the Boussinesq approximation. Hence one arrives at

$$\nabla \cdot \mathbf{u} = 0 \tag{11.16}$$

The velocity field is non-divergent or solenoidal. Material volumes are conserved. The flow, as opposed to the fluid, is incompressible. This condition is usually referred to as the "incompressibility condition."

The condition $H \ll \tilde{D}_{\mathrm{a}}$ implies that the reference density across the oceanic depth range does not differ much from a constant value ρ_0. Hence $\rho_{\mathrm{d}} \ll \rho_{\mathrm{r}} \approx \rho_0$ and one can replace ρ by a constant density ρ_0 wherever it occurs as a factor. The momentum equation, with the hydrostatically balanced reference state subtracted out, thus takes the form

$$\rho_0 \frac{\mathrm{D}}{\mathrm{D}t}\mathbf{u}_{\mathrm{d}} = -\nabla p_{\mathrm{d}} - \rho_0 2\boldsymbol{\Omega} \times \mathbf{u}_{\mathrm{d}} - \rho_{\mathrm{d}} \nabla \phi + \nabla \cdot \boldsymbol{\sigma}^{\mathrm{mol}} \tag{11.17}$$

The basic results of these approximations to the density field are that mass conservation is replaced by volume conservation and that density-induced nonlinearities are eliminated from the inertial terms in the momentum equation. These approximations simplify the equations. No additional waves are eliminated.

11.3 Equations

Implementing the above approximations one arrives at the equations of motion in the Boussinesq approximation. We write these equations in terms of the original variables. The reference state was only introduced to facilitate the approximations. Thus

$$\rho_0 \frac{\mathrm{D}}{\mathrm{D}t}\mathbf{u} = -\nabla p - \rho_0 2\boldsymbol{\Omega} \times \mathbf{u} - \rho \nabla \phi + \nabla \cdot \boldsymbol{\sigma}^{\mathrm{mol}} \tag{11.18}$$

$$\nabla \cdot \mathbf{u} = 0 \tag{11.19}$$

$$\frac{\mathrm{D}}{\mathrm{D}t}\theta = D_\theta^{\mathrm{mol}} \tag{11.20}$$

$$\frac{\mathrm{D}}{\mathrm{D}t}S = D_S^{\mathrm{mol}} \tag{11.21}$$

and

$$\rho = \rho(p_{\mathrm{r}}, \theta, S) \tag{11.22}$$

Since $p_{\mathrm{r}} = p_{\mathrm{r}}(z)$ this can also be written as

$$\rho = \rho(z, \theta, S) \tag{11.23}$$

Note that ρ multiplies the gravitational acceleration, not ρ_0. Also, the reference state or any other hydrostatically balanced background state can be subtracted out of the momentum balance.

The above equations contain no explicit equation for the pressure field. It has to be calculated from the constraint on the velocity field. Though physically transparent the equations are thus not very amenable to solution algorithms. A mathematically better posed set of equations is obtained by taking the divergence of the momentum equation. This yields a three-dimensional Poisson equation for the pressure

$$\Delta p = -\rho_0 \nabla \cdot (\mathbf{u} \cdot \nabla \mathbf{u}) + \rho_0 2\boldsymbol{\Omega} \cdot (\nabla \times \mathbf{u}) - \nabla \cdot (\rho \nabla \phi) + \nabla \cdot (\nabla \cdot \boldsymbol{\sigma}^{\text{mol}}) \tag{11.24}$$

This equation can be used to replace one component of the momentum equation, for example the third one. One then has explicit prognostic equations for (u_1, u_2, θ, S) and explicit diagnostic equations for (u_3, ρ, p).

11.4 Theorems

Equations for other quantities can be derived in a straightforward manner, using the equations in the Boussinesq approximation.

Density The equation for the density follows from differentiating (11.22) and is given by

$$\frac{\mathrm{D}}{\mathrm{D}t}\rho = -c^{-2}\rho_0 \mathbf{u} \cdot \nabla \phi - \rho_0 \tilde{\alpha} D_\theta^{\text{mol}} + \rho_0 \tilde{\beta} D_S^{\text{mol}} \tag{11.25}$$

The thermodynamic coefficients are evaluated at (p_r, θ, S). As already pointed out mass is no longer conserved. Molecular diffusion processes and changes in (background) pressure lead to density changes. In the basic equations for oceanic motions, these changes were balanced by a change in volume, so that total mass was conserved. This change of volume is suppressed in the Boussinesq approximation and mass is no longer conserved. The consequences of this violation need to be considered carefully. The terms on the right-hand side are important for processes such as water mass formation. If these are neglected one arrives at the material conservation of density, $\mathrm{D}\rho/\mathrm{D}t = 0$. It is emphasized that material conservation of density is not a consequence of mass conservation $\partial_t \rho + \nabla \cdot (\rho \mathbf{u}) = 0$, which is replaced by volume conservation $\nabla \cdot \mathbf{u} = 0$ in the Boussinesq approximation but a consequence of assuming that the adiabatic and diabatic processes on the right-hand side of (11.25) are neglected.

Temperature The temperature in the Boussinesq approximation is given by $T = T(p_r, \theta, S)$. Hence

$$\rho_0 c_p \frac{\mathrm{D}}{\mathrm{D}t} T = -\rho_0 c_p \Gamma \rho_0 \mathbf{u} \nabla \phi + D_h^{\mathrm{mol}} \tag{11.26}$$

Again all the thermodynamic coefficients are evaluated at (p_r, θ, S).

Vorticity The vorticity equation in the Boussinesq approximation becomes

$$\rho_0 \frac{\mathrm{D}}{\mathrm{D}t} \omega = \rho_0 (2\boldsymbol{\Omega} + \omega) \cdot \nabla \mathbf{u} - \nabla \rho \times \nabla \phi + \nabla \times \mathbf{F}^{\mathrm{mol}} \tag{11.27}$$

The divergence term vanishes.

Potential vorticity Ertel's potential vorticity

$$q_\psi = \frac{1}{\rho_0} (\omega + 2\boldsymbol{\Omega}) \cdot \nabla \psi \tag{11.28}$$

is governed by

$$\partial_t (\rho_0 q_\psi) + \nabla \cdot (\rho_0 q_\psi \mathbf{u} + \mathbf{I}_q) = 0 \tag{11.29}$$

where

$$\mathbf{I}_q := \frac{\rho \nabla \phi \times \nabla \psi}{\rho_0} - \frac{1}{\rho_0} \mathbf{F}^{\mathrm{mol}} \times \nabla \psi - (\omega + 2\boldsymbol{\Omega}) \dot{\psi} \tag{11.30}$$

and $\dot{\psi} := \mathrm{D}\psi/\mathrm{D}t$. Again, Ertel's potential vorticity is conserved integrally and becomes materially conserved when the divergence of $\mathbf{I}_q$ vanishes.

Energy The equation for the kinetic energy density $\rho_0 u_i u_i / 2$ is given by

$$\partial_t \left(\frac{1}{2} \rho_0 u_i u_i \right) + \partial_j \left(\frac{1}{2} \rho_0 u_i u_i u_j + p u_j - u_i \sigma_{ij}^{\mathrm{mol}} \right) = -\rho u_i \partial_i \phi - \sigma_{ij}^{\mathrm{mol}} D_{ij} \tag{11.31}$$

where the first term on the right describes the exchange with the potential energy and the second term the irreversible dissipation into internal energy by molecular friction. The potential energy density $\rho\phi$ satisfies

$$\partial_t (\rho \phi) + \partial_j (\rho \phi u_j) = \rho u_i \partial_i \phi + \phi \frac{\mathrm{D}}{\mathrm{D}t} \rho \tag{11.32}$$

There appears a new source term on the right-hand side that is again due to the fact that density changes are not compensated for by volume changes to conserve mass. Again, the implications need to be considered carefully.

11.5 Dynamical significance of two-component structure

Sea water is a two-component system. Both potential temperature (or temperature) and salinity are required to specify:

- the density $\rho(p_r, \theta, S)$;
- thermodynamic coefficients like the specific heat $c_p(p_r, \theta, S)$;
- molecular diffusion coefficients like $\lambda(p_r, \theta, S)$;
- the molecular heat and salt fluxes, since they depend on the gradients of T and S; and
- the boundary conditions.

The two-component structure manifests itself in specific phenomena such as double diffusion, which relies on the different molecular diffusion coefficients for heat and salt (see Section 21.2). One can ask under what circumstances sea water behaves as a one-component system. Sufficient conditions are:

- the neglect of the molecular diffusion of heat and salt;
- a non-thermobaric equation of state

$$\tilde{\alpha}\partial c^{-2}/\partial S + \tilde{\beta}\partial c^{-2}/\partial\theta = 0 \tag{11.33}$$

- boundary conditions that do not require θ and S.

In this case, the density is given by (see Table 2.1)

$$\rho = \rho(p_r, \rho_{pot}) \tag{11.34}$$

and the two prognostic equations for θ and S can be replaced by

$$\frac{D}{Dt}\rho_{pot} = 0 \tag{11.35}$$

The temperature–salinity mode of motion is eliminated.

Instead of assuming a non-thermobaric equation of state, one often makes the stronger assumption $c^{-2} = c^{-2}(p_r, \theta_r, S_r)$ or

$$\partial c^{-2}/\partial\theta = 0 \tag{11.36}$$

$$\partial c^{-2}/\partial S = 0 \tag{11.37}$$

Both terms in (11.33) vanish separately. Then

$$\frac{D}{Dt}\rho = -c_r^{-2}\rho_0 \mathbf{u} \cdot \nabla\phi \tag{11.38}$$

where $c_r = c(p_r, \theta_r, S_r)$. The density is in this case determined by a prognostic (evolution) equation, not by a (diagnostic) equation of state. The density equation (11.38) together with the momentum equation (11.18) and the divergence equation (11.19) form a closed set of equations for $(\mathbf{u}, p, \rho)$. No equation of state is needed. The deviation density ρ_d is materially conserved, $D\rho_d/Dt = 0$, and equal to ρ_d^{pot}.

We will generally employ one-component dynamics by assuming $c^{-2} = \tilde{c}^{-2} = c^{-2}(p_{\mathrm{r}}, \tilde{\theta}, \tilde{S})$ where the variables with the tilde represent a hydrostatically balanced background state. Then the density is determined by the dynamical equation

$$\frac{\mathrm{D}}{\mathrm{D}t}\rho = -\tilde{c}^{-2}\rho_0 w \tag{11.39}$$

and the deviation density by

$$\frac{\mathrm{D}}{\mathrm{D}t}\delta\rho - w\frac{\rho_0}{g_0}N^2 = 0 \tag{11.40}$$

where $N^2 := -\frac{g_0}{\rho_0}\frac{\mathrm{d}\tilde{\rho}}{\mathrm{d}z} - \frac{g_0^2}{\tilde{c}^2}$ is the buoyancy frequency.

12

Large-scale motions

The equations in the Boussinesq approximation are assumed to describe all oceanic motions except sound waves. The further approximations of the Boussinesq equations are based on a triple decomposition into large-, medium-, and small-scale motions accomplished by two Reynolds decompositions. One Reynolds decomposition separates large- from medium-scale motions and the second one medium- from small-scale motions. The equations for large-scale motions have to parametrize the eddy fluxes of the medium- and small-scale motions. The equations for medium-scale motions have to specify the large-scale background and the eddy fluxes by the small-scale motions. The equations for the small-scale motions have to specify the large- and medium-scale background. Their subgridscale fluxes are the molecular fluxes.

In this chapter we derive the equations for large-scale motions obtained by Reynolds averaging the Boussinesq equations. We introduce the standard parametrization of eddy fluxes in terms of eddy diffusion and eddy viscosity coefficients, and the standard boundary conditions. The final set of equations is given in pseudo-spherical coordinates.

12.1 Reynolds average of Boussinesq equations

Equations Reynolds decomposition and averaging of the Boussinesq equations lead to

$$\rho_0 \left(\partial_t + \overline{\mathbf{u}} \cdot \boldsymbol{\nabla}\right) \overline{\mathbf{u}} + \rho_0 2\boldsymbol{\Omega} \times \overline{\mathbf{u}} = -\boldsymbol{\nabla}\overline{p} - \overline{\rho}\boldsymbol{\nabla}\phi + \boldsymbol{\nabla} \cdot \boldsymbol{\sigma}^{\text{eddy}} \tag{12.1}$$

$$\boldsymbol{\nabla} \cdot \overline{\mathbf{u}} = 0 \tag{12.2}$$

$$\left(\partial_t + \overline{\mathbf{u}} \cdot \boldsymbol{\nabla}\right) \overline{\theta} + \boldsymbol{\nabla} \cdot \mathbf{I}_\theta^{\text{eddy}} = 0 \tag{12.3}$$

$$\left(\partial_t + \overline{\mathbf{u}} \cdot \boldsymbol{\nabla}\right) \overline{S} + \boldsymbol{\nabla} \cdot \mathbf{I}_S^{\text{eddy}} = 0 \tag{12.4}$$

$$\overline{\rho} = \rho(p_{\mathrm{r}}, \overline{\theta}, \overline{S}) + \frac{1}{2}\frac{\partial^2 \rho}{\partial \theta^2}\overline{\theta'\theta'} + \frac{\partial^2 \rho}{\partial \theta \partial S}\overline{\theta' S'} + \frac{1}{2}\frac{\partial^2 \rho}{\partial S^2}\overline{S'S'} + \dots \tag{12.5}$$

where

$$\boldsymbol{\sigma}^{\text{eddy}} := -\rho_0 \overline{\mathbf{u}'\mathbf{u}'} \quad \text{Reynolds stress tensor} \tag{12.6}$$

$$\mathbf{I}_\theta^{\text{eddy}} := \overline{\mathbf{u}'\theta'} \quad \text{eddy temperature flux} \tag{12.7}$$

$$\mathbf{I}_S^{\text{eddy}} := \overline{\mathbf{u}'S'} \quad \text{eddy salinity flux} \tag{12.8}$$

are the eddy induced fluxes. The molecular fluxes are neglected. The viscous stress tensor $\boldsymbol{\sigma}^{\text{mol}}$ is thus replaced by the Reynolds stress tensor $\boldsymbol{\sigma}^{\text{eddy}}$. The molecular source terms D_θ^{mol} and D_S^{mol} are replaced by the divergence of the eddy temperature and eddy salinity fluxes $\mathbf{I}_\theta^{\text{eddy}}$ and $\mathbf{I}_S^{\text{eddy}}$. Usually, the eddy contributions to the mean density are neglected.

Reynolds averaging also needs to be applied to theorems and any other equations of interest.

Density If $\overline{\rho} = \rho(p_r, \overline{\theta}, \overline{S})$ is assumed, the evolution of the mean density is given by

$$(\partial_t + \overline{\mathbf{u}} \cdot \boldsymbol{\nabla})\, \overline{\rho} = -\overline{c}^{-2} \rho_0 \overline{\mathbf{u}} \cdot \boldsymbol{\nabla}\phi + D_\rho^{\text{eddy}} \tag{12.9}$$

where

$$D_\rho^{\text{eddy}} := \overline{\rho}\overline{\alpha} \boldsymbol{\nabla} \cdot \mathbf{I}_\theta^{\text{eddy}} - \overline{\rho}\overline{\beta} \boldsymbol{\nabla} \cdot \mathbf{I}_S^{\text{eddy}} \tag{12.10}$$

All coefficients are evaluated at $(p_r, \overline{\theta}, \overline{S})$. Note that the right-hand side of the density equation is not the divergence of a flux.

Tracer The concentration c_{A} of a conservative tracer A is materially conserved, $\text{D}c_{\text{A}}/\text{D}t = 0$. Reynolds averaging results in

$$(\partial_t + \overline{\mathbf{u}} \cdot \boldsymbol{\nabla})\, \overline{c}_{\text{A}} = -\boldsymbol{\nabla} \cdot \left(\overline{\mathbf{u}'c'_{\text{A}}}\right) \tag{12.11}$$

The mean concentration is not materially conserved under advection with the mean velocity. The divergence of the eddy flux $\overline{\mathbf{u}'c'_{\text{A}}}$ appears as a source term.

Energy The mean kinetic energy density consists of two contributions, $\rho_0 \overline{u_i u_i}/2 = \rho_0 \overline{u}_i \overline{u}_i/2 + \rho_0 \overline{u'_i u'_i}/2$, which are governed by

$$\partial_t \left(\tfrac{1}{2}\rho_0 \overline{u}_i \overline{u}_i\right) + \partial_j \left(\tfrac{1}{2}\rho_0 \overline{u}_i \overline{u}_i \overline{u}_j + \overline{p}\overline{u}_j - \overline{u}_i \sigma_{ij}^{\text{eddy}}\right) = -\overline{\rho}\overline{u}_i \partial_i \phi - \sigma_{ij}^{\text{eddy}} D_{ij} \tag{12.12}$$

$$\partial_t \left(\tfrac{1}{2}\rho_0 \overline{u'_i u'_i}\right) + \partial_j \left(\tfrac{1}{2}\rho_0 \overline{u'_i u'_i}\,\overline{u}_j + \tfrac{1}{2}\rho_0 \overline{u'_i u'_i u'_j} + \overline{p' u'_j}\right) = -\overline{\rho' u'_i}\partial_i \phi + \sigma_{ij}^{\text{eddy}} D_{ij} \tag{12.13}$$

The "energy dissipation" term $\boldsymbol{\sigma}^{\text{eddy}} : \mathbf{D}$ describes the exchange between the two components of the mean kinetic energy density.

12.2 Parametrization of eddy fluxes

In the interior of the ocean the eddy fluxes are often parametrized in analogy to the molecular fluxes. We treat the eddy fluxes of scalars and momentum separately.

Eddy diffusivities For scalars $\psi = \theta, S, \ldots$ one assumes

$$\overline{\psi' u_i'} = -K_{ij} \partial_j \overline{\psi} \tag{12.14}$$

Here $\mathbf{K}$ is the eddy or turbulent diffusion tensor. It is a property of the flow field, not a property of the fluid. Heuristically, this diffusion parametrization is based on the *mixing length theory*. It asserts that a fluid parcel maintains its properties over a distance L when displaced from its original position. Then it assumes the value of its surroundings. The flux is thus given by

$$\overline{\psi' \mathbf{u}'} = -L (\overline{\mathbf{u}' \cdot \mathbf{u}'})^{1/2} \frac{\partial \overline{\psi}}{\partial \mathbf{x}} \tag{12.15}$$

and the diffusion coefficient by

$$K = L (\overline{\mathbf{u}' \cdot \mathbf{u}'})^{1/2} \tag{12.16}$$

The mixing length L is the analog of the mean free path of a molecule in statistical mechanics. It is, however, stressed that there is no general theory that justifies the diffusion form and allows the calculation of the diffusion tensor $\mathbf{K}$. The diffusion tensor does not need to be constant. Nor does it need to be positive definite. There can be up-gradient fluxes. No general values can be assigned to the components of $\mathbf{K}$. The value will depend on the parametrized flow field and on the averaging scale.

The eddy diffusion tensor can be decomposed into its symmetric and antisymmetric parts

$$K_{ij}^{\text{s}} := \frac{1}{2}(K_{ij} + K_{ji}) \tag{12.17}$$

$$K_{ij}^{\text{a}} := \frac{1}{2}(K_{ij} - K_{ji}) \tag{12.18}$$

The antisymmetric part describes a "skew flux" normal to the mean gradient of the scalar. Part of this flux is non-divergent and does not affect the mean scalar field. The remaining divergent part may be expressed as a simple advection

$$\partial_i \left(K_{ij}^{\text{a}} \partial_j \overline{\psi} \right) = u_j^* \partial_j \overline{\psi} \tag{12.19}$$

with a skew velocity

$$u_j^* = \partial_i K_{ij}^{\mathrm{a}} \tag{12.20}$$

The symmetric part of the diffusion tensor can be diagonalized by a suitable coordinate rotation. Different assumptions are made about the orientation of the principal axes:

1. Principal axes in horizontal and vertical directions

$$\mathbf{K}^{\mathrm{s}} = \begin{pmatrix} K_{\mathrm{h}} & 0 & 0 \\ 0 & K_{\mathrm{h}} & 0 \\ 0 & 0 & K_{\mathrm{v}} \end{pmatrix} \tag{12.21}$$

   where K_{h} is the horizontal and K_{v} the vertical diffusion coefficient.
2. Horizontal diffusion causes fluxes across isopycnal surfaces if these are inclined from the horizontal. The physical processes causing epipycnal (i.e., diffusion within an isopycnal surface) are often assumed to be different from those causing diapycnal diffusion (i.e., diffusion across an isopycnal surface). One then assumes a diffusion tensor that is diagonal in a rotated coordinate system aligned with isopycnals, with an epipycnal diffusion coefficient K_{e} and a diapycnal diffusion coefficient K_{d}. In the horizontal/vertical coordinate system this tensor has the form

$$\mathbf{K}^{\mathrm{s}} = \frac{1}{|\nabla\rho|^2} \begin{pmatrix} \begin{matrix} K_{\mathrm{d}}\overline{\rho}_x^2 + \\ K_{\mathrm{e}}(\overline{\rho}_y^2 + \overline{\rho}_z^2) \end{matrix} & (K_{\mathrm{d}} - K_{\mathrm{e}})\overline{\rho}_x\,\overline{\rho}_y & (K_{\mathrm{d}} - K_{\mathrm{e}})\overline{\rho}_x\,\overline{\rho}_z \\ (K_{\mathrm{d}} - K_{\mathrm{e}})\overline{\rho}_x\,\overline{\rho}_y & \begin{matrix} K_{\mathrm{d}}\overline{\rho}_y^2 + \\ K_{\mathrm{e}}(\overline{\rho}_x^2 + \overline{\rho}_z^2) \end{matrix} & (K_{\mathrm{d}} - K_{\mathrm{e}})\overline{\rho}_y\,\overline{\rho}_z \\ (K_{\mathrm{d}} - K_{\mathrm{e}})\overline{\rho}_x\,\overline{\rho}_z & (K_{\mathrm{d}} - K_{\mathrm{e}})\overline{\rho}_y\,\overline{\rho}_z & \begin{matrix} K_{\mathrm{e}}(\overline{\rho}_x^2 + \overline{\rho}_y^2) \\ +K_{\mathrm{d}}\overline{\rho}_z^2 \end{matrix} \end{pmatrix} \tag{12.22}$$

   If one assumes $K_{\mathrm{d}} \ll K_{\mathrm{e}}$ and small isopycnal slopes $\frac{\overline{\rho}_x^2+\overline{\rho}_y^2}{\overline{\rho}_z^2} \ll 1$ then

$$\mathbf{K}^{\mathrm{s}} = \begin{pmatrix} K_{\mathrm{e}} & 0 & -K_{\mathrm{e}}\frac{\overline{\rho}_x}{\overline{\rho}_z} \\ 0 & K_{\mathrm{e}} & -K_{\mathrm{e}}\frac{\overline{\rho}_y}{\overline{\rho}_z} \\ -K_{\mathrm{e}}\frac{\overline{\rho}_x}{\overline{\rho}_z} & -K_{\mathrm{e}}\frac{\overline{\rho}_y}{\overline{\rho}_z} & -K_{\mathrm{e}}\frac{\overline{\rho}_x^2+\overline{\rho}_y^2}{\overline{\rho}_z} + K_{\mathrm{d}} \end{pmatrix} \tag{12.23}$$

3. A third choice of the orientation is motivated by an energetic argument. The exchange of particles along isopycnals does not require any work since the particles have the same density. The exchange of particles along geopotential surfaces does not require any work since gravity acts perpendicular to these surfaces. When particles are exchanged within the wedge between the horizontal and isopycnal surface potential energy can be released. Formally, the Archimedean work done on a parcel displaced by an infinitesimal distance $\mathrm{d}\mathbf{x}$ is given by

$$\mathrm{d}W = -[\nabla\rho \cdot \mathrm{d}\mathbf{x}](\nabla\phi \cdot \mathrm{d}\mathbf{x}) \tag{12.24}$$

This form suggests a diffusion tensor with an orientation as follows: a large value along the axis halfway between the isopycnal and geopotential surface where the exchange of water parcels might release potential energy, a medium value along the intersection of isopycnal and geopotential surface where no work is required, and a small value in a direction perpendicular to these two axes where exchange of water parcels requires work.

Irreversibility The second law of thermodynamics states that any closed system irreversibly approaches thermodynamic equilibrium. During the approach the entropy increases monotonically until it is maximal at equilibrium. The law implies that the molecular diffusion coefficients are positive. There is no analogous law that implies that eddy diffusion coefficients have to be positive as well. Here we demonstrate that **if** the diffusion tensor is positive definite then eddy diffusion causes similar irreversible tendencies. For this we consider the diffusion equation

$$\partial_t \rho_{\rm A} = \partial_i K_{ij} \partial_j \rho_{\rm A} \tag{12.25}$$

which describes the diffusion of a substance A with density $\rho_{\rm A}$, diffusion tensor $\mathbf{K}$ and diffusive flux

$$I_i := -K_{ij} \partial_j \rho_{\rm A} \tag{12.26}$$

Consider a fixed volume V enclosed by a surface δV. The diffusion equation has three important integral properties:

1. If there is no net flux through the boundary ($\iint_{\delta V} \mathrm{d}^2x \, n_i I_i = 0$) then the total amount of the substance in the volume does not change since

$$\frac{\mathrm{d}}{\mathrm{d}t} \iiint_{\rm V} \mathrm{d}^3x \; \rho_{\rm A} = - \iint_{\delta V} \mathrm{d}^2x \; n_i I_i = 0 \tag{12.27}$$

after applying Reynolds, transport theorem, the diffusion equation, and Gauss' theorem.

2. Diffusion is the prototype of an irreversible process. Irreversibility requires that the symmetric part of the diffusion tensor is positive definite. There are two types of stationary solutions. One is the no-flux solution $\rho_{\rm A} = \mathit{constant}$ and the other the constant flux solution $\partial_i I_i = 0$. The conditions for evolution towards the first solution can be inferred from considering the evolution of the total variance of the density field

$$\frac{\mathrm{d}}{\mathrm{d}t} \iiint_{\rm V} \mathrm{d}^3x \; \rho_{\rm A}^2 = -2 \iiint_{\rm V} \mathrm{d}^3x \; K_{ij} (\partial_i \rho_{\rm A})(\partial_j \rho_{\rm A}) - 2 \iint_{\delta V} \mathrm{d}^2x \; \rho_{\rm A} n_i I_i \tag{12.28}$$

The second term on the right-hand side vanishes for an isolated system where $n_i I_i = 0$ everywhere on the surface δV. If the symmetric part of $\mathbf{K}$ is positive definite then the first term is non-positive and zero for $\boldsymbol{\nabla} \rho_{\rm A} = 0$ or $\rho_{\rm A} = \mathit{constant}$. The total variance of the density field of an isolated system thus decreases monotonically until the solution $\rho_{\rm A} = \mathit{constant}$ is reached. This is the "thermodynamic" equilibrium solution.

3. The conditions for evolution towards the second type of solution can be inferred by considering the density gradient $q_j = \partial_j \rho_A$, which is governed by

$$\partial_t q_j = \partial_j \partial_i K_{ik} q_k \tag{12.29}$$

Its total weighted variance changes according to

$$\frac{d}{dt} \iiint_V d^3x \; K_{ij} q_i q_j = \iiint_V d^3x \; q_i q_j \partial_t K_{ij} - 2 \iiint_V d^3x \; (\partial_i I_i)(\partial_j I_j) - 2 \iint_{\delta V} d^2x \; n_j I_j \partial_t \rho_A \tag{12.30}$$

Consider the case of a time-independent diffusion tensor. Then the first term on the right-hand side vanishes. The second term is non-positive and vanishes for $\partial_i I_i = 0$. The third term vanishes if $\partial_t \rho_A = 0$ everywhere on the boundary. If the diffusion tensor and the density ρ_A on the surface are constant in time then the weighted variance of $\mathbf{q}$ decreases monotonically until $\partial_i I_i = 0$ everywhere in V. This is the constant flux solution.

Eddy viscosity The Reynolds stress tensor is decomposed into

$$\sigma_{ij}^{\text{eddy}} = -\rho_0 \overline{u_i' u_j'} = -\rho_0 \frac{\overline{u_k' u_k'}}{3} \delta_{ij} + \tau_{ij}^{\text{eddy}} \tag{12.31}$$

The first term on the right-hand side can be absorbed into the mean pressure. The second term has trace zero and is parametrized by

$$\tau_{ij}^{\text{eddy}} = 2\rho_0 A_{ijkl} D_{kl} \tag{12.32}$$

where $\mathbf{D}$ is the rate of deformation tensor and A_{ijkl} the 81 components of a fourth-order eddy or turbulent viscosity tensor. Only 30 components are independent since

1. A_{ijkl} is symmetric in the indices (i, j) and (k, l) because of its definition
2. $A_{iikl} = 0$ because $\tau_{ii} = 0$.

   One further assumes

3. $A_{ijkl} = A_{klij}$ (see below)
4. axial symmetry about the "vertical" axis.

Then the eddy stress tensor takes the form

$$\begin{aligned}
\tau_{11}^{\text{eddy}} &= \rho_0 A_h (D_{11} - D_{22}) + \rho_0 A^* \left(\tfrac{D_{ii}}{3} - D_{33}\right) \\
\tau_{12}^{\text{eddy}} &= \tau_{21} = 2\rho_0 A_h D_{12} \\
\tau_{13}^{\text{eddy}} &= \tau_{31} = 2\rho_0 A_v D_{13} \\
\tau_{22}^{\text{eddy}} &= \rho_0 A_h (D_{22} - D_{11}) + \rho_0 A^* \left(\tfrac{D_{ii}}{3} - D_{33}\right) \\
\tau_{23}^{\text{eddy}} &= \tau_{32} = 2\rho_0 A_v D_{23} \\
\tau_{33}^{\text{eddy}} &= -2\rho_0 A^* \left(\tfrac{D_{ii}}{3} - D_{33}\right)
\end{aligned} \tag{12.33}$$

with only three independent coefficients: the horizontal eddy viscosity coefficient A_{h}, the vertical viscosity coefficient A_{v}, and the eddy viscosity coefficient A^*. In the Boussinesq approximation $D_{ii} = \nabla \cdot \overline{\mathbf{u}} = 0$. If we had assumed isotropy instead of axial symmetry we would have obtained

$$\tau_{ij}^{\text{eddy}} = 2\rho_0 A \left(D_{ij} - \frac{\nabla \cdot \overline{\mathbf{u}}}{3} \right) \tag{12.34}$$

with only one coefficient A. This differs from the molecular case since we did not require $\sigma_{ii}^{\text{mol}} = 0$.

The energy dissipation is given by

$$\begin{aligned} \tau_{ij}^{\text{eddy}} D_{ij} &= 2\rho_0 A_{ijkl} D_{ij} D_{kl} \\ &= A_{\text{h}}(D_{11} - D_{22})^2 + 4A_{\text{h}} D_{12}^2 + 3A^* D_{33}^2 + 4A_{\text{v}}\left(D_{13}^2 + D_{23}^2\right) \end{aligned} \tag{12.35}$$

and is positive if $A_{\text{h}}, A_{\text{v}}, A^* \geq 0$. Condition (3) above derives from this energy equation. It suppresses components of the eddy viscosity tensor that do not contribute to the dissipation of energy.

Warning It is reiterated that once we Reynolds average the equations we encounter the closure problem. The moments of the unresolved fluctuating fields that enter the averaged equations must be expressed in terms of resolved field variables. There is no general theory on how to do such a closure. Specifics will surely depend on the averaging scale, whether the equations are averaged over meters or hundreds of kilometers. In each specific case one has to assess carefully how the unresolved fluctuations affect the resolved mean flow. Parametrizations and boundary conditions need to be compatible with the basic conservation laws. They should also lead to a well-posed mathematical problem. The specific parametrizations given in this section and the boundary conditions given in the next section should only be viewed as examples of what is often done in certain circumstances. These parametrizations are physically sensible and lead to well-posed problems, but are by no means “correct.” Other parametrizations and boundary conditions might be more appropriate.

12.3 Boundary conditions

The Reynolds averaged equations need to be augmented by boundary conditions. We discussed in Section 10.5 that it is difficult to obtain these boundary conditions in any systematic manner. Here we describe some empirical laws that are often employed in the formulation of boundary conditions for large- and medium-scale motions.

Law of the wall In Section 10.5 we pointed out that the concept of a constant flux layer implies the continuity of fluxes across mean interfaces. Another important consequence is the law of the wall. Consider a fluid above a rigid horizontal surface and the Reynolds stress components $\boldsymbol{\tau}_\mathrm{h} := (\tau_{13}^{\mathrm{eddy}}, \tau_{23}^{\mathrm{eddy}})$ that describe the vertical, i.e., normal, flux of horizontal momentum. Within the turbulent part of the constant flux layer the velocity shear can only depend on the density, the stress, and, of course, the distance from the surface. Dimensional arguments then require the law of the wall with a logarithmic velocity profile and a drag law

$$\boldsymbol{\tau}_\mathrm{h} = \rho c_\mathrm{d} |\mathbf{u}_\mathrm{h}| \mathbf{u}_\mathrm{h} \tag{12.36}$$

where $\mathbf{u}_\mathrm{h} = (u_1, u_2)$ is the velocity at some prescribed height above the surface and c_d a dimensionless drag coefficient that depends on the bottom roughness. These drag laws are fairly well established for the atmosphere above the ocean and for the ocean above a flat bottom.

At the ocean surface, the velocity in expression (12.36) is $\mathbf{u}_\mathrm{h} = \overline{\mathbf{u}}_\mathrm{h}^\mathrm{a} - \overline{\mathbf{u}}_\mathrm{h}(0)$ where $\overline{\mathbf{u}}_\mathrm{h}^\mathrm{a}$ is the mean wind velocity at some prescribed height, usually 10 m or 2 m, and $\overline{\mathbf{u}}_\mathrm{h}(0)$ the mean surface drift velocity. If the drift velocity is neglected one arrives at the boundary condition

$$\boldsymbol{\tau}_\mathrm{h} = \boldsymbol{\tau}_\mathrm{h}^\mathrm{a} \tag{12.37}$$

where the mean atmospheric windstress

$$\boldsymbol{\tau}_\mathrm{h}^\mathrm{a} = c_\mathrm{d} \overline{\rho}_\mathrm{e} \left| \overline{\mathbf{u}}_\mathrm{h}^\mathrm{a} \right| \overline{\mathbf{u}}_\mathrm{h}^\mathrm{a} \tag{12.38}$$

is solely determined by the mean atmospheric wind velocity. The boundary condition (12.37) is thus a prescription of the oceanic Reynolds stress at the ocean surface in terms of the atmospheric windstress. The ocean is forced by the atmospheric windstress.

At the ocean bottom, the law of the wall becomes

$$\boldsymbol{\tau}_\mathrm{h} = \overline{\rho} c_\mathrm{d} |\overline{\mathbf{u}}_\mathrm{h}| \overline{\mathbf{u}}_\mathrm{h} \tag{12.39}$$

where $\overline{\mathbf{u}}_\mathrm{h}$ is the mean ocean velocity just outside the log layer. This is a prescription of the oceanic Reynolds stress at the ocean bottom in terms of the ocean velocity. The ocean is slowed down by *bottom friction*.

Evaporation Boundary layer theory and observations also suggest that the rate of evaporation $\overline{E}$ at the ocean surface is given by

$$\overline{E} = \overline{\rho}_\mathrm{a} c_E (\overline{q}_\mathrm{s} - \overline{q}) \left| \overline{\mathbf{u}}_\mathrm{h}^\mathrm{a} \right| \tag{12.40}$$

where $\overline{q}$ is the mean specific humidity at some prescribed height, $\overline{q}_\mathrm{s}$ the saturation humidity, $\overline{\mathbf{u}}_\mathrm{h}^\mathrm{a}$ the wind velocity at the same height, and c_E a dimensionless coefficient, often called the Dalton number. The mass and salt flux across the air–sea interface are thus prescribed to be

$$\overline{I}_{\text{mass}} = \overline{E} - \overline{P} \tag{12.41}$$

$$\rho_0 \mathbf{I}_S^{\text{eddy}} = -(\overline{E} - \overline{P})\overline{S} \tag{12.42}$$

where $\overline{P}$ is the mean precipitation rate.

Heat flux Similarly, the sensible heat flux in the boundary layer above the ocean surface is given by

$$\overline{Q}_S = \overline{\rho}_{\text{a}} \overline{c}_{\text{p}} c_{\text{H}} (\overline{T} - \overline{T}_{\text{a}}) \left| \overline{\mathbf{u}}_{\text{h}}^{\text{a}} \right| \tag{12.43}$$

where $\overline{T}$ is the mean sea surface temperature, $\overline{T}_{\text{a}}$ the mean air temperature at some prescribed height, and c_{H} a dimensionless coefficient, often called the Stanton number.

The oceanic heat flux at the sea surface is thus given by

$$\rho_0 c_{\text{p}} \mathbf{I}_\theta^{\text{eddy}} = Q^{\text{a}} \tag{12.44}$$

with

$$Q^{\text{a}} = \overline{Q}_S + \overline{Q}_L + \overline{Q}_{\text{SW}} + \overline{Q}_{\text{LW}} \tag{12.45}$$

where $\overline{Q}_S$ is the mean sensible heat flux, $\overline{Q}_L = \overline{L_{\text{v}} E}$ the mean latent heat flux, $\overline{Q}_{\text{SW}}$ the mean short-wave radiation absorbed in the surface, and $\overline{Q}_{\text{LW}}$ the mean long-wave radiation emitted from or absorbed at the surface.

12.4 Boussinesq equations in spherical coordinates

In pseudo-spherical coordinates (φ, θ, z) with metric $\text{d}s^2 = r_0^2 \cos^2\theta \text{d}\varphi^2 + r_0^2 \text{d}\theta^2 + \text{d}z^2$ the Reynolds averaged Boussinesq equations take the form (with the overbar now omitted):

$$\begin{aligned} &\partial_t u + \frac{u}{r_0 \cos\theta} \frac{\partial u}{\partial \varphi} + \frac{v}{r_0} \frac{\partial u}{\partial \theta} + w \frac{\partial u}{\partial z} - \frac{uv}{r_0} \tan\theta - 2\Omega \sin\theta\, v + 2\Omega \cos\theta\, w \\ &\quad = -\frac{1}{\rho_0 r_0 \cos\theta} \frac{\partial p}{\partial \varphi} + \frac{1}{\rho_0} F_\varphi^{\text{eddy}} \end{aligned} \tag{12.46}$$

$$\begin{aligned} &\partial_t v + \frac{u}{r_0 \cos\theta} \frac{\partial v}{\partial \varphi} + \frac{v}{r_0} \frac{\partial v}{\partial \theta} + w \frac{\partial v}{\partial z} + \frac{u^2}{r_0} \tan\theta + 2\Omega \sin\theta\, u \\ &\quad = -\frac{1}{\rho_0 r_0} \frac{\partial p}{\partial \theta} + \frac{1}{\rho_0} F_\theta^{\text{eddy}} \end{aligned} \tag{12.47}$$

$$\begin{aligned} &\partial_t w + \frac{u}{r_0 \cos\theta} \frac{\partial w}{\partial \varphi} + \frac{v}{r_0} \frac{\partial w}{\partial \theta} + w \frac{\partial w}{\partial z} - 2\Omega \cos\theta\, u \\ &\quad = -\frac{1}{\rho_0} \frac{\partial p}{\partial z} - \frac{\rho g_0}{\rho_0} + \frac{1}{\rho_0} F_z^{\text{eddy}} \end{aligned} \tag{12.48}$$

$$\frac{1}{r_0 \cos\theta} \frac{\partial u}{\partial \varphi} + \frac{1}{r_0 \cos\theta} \frac{\partial (v \cos\theta)}{\partial \theta} + \frac{\partial w}{\partial z} = 0 \tag{12.49}$$

$$\left(\partial_t + \frac{u}{r \cos\varphi} \frac{\partial}{\partial \varphi} + \frac{v}{r} \frac{\partial}{\partial \theta} + w \frac{\partial}{\partial z} \right) \psi = D_\psi^{\text{eddy}} \tag{12.50}$$

where

$$\begin{aligned}
\varphi &= \text{longitude} \\
\theta &= \text{latitude} \\
z &= \text{vertical distance} \\
u &= \text{mean zonal velocity component} \\
v &= \text{mean meridional velocity component} \\
w &= \text{mean vertical velocity component} \\
p &= \text{mean pressure} \\
\rho &= \text{mean density} \\
\mathbf{F}^{\text{eddy}} &:= \nabla \cdot \boldsymbol{\tau}^{\text{eddy}} \ \text{divergence of eddy stress tensor} \\
D_\psi^{\text{eddy}} &:= -\nabla \cdot \mathbf{I}_\psi^{\text{eddy}} \ \text{divergence of eddy flux} \\
\psi &= \begin{cases} \theta & \text{mean potential temperature} \\ S & \text{mean salinity} \\ c_{\text{A}} & \text{mean tracer concentration} \end{cases}
\end{aligned} \tag{12.51}$$

The general expressions for the divergence of the eddy stress tensor will not be given here but can be constructed from the formulae in Section C.4. Their shallow water limit will be given in Chapter 13 (formulae (13.25) to (13.27)). For the parametrizations (12.33) with constant eddy viscosity coefficients, the general expressions reduce to

$$\begin{aligned}
\frac{1}{\rho_0} F_\varphi^{\text{eddy}} = A_{\text{h}} \left(\Delta_{\text{h}} u + \frac{\cos 2\theta}{r_0^2 \cos^2\theta} u - \frac{2 \sin\theta}{r_0^2 \cos^2\theta} \frac{\partial v}{\partial \varphi} \right) + A_{\text{v}} \frac{\partial^2 u}{\partial z^2} \\
+ (A_{\text{v}} - A_*) \frac{1}{r_0 \cos\theta} \frac{\partial^2 w}{\partial \varphi \partial z}
\end{aligned} \tag{12.52}$$

$$\begin{aligned}
\frac{1}{\rho_0} F_\theta^{\text{eddy}} = A_{\text{h}} \left(\Delta_{\text{h}} v + \frac{\cos 2\theta}{r_0^2 \cos^2\theta} v + \frac{2 \sin\theta}{r_0^2 \cos^2\theta} \frac{\partial u}{\partial \varphi} \right) + A_{\text{v}} \frac{\partial^2 v}{\partial z^2} \\
+ (A_{\text{v}} - A_*) \frac{1}{r_0} \frac{\partial^2 w}{\partial \theta \partial z}
\end{aligned} \tag{12.53}$$

$$\frac{1}{\rho_0} F_z^{\text{eddy}} = A_{\text{v}} \left(\Delta_{\text{h}} + \frac{\partial^2}{\partial z^2} \right) w - 2 (A_{\text{v}} - A_*) \frac{\partial^2 w}{\partial z^2} \tag{12.54}$$

where

$$\Delta_{\mathrm{h}} = \frac{1}{r_0 \cos\theta} \left[\frac{\partial}{\partial \varphi} \left(\frac{1}{r_0 \cos\theta} \frac{\partial}{\partial \varphi} \right) + \frac{\partial}{\partial \theta} \left(\frac{\cos\theta}{r_0} \frac{\partial}{\partial \theta} \right) \right] \tag{12.55}$$

is the horizontal Laplace operator.

The general expression for the divergence of the eddy flux is given by

$$D_{\psi}^{\mathrm{eddy}} = -\left(\frac{1}{r_0 \cos\theta} \frac{\partial I_{\varphi}^{\mathrm{eddy}}}{\partial \varphi} + \frac{1}{r_0 \cos\theta} \frac{\partial \left(I_{\theta}^{\mathrm{eddy}} \cos\theta \right)}{\partial \theta} + \frac{\partial I_z^{\mathrm{eddy}}}{\partial z} \right) \tag{12.56}$$

which reduces to

$$D_{\psi}^{\mathrm{eddy}} = K_{\mathrm{h}} \Delta_{\mathrm{h}} \psi + K_{\mathrm{v}} \frac{\partial^2 \psi}{\partial z^2} \tag{12.57}$$

for the parametrization (12.21) with constant eddy diffusion coefficients.

Because of the vector identity $\mathbf{u} \cdot \boldsymbol{\nabla} \mathbf{u} = \boldsymbol{\nabla} (\mathbf{u} \cdot \mathbf{u}/2) + \boldsymbol{\omega} \times \mathbf{u}$ the advection terms in the momentum equation can also be written as

$$(\mathbf{u} \cdot \boldsymbol{\nabla} \mathbf{u})_{\varphi} = \frac{1}{r_0 \cos\theta} \frac{\partial}{\partial \varphi} \left(\frac{u^2 + v^2 + w^2}{2} \right) + \omega_{\theta} w - \omega_z v \tag{12.58}$$

$$(\mathbf{u} \cdot \boldsymbol{\nabla} \mathbf{u})_{\theta} = \frac{1}{r_0} \frac{\partial}{\partial \theta} \left(\frac{u^2 + v^2 + w^2}{2} \right) + \omega_z u - \omega_{\varphi} w \tag{12.59}$$

$$(\mathbf{u} \cdot \boldsymbol{\nabla} \mathbf{u})_z = \frac{\partial}{\partial z} \left(\frac{u^2 + v^2 + w^2}{2} \right) + \omega_{\varphi} v - \omega_{\theta} u \tag{12.60}$$

where the relative vorticity vector $\boldsymbol{\omega} = \boldsymbol{\nabla} \times \mathbf{u}$ has components

$$\omega_{\varphi} = \frac{1}{r_0} \frac{\partial w}{\partial \theta} - \frac{\partial v}{\partial z} \tag{12.61}$$

$$\omega_{\theta} = \frac{\partial u}{\partial z} - \frac{1}{r_0 \cos\theta} \frac{\partial w}{\partial \varphi} \tag{12.62}$$

$$\omega_z = \frac{1}{r_0 \cos\theta} \frac{\partial v}{\partial \varphi} - \frac{1}{r_0 \cos\theta} \frac{\partial (u \cos\theta)}{\partial \theta} \tag{12.63}$$

13
Primitive equations

In this chapter we assume that the aspect ratio of the motion is small. This assumption leads to the primitive equations that are supposed to govern most large-scale oceanic flows. The smallness of the aspect ratio is exploited in the shallow water approximation. Its main consequence is that the vertical momentum balance reduces to the hydrostatic balance. Both sound and gyroscopic waves are eliminated. A second consequence is that the meridional component of the planetary vorticity can be neglected. Gravity wins over rotation. We do not perform a complete asymptotic expansion with respect to the aspect ratio but settle for a simple scale analysis. We derive the primitive equations in height coordinates. Other representations of the vertical structure will be considered in the next chapter.

13.1 Shallow water approximation

Assume that the fluid motion under consideration has a vertical length scale D much smaller than its horizontal length scale L. Its aspect ratio

$$\delta := \frac{D}{L} \ll 1 \qquad (13.1)$$

is small. The smallness of the aspect ratio will be exploited by a scale analysis. The starting point is the Boussinesq equations (12.46) to (12.49) with the parametrizations (12.52) to (12.57). All overbars and "eddy" superscripts are omitted for compactness of notation.

The characteristic scales and values are denoted by

$$\begin{array}{ll} T & \text{time scale} \\ L & \text{horizontal scale} \\ D & \text{vertical scale} \\ U & \text{horizontal velocity} \\ W & \text{vertical velocity} \\ \Delta P & \text{deviation pressure} \\ \Delta R & \text{deviation density} \end{array} \tag{13.2}$$

These scales and values can be used to obtain estimates of the order of magnitude of the various terms in the equations of motion. Technically, one replaces all dependent variables by their characteristic values and derivatives by the inverse of the scale

$$\begin{aligned} \frac{\partial}{\partial t} &\to T^{-1} \\ \frac{1}{r_0 \cos\theta}\frac{\partial}{\partial \varphi} &\to L^{-1} \\ \frac{1}{r_0}\frac{\partial}{\partial \theta} &\to L^{-1} \\ \frac{\partial}{\partial z} &\to D^{-1} \end{aligned} \tag{13.3}$$

Scaling of the incompressibility condition leads to

$$\begin{array}{ccccccccc} \frac{1}{r_0\cos\theta}\frac{\partial u}{\partial \varphi} & + & \frac{1}{r_0}\frac{\partial v}{\partial \theta} & - & \frac{v}{r_0}\tan\theta & + & \frac{\partial w}{\partial z} & = 0 \\ \frac{U}{L} & & \frac{U}{L} & & \frac{U}{r_0} & & \frac{W}{D} & \\ 1 & & 1 & & \gamma & & \frac{W}{U}\delta^{-1} & \end{array} \tag{13.4}$$

where

$$\gamma := \frac{L}{r_0} \tag{13.5}$$

is the ratio of the horizontal length scale and the Earth's radius. The order of magnitude or characteristic value of each term is given in the second row. The third row is obtained by normalizing by U/L. Comparing the various terms in the equation an upper bound for the vertical velocity is obtained

$$\frac{W}{U}\delta^{-1} \leq 1 \quad \text{or} \quad W \leq \delta U \tag{13.6}$$

A larger value of W cannot be balanced.

Scaling of the zonal momentum balance leads to

$$\begin{array}{ccccccc}
\partial_t u & + & \frac{u}{r_0\cos\theta}\frac{\partial u}{\partial\varphi} & + & \frac{v}{r_0}\frac{\partial u}{\partial\theta} & + & w\frac{\partial u}{\partial z} \\
\frac{U}{T} & & \frac{U^2}{L} & & \frac{U^2}{L} & & \frac{U^2}{L} \\
\tilde{\mathrm{Ro}} & & \mathrm{Ro} & & \mathrm{Ro} & & \mathrm{Ro} \\
-\frac{uv}{r_0}\tan\theta & - & 2\Omega\sin\theta\, v & & +\,2\Omega\cos\theta\, w & = & -\frac{1}{\rho_0 r_0\cos\theta}\frac{\partial p}{\partial\varphi} \\
\frac{U^2}{r_0} & & fU & & \tilde{f}\delta U & & \frac{\Delta P}{\rho_0 L} \\
\mathrm{Ro}\gamma & & 1 & & \delta & & \frac{\Delta P}{\rho_0 L} \\
+A_{\mathrm{h}}\Big(\Delta_{\mathrm{h}} u & + & \frac{\cos 2\theta}{r_0^2\cos^2\theta}u & - & \frac{2\sin\theta}{r_0^2\cos^2\theta}\frac{\partial v}{\partial\varphi}\Big) & + & A_{\mathrm{v}}\frac{\partial^2 u}{\partial z^2} \\
\frac{A_{\mathrm{h}}U}{L^2} & & \frac{A_{\mathrm{h}}U}{r_0^2} & & \frac{A_{\mathrm{h}}U}{r_0 L} & & \frac{A_{\mathrm{v}}U}{D^2} \\
E_{\mathrm{h}} & & \gamma^2 E_{\mathrm{h}} & & \gamma E_{\mathrm{h}} & & E_{\mathrm{v}} \\
+(A_{\mathrm{v}} & - & A_*)\frac{1}{r_0\cos\theta}\frac{\partial^2 w}{\partial\varphi\partial z} & & & & \\
\frac{A_{\mathrm{v}}U}{L^2} & & \frac{A_* U}{L^2} & & & & \\
\delta^2 E_{\mathrm{v}} & & \frac{A_*}{A_{\mathrm{v}}}\delta^2 E_{\mathrm{v}} & & & &
\end{array} \tag{13.7}$$

where

$$f = 2\Omega\sin\theta \quad \text{vertical component of planetary vorticity} \tag{13.8}$$

$$\tilde{f} = 2\Omega\cos\theta \quad \text{meridional component of planetary vorticity} \tag{13.9}$$

and

$$\tilde{\mathrm{Ro}} = \frac{1}{T|f|} \quad \text{temporal Rossby number} \tag{13.10}$$

$$\mathrm{Ro} = \frac{U}{fL} \quad \text{advective Rossby number} \tag{13.11}$$

$$E_{\mathrm{h}} = \frac{A_{\mathrm{h}}}{fL^2} \quad \text{horizontal Ekman number} \tag{13.12}$$

$$E_{\mathrm{v}} = \frac{A_{\mathrm{v}}}{fH^2} \quad \text{vertical Ekman number} \tag{13.13}$$

The scaling is only valid for midlatitudes where $O(\sin\theta) = O(\cos\theta) = O(\tan\theta) = O(1)$.

The scaling of the meridional momentum balance yields identical results. The scaling of the vertical momentum balance results in

$$\begin{array}{ccccc}
\partial_t w & +\frac{u}{r_0 \cos\theta}\frac{\partial w}{\partial\varphi} + & \frac{v}{r_0}\frac{\partial w}{\partial\theta} & + \quad w\frac{\partial w}{\partial z} & -2\Omega\cos\theta u \\
\frac{W}{T} & \frac{U}{L}W & \frac{U}{L}W & \frac{W}{D}W & \tilde{f}U \\
\tilde{\mathrm{Ro}}\delta^2 & \mathrm{Ro}\delta^2 & \mathrm{Ro}\delta^2 & \mathrm{Ro}\delta^2 & \delta \\
= -\frac{1}{\rho_0 r_0}\frac{\partial p}{\partial z} - & \frac{\rho}{\rho_0}g_0 & + A_\mathrm{v}\Delta_\mathrm{h} w & + 2A_*\frac{\partial^2 w}{\partial z^2} - & A_\mathrm{v}\frac{\partial^2 w}{\partial z^2} \\
\frac{\Delta P}{\rho_0 D} & \frac{\Delta R g_0}{\rho_0} & \frac{A_\mathrm{v} W}{L^2} & A_*\frac{W}{D^2} & A_\mathrm{v}\frac{W}{D^2} \\
\frac{\Delta P}{\rho_0 f L U} & \frac{\Delta R g_0 D}{\rho_0 f L U} & \delta^4 E_\mathrm{v} & \frac{A_*}{A_\mathrm{v}}\delta^2 E_\mathrm{v} & \delta^2 E_\mathrm{v}
\end{array} \tag{13.14}$$

The scaling of the pressure and gravitational terms assume that the reference state (or a background state) has been subtracted out.

Formally, one would now proceed with an asymptotic expansion with respect to the aspect ratio δ. We will not go through this procedure. We implement the shallow water approximation by simply neglecting all terms of $O(\delta)$ and smaller in the above equations, being aware that the resulting equations are the lowest order of an asymptotic expansion.

Besides δ, the above scale analysis also produced the non-dimensional parameters γ, $\tilde{\mathrm{Ro}}$, Ro, E_h, E_v. Expansions with respect to these parameters will be considered in subsequent chapters.

13.2 Primitive equations in height coordinates

Equations Neglecting all terms of $O(\delta)$ one obtains the primitive equations

$$\partial_t u + \frac{u}{r_0\cos\theta}\frac{\partial u}{\partial\varphi} + \frac{v}{r_0}\frac{\partial u}{\partial\theta} + w\frac{\partial u}{\partial z} - \frac{uv}{r_0}\tan\theta - fv = -\frac{1}{\rho_0 r_0\cos\theta}\frac{\partial p}{\partial\varphi} + \frac{1}{\rho_0}F_\varphi \tag{13.15}$$

$$\partial_t v + \frac{u}{r_0\cos\theta}\frac{\partial v}{\partial\varphi} + \frac{v}{r_0}\frac{\partial v}{\partial\theta} + w\frac{\partial v}{\partial z} + \frac{u^2}{r_0}\tan\theta + fu = -\frac{1}{\rho_0 r_0}\frac{\partial p}{\partial\theta} + \frac{1}{\rho_0}F_\theta \tag{13.16}$$

$$0 = \frac{\partial p}{\partial z} + \rho g_0 \tag{13.17}$$

$$\frac{1}{r_0\cos\theta}\frac{\partial u}{\partial\varphi} + \frac{1}{r_0\cos\theta}\frac{\partial(v\cos\theta)}{\partial\theta} + \frac{\partial w}{\partial z} = 0 \tag{13.18}$$

$$\left(\partial_t + \frac{u}{r\cos\varphi}\frac{\partial}{\partial\varphi} + \frac{v}{r_0}\frac{\partial}{\partial\theta} + w\frac{\partial}{\partial z}\right)\psi = D_\psi \tag{13.19}$$

$$\rho = \rho(z, \theta, S) \tag{13.20}$$

where $\psi = \theta, S$, or any other tracer. The divergence D_ψ of the eddy flux of ψ is given by (12.56) and (12.57). The divergences $F_{\varphi,\theta}$ of the eddy stress tensor are given further below. The variables are all Reynolds averaged variables. Eddy terms in the equation of state are neglected. Instead of the pressure p and density ρ one can also substitute the deviations from a reference or background state.

The first main simplification is that the pressure and density are in hydrostatic balance. This is the *hydrostatic approximation*. The three-dimensional Poisson equation (11.24) for the pressure reduces to the first order ordinary differential equation (13.17). The second main simplification is that the Coriolis terms containing the meridional component of the planetary vorticity drop out of the momentum balance. This is the *traditional approximation*. The traditional approximation needs to be applied in conjunction with the hydrostatic balance. Otherwise the Coriolis force would do work.

The above primitive equations are derived and given in pseudo-spherical coordinates. They can also be written in three-dimensional vector notation. However, since the primitive equations treat the horizontal and vertical dimensions differently it is more convenient to introduce horizontal vectors and differential operators, denoted by a subscript "h." The rules for these two-dimensional vectors and operators are given in Section B.3 and their representation in spherical coordinates in Section C.5. In this two-dimensional vector notation the primitive equations take the form

$$\frac{\mathrm{D}}{\mathrm{D}t}\mathbf{u}_\mathrm{h} + f\hat{\mathbf{u}}_\mathrm{h} = -\frac{1}{\rho_0}\boldsymbol{\nabla}_\mathrm{h} p + \frac{1}{\rho_0}\mathbf{F}_\mathrm{h} \tag{13.21}$$

$$0 = \frac{\partial p}{\partial z} + \rho g_0 \tag{13.22}$$

$$\boldsymbol{\nabla}_\mathrm{h} \cdot \mathbf{u}_\mathrm{h} + \frac{\partial w}{\partial z} = 0 \tag{13.23}$$

$$\frac{\mathrm{D}}{\mathrm{D}t}\psi = -\boldsymbol{\nabla}_\mathrm{h} \cdot \mathbf{I}_\mathrm{h}^\psi - \frac{\partial}{\partial z} I_z^\psi \tag{13.24}$$

where $\mathrm{D}/\mathrm{D}t = \partial_t + \mathbf{u}_\mathrm{h} \cdot \boldsymbol{\nabla}_\mathrm{h} + w\partial_z$. A hat on a horizontal vector denotes the vector that is obtained by a 90° anticlockwise rotation, e.g., $\hat{\mathbf{u}}_\mathrm{h} = (-v, u)$. In (13.21) one can also employ the vector identity $\mathbf{u}_\mathrm{h} \cdot \boldsymbol{\nabla}_\mathrm{h}\mathbf{u}_\mathrm{h} = \boldsymbol{\nabla}_\mathrm{h}(\mathbf{u}_\mathrm{h} \cdot \mathbf{u}_\mathrm{h}/2) + \omega_z\hat{\mathbf{u}}_\mathrm{h}$ where $\omega_z = \hat{\boldsymbol{\nabla}}_\mathrm{h} \cdot \mathbf{u}_\mathrm{h}$ is the vertical component of the vorticity.

Eddy fluxes It is convenient to consider the horizontal and vertical divergences of the eddy stress tensor separately. We thus write $\mathbf{F}_\mathrm{h} = \mathbf{F}^\mathrm{h}_\mathrm{h} + \mathbf{F}^\mathrm{v}_\mathrm{h}$ where

$$\begin{aligned} F^\mathrm{h}_\varphi &= \frac{1}{r_0 \cos\theta}\frac{\partial \tau_{\varphi\varphi}}{\partial \varphi} + \frac{1}{r_0 \cos^2\theta}\frac{\partial}{\partial \theta}\left(\cos^2\theta\, \tau_{\varphi\theta}\right) \\ &= \rho_0 A_\mathrm{h}\left(\Delta_\mathrm{h} u + \frac{\cos 2\theta}{r_0^2 \cos^2\theta} u - \frac{2\sin\theta}{r_0^2\cos^2\theta}\frac{\partial v}{\partial \varphi}\right) \end{aligned} \tag{13.25}$$

$$\begin{aligned} F^\mathrm{h}_\theta &= \frac{1}{r_0 \cos\theta}\frac{\partial \tau_{\theta\varphi}}{\partial \varphi} + \frac{\tan\theta}{r_0}\tau_{\varphi\varphi} + \frac{1}{r_0 \cos\theta}\frac{\partial}{\partial \theta}(\cos\theta\, \tau_{\theta\theta}) \\ &= \rho_0 A_\mathrm{h}\left(\Delta_\mathrm{h} v + \frac{\cos 2\theta}{r_0^2 \cos^2\theta} v + \frac{2\sin\theta}{r_0^2\cos^2\theta}\frac{\partial u}{\partial \varphi}\right) \end{aligned} \tag{13.26}$$

$$\begin{aligned} \mathbf{F}^\mathrm{v}_\mathrm{h} &= \frac{\partial}{\partial z}\boldsymbol{\tau}_\mathrm{h} \\ &= \rho_0 A_\mathrm{v} \partial_z \partial_z \mathbf{u}_\mathrm{h} \end{aligned} \tag{13.27}$$

The horizontal vector $\boldsymbol{\tau}_\mathrm{h}$ has components $(\tau_{\varphi z}, \tau_{\theta z})$. The second equality sign holds for the eddy viscosity parametrization (12.33) and the shallow water limit. The eddy coefficient A_* is eliminated.

Boundary conditions The primitive equations have to be augmented by boundary conditions.

For large-scale flows one usually assumes that the surface elevation η is much smaller than the vertical scale D of the motion or the depth H of the ocean. Expansion of the boundary conditions about $z = 0$ then results in the air–sea boundary conditions

$$(\partial_t + \mathbf{u}_\mathrm{h} \cdot \boldsymbol{\nabla}_\mathrm{h})\eta = w - \frac{E - P}{\rho_0} \tag{13.28}$$

$$\boldsymbol{\tau}_\mathrm{h} = \boldsymbol{\tau}^\mathrm{a}_\mathrm{h} \tag{13.29}$$

$$p = \rho_0 g_0 \eta + p_\mathrm{a} \tag{13.30}$$

$$\rho_0 I^S_z = -(E - P)S \tag{13.31}$$

$$\rho_0 c_p I^\theta_z = Q^\mathrm{a} \tag{13.32}$$

where E is the rate of evaporation, P the rate of precipitation, $\boldsymbol{\tau}^\mathrm{a}_\mathrm{h}$ the atmospheric windstress, p_a the atmospheric pressure, and Q^a the atmospheric heat flux. For our standard parametrizations (12.21) and (12.33) $\boldsymbol{\tau}_\mathrm{h} = A_\mathrm{v}\partial_z \mathbf{u}_\mathrm{h}$, $I^S_z = -K_\mathrm{v}\partial_z S$, and $I^\theta_z = -K_\mathrm{v}\partial_z\theta$. All conditions are taken at $z = 0$. The boundary condition (13.28) is called the kinematic boundary condition. Equations (13.29) and (13.30) are called the dynamic boundary condition. The pressure condition is the continuity condition for the component $\Pi_{zz} = -p + \tau_{zz}$ of the stress tensor, with the term τ_{zz} being neglected. The additional term $\rho_0 g_0 \eta$ arises from an expansion of the boundary condition $p = p_{\mathrm{a}}$ at $z = \eta$. If p_{a} were the only forcing term and time independent the surface elevation would adjust to $\eta_{\mathrm{a}} = -p_{\mathrm{a}}/\rho_0 g_0$. This is the *inverse barometer effect*. Often one uses η_{a} as a variable instead of p_{a}. Then the pressure boundary condition reads $p = \rho_0 g_0(\eta - \eta_{\mathrm{a}})$. In the kinematic boundary condition (13.28) the mass flux $E - P$ describes the addition or subtraction of mass to or from the ocean. In the boundary condition (13.31) $E - P$ determines the salinity flux $(E - P)S$ that causes changes in the salinity and hence the density of the ocean.

At the bottom $z = -H(\varphi, \theta)$ with normal vector $\mathbf{n}$ one typically employs the kinematic boundary condition

$$\mathbf{u} \cdot \mathbf{n} = 0 \tag{13.33}$$

dynamic boundary conditions such as

$$\begin{aligned} \partial_z \mathbf{u}_{\mathrm{h}} &= 0 && \text{free slip} \\ \mathbf{u}_{\mathrm{h}} &= 0 && \text{no slip} \\ \boldsymbol{\tau}_{\mathrm{h}}^{\mathrm{b}} &= \rho_0 H \mathbf{r} \cdot \mathbf{u}_{\mathrm{h}} && \text{“linear” bottom friction} \\ \boldsymbol{\tau}_{\mathrm{h}}^{\mathrm{b}} &= \rho_0 c_{\mathrm{d}} |\mathbf{u}_{\mathrm{h}}| \mathbf{u}_{\mathrm{h}} && \text{“quadratic” bottom friction} \end{aligned} \tag{13.34}$$

and flux conditions

$$\mathbf{I}_S \cdot \mathbf{n} = 0 \tag{13.35}$$

$$\mathbf{I}_\theta \cdot \mathbf{n} = Q^{\mathrm{b}} \tag{13.36}$$

Here $\mathbf{r}$ is a bottom friction tensor, c_{d} the drag coefficient, and Q^{b} the geothermal heat flux through the bottom. The dynamic boundary condions are formulated for a flat bottom.

Structure The primitive equations together with the above boundary condition constitute a well-defined problem with:

- prognostic variables (u, v, θ, S, η);
- diagnostic variables (w, p, ρ); and
- external forcing fields $(E - P, p^{\mathrm{a}}, \boldsymbol{\tau}_{\mathrm{h}}^{\mathrm{a}}, Q^{\mathrm{a,b}})$.

The dynamic boundary condition (13.29) at the surface and the dynamic boundary condition at the bottom are needed to solve the prognostic equation for $\mathbf{u}_{\mathrm{h}}$. The prescribed heat and salt fluxes at the surface and bottom are needed to solve the prognostic equations for θ and S. The surface mass flux enters the prognostic equation for η. Often the primitive equations neglect the nonlinear advection terms in the momentum balance.

The diagnostic variables can be determined explicitly as follows. Integration of the incompressibility condition upwards from the bottom yields the vertical velocity

$$w(z) = w(-H) - \int_{-H}^{z} \mathrm{d}z\, \nabla_{\mathrm{h}} \cdot \mathbf{u}_{\mathrm{h}} = -\nabla_{\mathrm{h}} \cdot \int_{-H}^{z} \mathrm{d}z\, \mathbf{u}_{\mathrm{h}} \tag{13.37}$$

where the kinematic bottom boundary condition $w = -\mathbf{u}_{\mathrm{h}} \cdot \nabla_{\mathrm{h}} H$ at $z = -H$ has been used. Integration of the hydrostatic balance downwards from the surface yields the pressure

$$p(z) = \rho_0 g_0 \eta + p_{\mathrm{a}} + \int_{z}^{0} \mathrm{d}z\, \rho g_0 \tag{13.38}$$

where the surface boundary condition (13.30) has been used. The density is determined by the equation of state.

The main structural change accomplished by the shallow water approximation is the considerable simplification of the diagnostic equation for the pressure. Instead of a three-dimensional Poisson equation one now has to solve a first order ordinary differential equation in z. Its explicit solution is given by (13.38) above.

The vertical eigenvalue problem of the primitive equations is discussed in Section 14.6. It does not contain the internal modes of the second kind. Thus, sound and gyroscopic waves are eliminated. Rossby and gravity waves are retained in their low-frequency limit $\omega^2 \ll N^2$.

13.3 Vorticity equations

In the shallow water approximation, the vorticity vector can be approximated by

$$\tilde{\omega}_{\mathrm{h}} = \frac{\partial}{\partial z} \hat{\mathbf{u}}_{\mathrm{h}} = \left(-\frac{\partial v}{\partial z}, \frac{\partial u}{\partial z} \right) \tag{13.39}$$

$$\omega_z = \hat{\nabla}_{\mathrm{h}} \cdot \mathbf{u}_{\mathrm{h}} = \partial_1 u_2 - \partial_2 u_1 \tag{13.40}$$

The horizontal components represent the vertical shear only. In concordance, the vorticity equation takes the form

$$\frac{\mathrm{D}}{\mathrm{D}t} \tilde{\omega}_{\mathrm{h}} = \tilde{\omega}_{\mathrm{h}} \cdot \nabla_{\mathrm{h}} \mathbf{u}_{\mathrm{h}} + (\omega_z + f) \partial_z \mathbf{u}_{\mathrm{h}} + \frac{g_0}{\rho_0} \hat{\nabla}_{\mathrm{h}} \rho + \frac{1}{\rho_0} \partial_z \hat{\mathbf{F}}_{\mathrm{h}} \tag{13.41}$$

$$\frac{\mathrm{D}}{\mathrm{D}t} (\omega_z + f) = \tilde{\omega}_{\mathrm{h}} \cdot \nabla_{\mathrm{h}} w + (\omega_z + f) \partial_z w + \frac{1}{\rho_0} \hat{\nabla}_{\mathrm{h}} \cdot \mathbf{F}_{\mathrm{h}} \tag{13.42}$$

Ertel's potential vorticity is given by

$$q_\psi = \tilde{\omega}_{\mathrm{h}} \cdot \nabla_{\mathrm{h}} \psi + (\omega_z + f) \frac{\partial \psi}{\partial z} \tag{13.43}$$

and governed by

$$\partial_t q_\psi + \nabla_\mathrm{h} \cdot \left(\mathbf{u}_\mathrm{h} q_\psi + \mathbf{I}^q_\mathrm{h}\right) + \partial_z \left(w q_\psi + I^q_z\right) = 0 \tag{13.44}$$

where

$$\mathbf{I}^q_\mathrm{h} = \frac{\rho g}{\rho_0} \hat{\nabla}_\mathrm{h} \psi + \frac{1}{\rho_0} \hat{\mathbf{F}}_\mathrm{h} \partial_z \psi - \tilde{\omega}_\mathrm{h} \dot{\psi} \tag{13.45}$$

$$I^q_z = -\frac{1}{\rho_0} \hat{\mathbf{F}}_\mathrm{h} \cdot \nabla_\mathrm{h} \psi - (\omega_z + f) \dot{\psi} \tag{13.46}$$

For $\psi = \rho$ the divergence of the first term of $\mathbf{I}^q_\mathrm{h}$ vanishes.

13.4 Rigid lid approximation

Often one introduces the surface pressure $p_\mathrm{s} = p(z = 0)$ as a prognostic variable instead of η. Time differentiation of p_s yields

$$(\partial_t + \mathbf{u}_\mathrm{h} \cdot)(p_\mathrm{s} - p_\mathrm{a}) = \rho_0 g_0 \left(w - \frac{E - P}{\rho_0} \right) \tag{13.47}$$

which acts as a prognostic boundary condition for the pressure field. The primitive equations then do not contain the surface elevation any more. It can be calculated a posteriori from the surface pressure

$$\eta = \frac{p_\mathrm{s} - p_\mathrm{a}}{\rho_0 g_0} \tag{13.48}$$

The surface pressure adjusts through the propagation of surface waves. If this adjustment is assumed to be instantaneous ($g_0 = \infty$), then the equation for the surface pressure reduces to its steady or diagnostic limit

$$w = \frac{E - P}{\rho_0} \quad \text{at} \quad z = 0 \tag{13.49}$$

This is the rigid lid approximation. The surface pressure p_s must now be calculated from the diagnostic constraint

$$\nabla_\mathrm{h} \cdot \int_{-H}^{0} \mathrm{d}z \, \mathbf{u}_\mathrm{h} = -\frac{E - P}{\rho_0} \tag{13.50}$$

The surface elevation is still determined by (13.48). The rigid lid approximation thus does not imply that there is no surface elevation, but only that it adjusts instantaneously. The rigid lid approximation eliminates surface gravity waves.

A real rigid lid would imply $w = 0$ and $\eta = 0$ and no forcing by a freshwater flux or the atmospheric pressure would be possible.

13.5 Homogeneous ocean

For a homogeneous ocean $\rho = \rho_0 =$ constant and the primitive equations become the *shallow water equations*. The shallow water equations are not only used to describe large-scale flows but also tidal flows and long surface gravity waves (see Chapter 17 and Section 21.3). For these latter motions $|\eta|$ is not necessarily smaller than H and one has to evaluate the boundary conditions at $z = \eta$. We will do so in this section in order to have a common reference. The assumption of a constant density simplifies the primitive equations considerably. First, forcing by heat and salt fluxes must be discarded. Second, the pressure is given by (see (13.38))

$$p(z) = \rho_0 g_0 \eta + p_a - \rho_0 g_0 z \tag{13.51}$$

and the horizontal pressure gradient by

$$\nabla_h p = \rho_0 g_0 \nabla_h \eta + \nabla_h p_a = \rho_0 g_0 \nabla_h (\eta - \eta_a) \tag{13.52}$$

The pressure force is hence independent of depth. For consistency, one requires that the eddy term $\mathbf{F}_h = \mathbf{F}_h^h + \mathbf{F}_h^v$ with $\mathbf{F}_h^v = \partial_z \boldsymbol{\tau}_h$ is depth-independent as well. Then the horizontal velocity $\mathbf{u}_h$ may be assumed to be depth-independent. The internal modes that have a z-dependent structure are thus eliminated and only the external mode is retained. Specifically, the stress $\boldsymbol{\tau}_h$ must be a linear function of z such that

$$\partial_z \boldsymbol{\tau}_h = \frac{\boldsymbol{\tau}_h^s - \boldsymbol{\tau}_h^b}{H + \eta} \tag{13.53}$$

For large-scale motions, the surface stress is generally assumed to be the prescribed atmospheric windstress, $\boldsymbol{\tau}_h^s = \boldsymbol{\tau}_h^a$, and the bottom stress is generally parametrized by linear or quadratic bottom friction (see (13.34)).

Equations The primitive equation for a homogeneous ocean or the shallow water equations thus take the form

$$(\partial_t + \mathbf{u}_h \cdot \nabla_h)\mathbf{u}_h + f\hat{\mathbf{u}}_h = -g_0 \nabla_h (\eta - \eta_a) + \frac{1}{\rho_0}\mathbf{F}_h \tag{13.54}$$

$$\partial_t \eta + \nabla_h \cdot [(H + \eta)\mathbf{u}_h] = -\frac{E - P}{\rho_0} \tag{13.55}$$

The second equation is the depth integrated incompressibility condition and explicitly expresses volume conservation. The equations are prognostic equations for the horizontal velocity $\mathbf{u}_h$ and surface elevation η. The pressure and vertical velocity do not enter but can be calculated from (13.51) and (see (13.37))

$$w(z) = w(-H) - (z + H)\nabla_h \cdot \mathbf{u}_h \tag{13.56}$$

The volume conservation equation becomes linear in η and $\mathbf{u}_h$ if $|\eta| \ll H$ is assumed. In terms of the thickness $h := H + \eta$ of the water column it takes the form

$$\partial_t h + \nabla_h \cdot (h\mathbf{u}_h) = -\frac{E-P}{\rho_0} \tag{13.57}$$

If the depth integrated volume transport $\mathbf{M}_h := h\mathbf{u}_h$ is introduced, instead of $\mathbf{u}_h$, then the volume equation becomes linear in h and $\mathbf{M}_h$ regardless of the magnitude of η.

Vorticity By taking the curl of the momentum equation (13.54), i.e., by applying the operator $\hat{\nabla}_h \cdot$, one obtains the vorticity equation

$$(\partial_t + \mathbf{u}_h \cdot \nabla_h)\omega_z = -(\omega_z + f)\nabla_h \cdot \mathbf{u}_h + \frac{1}{\rho_0}\hat{\nabla}_h \cdot \mathbf{F}_h \tag{13.58}$$

Consider the relative height of a fluid particle

$$\sigma = \frac{z-\eta}{H+\eta} \tag{13.59}$$

It is $\sigma = 0$ at $z = \eta$ and $\sigma = -1$ at $z = -H$. It is materially conserved except for the effect of a freshwater flux

$$\frac{\mathrm{D}}{\mathrm{D}t}\sigma = \frac{z+H}{(H+\eta)^2}\frac{E-P}{\rho_0} \tag{13.60}$$

Consider next Ertel's potential vorticity

$$q_\sigma := (\omega_z + f)\,\partial_z \sigma = \frac{\omega_z + f}{h} \tag{13.61}$$

Its evolution is governed by

$$\partial_t q_\sigma + \nabla_h \cdot \left(\mathbf{u}_h q_\sigma + \mathbf{I}_h^q\right) + \partial_z\left(w q_\sigma + I_z^q\right) = 0 \tag{13.62}$$

where

$$\mathbf{I}_h^q := \frac{1}{\rho_0}\hat{\mathbf{F}}_h \partial_z \sigma \tag{13.63}$$

$$I_z^q := -\frac{1}{\rho_0}\hat{\mathbf{F}}_h \cdot \nabla_h \sigma - (\omega_z + f)\dot{\sigma} \tag{13.64}$$

which becomes

$$(\partial_t + \mathbf{u}_h \cdot \nabla_h)\, q_\sigma = \frac{1}{\rho_0 h}\hat{\nabla}_h \cdot \mathbf{F}_h + \frac{\omega_z + f}{h^2}\frac{E-P}{\rho_0} \tag{13.65}$$

If the curl of the eddy stress and the freshwater flux vanish then the potential vorticity $(\omega_z + f)/h$ is materially conserved. The atmospheric pressure p_a does not affect the potential vorticity. This simple form of the potential vorticity and potential vorticity equation is exploited in many interpretations of oceanographic phenomena.[1]

Rigid lid If the rigid lid approximation is made and the freshwater flux neglected the volume transport $H\mathbf{u}_h$ becomes non-divergent, $\nabla_h \cdot (H\mathbf{u}_h) = 0$, and one can introduce a transport streamfunction Ψ such that

$$\mathbf{u}_h = \frac{1}{H}\hat{\nabla}_h \Psi \tag{13.66}$$

The streamfunction is related to the relative vorticity by the Poisson equation

$$\nabla_h \cdot \left(\frac{1}{H} \nabla_h \Psi \right) = \omega_z \tag{13.67}$$

The evolution of ω_z is governed by the potential vorticity equation

$$(\partial_t + \mathbf{u}_h \cdot \nabla_h) \left(\frac{\omega_z + f}{H} \right) = \frac{1}{\rho_0} \frac{1}{H} \hat{\nabla}_h \cdot \mathbf{F}_h \tag{13.68}$$

This prognostic equation for ω_z and the diagnostic equations for Ψ and $\mathbf{u}_h$ constitute a closed set of equations for the homogeneous ocean. These equations describe the vorticity carrying or vortical mode of motion. The gravity mode of motion has been eliminated by the rigid lid approximation. We discuss this case further in Chapter 16 in conjunction with the geostrophic approximation. The pressure can be calculated from the momentum equation if needed, which might require integrability conditions (see Section 16.2).

[1] Often the reference to Ertel is suppressed since this form of the potential vorticity equation was derived independently and only later recognized as a special case of Ertel's general potential vorticity theorem.

14

Representation of vertical structure

The vertical structure of the flow field has been represented so far by specifying all field variables as a function of the vertical coordinate z. This is not the only possibility. In this chapter we describe other representations that are used commonly in oceanography:

1. the decomposition into barotropic and baroclinic flow components;
2. isopycnal coordinates;
3. sigma coordinates;
4. layer models; and
5. projection onto vertical normal modes.

All these representations recover the full vertical structure, except the layer models that are discrete representations. No additional assumptions need to be made for these representations, although their specific form often invites ancillary assumptions.

14.1 Decomposition into barotropic and baroclinic flow components

Decomposition As discussed in Chapter 8 on waves, the external or barotropic modes are characterized by the parameter gH_0, which is much larger than the parameter $N_0^2 H_0^2$ characterizing the internal or baroclinic modes. To separate the different underlying physics associated with these two modes one decomposes the flow field into its barotropic and baroclinic components. The barotropic and baroclinic components of the horizontal velocity are defined by

$$\mathbf{u}_\mathrm{h}^\mathrm{bt} := \frac{1}{H} \int_{-H}^{0} \mathrm{d}z \, \mathbf{u}_\mathrm{h} \tag{14.1}$$

$$\mathbf{u}_\mathrm{h}^\mathrm{bc} := \mathbf{u}_\mathrm{h} - \mathbf{u}_\mathrm{h}^\mathrm{bt} \tag{14.2}$$

Note that the integral extends to $z = 0$ only. The small contribution from the interval $(0, \eta)$ is neglected, as is appropriate for large-scale phenomena. The barotropic component is the depth average. It carries the integrated horizontal volume transport

$$\mathbf{M}_\mathrm{h} := \int_{-H}^{0} \mathrm{d}z\, \mathbf{u}_\mathrm{h} = H\mathbf{u}_\mathrm{h}^\mathrm{bt} \tag{14.3}$$

The baroclinic part does not contribute to the volume transport

$$\int_{-H}^{0} \mathrm{d}z\, \mathbf{u}_\mathrm{h}^\mathrm{bc} = 0 \tag{14.4}$$

Its depth integral is zero. Integration of the incompressibility condition yields

$$\partial_t \eta + \nabla_\mathrm{h} \cdot \mathbf{M}_\mathrm{h} = -\frac{E - P}{\rho_0} \tag{14.5}$$

The barotropic and baroclinic parts of the vertical velocity are defined by

$$w_\mathrm{bt} := -\nabla_\mathrm{h} \cdot \int_{-H}^{z} \mathrm{d}z'\, \mathbf{u}_\mathrm{h}^\mathrm{bt} \tag{14.6}$$

$$w_\mathrm{bc} := -\nabla_\mathrm{h} \cdot \int_{-H}^{z} \mathrm{d}z'\, \mathbf{u}_\mathrm{h}^\mathrm{bc} \tag{14.7}$$

which implies

$$w_\mathrm{bc}(z = 0) = 0 \tag{14.8}$$

Note that the barotropic part of the vertical velocity is not the depth averaged part.

The barotropic and baroclinic pressures are defined by

$$p_\mathrm{bt} := \rho_0 g \eta + p_\mathrm{a} \tag{14.9}$$

$$p_\mathrm{bc} := \int_{z}^{0} \mathrm{d}z'\, \rho g \tag{14.10}$$

Again, p_bt is not the depth averaged pressure, which is given by

$$\frac{1}{H}\int_{-H}^{0} \mathrm{d}z\, p = \frac{1}{H}\int_{-H}^{0} (\mathrm{d}(zp) - z\,\mathrm{d}p) = p(-H) + \frac{1}{H}\int_{-H}^{0} \mathrm{d}z\, \rho g z \tag{14.11}$$

One now defines:

- the barotropic flow field by $(\mathbf{u}_\mathrm{h}^\mathrm{bt}, w^\mathrm{bt}, p_\mathrm{bt}, \eta)$ or $(\mathbf{M}_\mathrm{h}, w^\mathrm{bt}, p_\mathrm{bt}, \eta)$; and
- the baroclinic flow field by $(\mathbf{u}_\mathrm{h}^\mathrm{bc}, w^\mathrm{bc}, p_\mathrm{bc}, \theta, S, \rho)$.

Equations for the barotropic flow field The equation for the barotropic horizontal velocity is obtained by depth averaging the horizontal momentum balance. The resulting expression is somewhat complicated because depth averaging does not commute with taking the horizontal gradient if the depth varies in space. This property complicates the evaluation of the nonlinear and horizontal friction terms. If these terms are disregarded[1] one obtains

$$\partial_t \mathbf{M}_\mathrm{h} + f\hat{\mathbf{M}}_\mathrm{h} = -\frac{H}{\rho_0}\nabla_\mathrm{h} p_\mathrm{bt} - \mathbf{T} + \frac{1}{\rho_0}\left[\boldsymbol{\tau}_\mathrm{h}^\mathrm{a} - \boldsymbol{\tau}_\mathrm{h}^\mathrm{b}\right] \tag{14.12}$$

where we have used the boundary conditions $\boldsymbol{\tau}_\mathrm{h} = \boldsymbol{\tau}_\mathrm{h}^{\mathrm{a,b}}$ at $z = 0, -H$ and where the force

$$\mathbf{T} := \frac{1}{\rho_0}\int_{-H}^{0} \mathrm{d}z\, \nabla_\mathrm{h} p_\mathrm{bc} \tag{14.13}$$

describes the influence of the baroclinic field. This term can be rewritten as

$$\mathbf{T} = \frac{1}{\rho_0}\left[\nabla_\mathrm{h}\int_{-H}^{0} \mathrm{d}z\, p_\mathrm{bc} - p_\mathrm{bc}(-H)\nabla_\mathrm{h} H\right] \tag{14.14}$$

Using

$$\mathrm{d}z\, p_\mathrm{bc} = \mathrm{d}(z p_\mathrm{bc}) - z\,\mathrm{d}p_\mathrm{bc} = \mathrm{d}(z p_\mathrm{bc}) + z\rho g\,\mathrm{d}z \tag{14.15}$$

one finds

$$\mathbf{T} = \frac{1}{\rho_0}\left[\nabla_\mathrm{h} E_\mathrm{pot} + H\nabla_\mathrm{h} p_\mathrm{bc}(-H)\right] \tag{14.16}$$

where

$$E_\mathrm{pot} := \int_{-H}^{0} \mathrm{d}z\, \rho g z \tag{14.17}$$

is the potential energy.

The evolution of the barotropic pressure is governed by

$$(\partial_t + \mathbf{u}_\mathrm{h}\cdot\nabla_\mathrm{h})(p_\mathrm{bt} - p_\mathrm{a}) = -\rho_0 g\left(\nabla_\mathrm{h}\cdot\mathbf{M}_\mathrm{h} + \frac{E-P}{\rho_0}\right) \tag{14.18}$$

The equations for $\mathbf{M}_\mathrm{h}$ and p_bt constitute the prognostic equations for the barotropic flow. Alternatively, one could have used $\mathbf{u}_\mathrm{h}^\mathrm{bt}$ and η as prognostic variables. If the term $\mathbf{T}$ is neglected then the equations for the barotropic flow component are identical to the equations for a homogeneous fluid, discussed in Section 13.5. The equations for the barotropic flow field are forced by:

- the atmospheric pressure p_a;
- the atmospheric windstress $\boldsymbol{\tau}_\mathrm{h}^\mathrm{a}$;
- the freshwater flux $E - P$; and
- the baroclinic flow through $\mathbf{T}$.

[1] In Section 13.1 we have seen that these terms are indeed small if $\mathrm{Ro} \ll 1$ and $E_\mathrm{h} \ll 1$.

The bottom stress $\boldsymbol{\tau}_{\rm h}^{\rm b}$ may also depend on the baroclinic flow, as is the case if one assumes linear or nonlinear bottom friction or a no-slip bottom boundary condition since these conditions depend on total $\mathbf{u}_{\rm h}$ and not just the barotropic component. Tidal forcing also affects the barotropic component.

The surface elevation and vertical velocity are given by

$$\eta = \frac{p_{\rm bt} - p_{\rm a}}{\rho_0 g} \tag{14.19}$$

$$w_{\rm bt}(z) = -\boldsymbol{\nabla}_{\rm h} \left[\left(\frac{z}{H} + 1 \right) \mathbf{M}_{\rm h} \right] \tag{14.20}$$

As discussed in Section 13.5 for the homogeneous ocean, one can apply the rigid lid approximation to eliminate surface gravity waves and focus on the potential vorticity carrying or vortical mode of motion.

Equations for the baroclinic flow field The equations for the baroclinic flow field ($\mathbf{u}_{\rm h}^{\rm bc}$, $w^{\rm bc}$, $p_{\rm bc}$, θ, S, ρ) are given by prognostic equations for $\mathbf{u}_{\rm h}^{\rm bc}$, θ, and S

$$\partial_t \mathbf{u}_{\rm h}^{\rm bc} + f \hat{\mathbf{u}}_{\rm h}^{\rm bc} = -\frac{1}{\rho_0} \boldsymbol{\nabla}_{\rm h} p^{\rm bc} + \frac{\mathbf{T}}{H} + \frac{1}{\rho_0} \partial_z \boldsymbol{\tau}_{\rm h} - \frac{1}{\rho_0 H} \left[\boldsymbol{\tau}_{\rm h}^{\rm a} - \boldsymbol{\tau}_{\rm h}^{\rm b} \right] \tag{14.21}$$

$$\left[\partial_t + \left(\mathbf{u}_{\rm h}^{\rm bt} + \mathbf{u}_{\rm h}^{\rm bc} \right) \cdot \boldsymbol{\nabla}_{\rm h} + \left(w^{\rm bt} + w^{\rm bc} \right) \partial_z \right] \begin{pmatrix} \theta \\ S \end{pmatrix} = - \begin{pmatrix} \boldsymbol{\nabla} \cdot \mathbf{I}_\theta \\ \boldsymbol{\nabla} \cdot \mathbf{I}_S \end{pmatrix} \tag{14.22}$$

and by diagnostic equations for $w^{\rm bc}$, $p^{\rm bc}$, and ρ

$$w^{\rm bc}(z) = \boldsymbol{\nabla}_{\rm h} \int_{-H}^{z} {\rm d}z' \, \mathbf{u}_{\rm h}^{\rm bc} \tag{14.23}$$

$$p^{\rm bc}(z) = \int_z^0 {\rm d}z' \, \rho g \tag{14.24}$$

$$\rho = \rho(z, \theta, S) \tag{14.25}$$

The baroclinic flow is forced by:

- the atmospheric windstress $\boldsymbol{\tau}_{\rm h}^{\rm a}$;
- the atmospheric and geothermal heat fluxes $Q^{\rm a,b}$;
- the surface salinity flux $(E - P)S$; and
- the barotropic velocity $\mathbf{u}_{\rm h}^{\rm bt}$.

The salinity and heat fluxes enter through the boundary conditions $\rho_0 I_z^S = -(E - P)S$ at the surface and $\rho_0 c_{\rm p} I_z^\theta = Q^{\rm a,b}$ at the surface and bottom. The barotropic velocity enters the advection operator in the temperature and salinity equation and the dynamic bottom boundary condition if this condition depends on total $\mathbf{u}_{\rm h}$ and not just the baroclinic component.

If nonlinear momentum advection and horizontal friction were included then additional coupling occurs between the barotropic and baroclinic components.

14.2 Generalized vertical coordinates

The fluid dynamic equations can be expressed in any coordinate system. The physical laws expressed by these equations are independent of the chosen coordinate system. Indeed, for theoretical purposes one usually strives to express these laws in coordinate invariant form. Nevertheless, expressing the equations in certain specific coordinate systems may have computational and conceptual advantages. One class of coordinate systems is obtained by replacing the vertical coordinate z by a generalized vertical coordinate

$$s = s(x_1, x_2, z, t) \tag{14.26}$$

Then $z = z(x_1, x_2, s, t)$ becomes a dependent variable. Such a replacement is possible if s is a strictly monotonic function z. The most common choices in oceanography are

1. isopycnal coordinates $s = \rho, \rho_{\text{pot}}$
2. sigma-coordinate system $s = (z - \eta)/(H + \eta)$

In meteorology, one also uses the isentropic ($s = \eta$), the isobaric ($s = p$), and the normalized pressure ($s = p/p_{\text{surface}}$) coordinate systems. Three points need to be noted.

1. The transformation from (x_1, x_2, z, t) to (x_1, x_2, s, t) is a time-dependent transformation.
2. The curvilinear coordinates (x_1, x_2, s) are not orthogonal. One might be tempted to introduce orthogonal curvilinear coordinates $(q_1, q_2, q_3 = s)$ and associated velocity components normal and tangential to the isopycnal surface, but then the tangential velocity components would be affected by gravity. For this reason one stays with the (x_1, x_2, s)-coordinate system and the horizontal velocities.
3. Because of the above two points we cannot use the coordinate transformation formulae for orthogonal curvilinear coordinates listed in Appendix C. The appropriate transformation formulae are listed next.

Transformation formulae For any function $f(x_1, x_2, z, t) = f(x_1, x_2, s(x_1, x_2, z, t), t)$ the derivatives at constant z and s are related by

$$\left.\frac{\partial f}{\partial a}\right|_z = \left.\frac{\partial f}{\partial a}\right|_s + \frac{\partial f}{\partial s}\frac{\partial s}{\partial a} \tag{14.27}$$

where a can be any of the independent variables x_1, x_2, and t. Application of this formula to $f = z$ implies

$$\frac{\partial s}{\partial a} = -\frac{\partial z/\partial a}{\partial z/\partial s} \tag{14.28}$$

and hence the transformation formulae

$$\left.\frac{\partial f}{\partial t}\right|_z = \left.\frac{\partial f}{\partial t}\right|_s - \frac{\partial f/\partial s}{\partial z/\partial s}\left.\frac{\partial z}{\partial t}\right|_s \tag{14.29}$$

$$\nabla_{\mathrm{h}}^z f = \nabla_{\mathrm{h}}^s f - \frac{\partial f/\partial s}{\partial z/\partial s}\nabla_{\mathrm{h}}^s z \tag{14.30}$$

$$\frac{\partial f}{\partial z} = \frac{1}{\partial z/\partial s}\frac{\partial f}{\partial s} \tag{14.31}$$

where ∇_{h}^s is the horizontal gradient operator at constant s. On the left-hand side f is a function of (x_1, x_2, z, t) and on the right-hand side a function of (x_1, x_2, s, t). In the following we do not indicate the variables that are held constant during differentiation, when this should be evident from the context.

Vorticity Using the above transformation formulae one finds that the vorticity components in height coordinates are given by

$$\tilde{\omega}_{\mathrm{h}} := \partial_z \hat{\mathbf{u}}_{\mathrm{h}} = \frac{1}{\partial z/\partial s}\tilde{\omega}_{\mathrm{h}}^s \tag{14.32}$$

$$\omega_z := \hat{\nabla}_{\mathrm{h}}^z \cdot \mathbf{u}_{\mathrm{h}} = \omega_s + \frac{1}{\partial z/\partial s}\tilde{\omega}_{\mathrm{h}}^s \cdot \nabla_{\mathrm{h}}^s z \tag{14.33}$$

where

$$\tilde{\omega}_{\mathrm{h}}^s := \partial_s \hat{\mathbf{u}}_{\mathrm{h}} \tag{14.34}$$

$$\omega_s := \hat{\nabla}_{\mathrm{h}}^s \cdot \mathbf{u}_{\mathrm{h}} \tag{14.35}$$

Vertical velocity When transforming to a general vertical coordinate s one also eliminates the vertical velocity $w = \mathrm{D}z/\mathrm{D}t$, which is the fluid velocity through a $z =$ constant surface. Instead one introduces the fluid velocity through a $s =$ constant surface, which is given by

$$\tilde{w} = (\mathbf{u} - \mathbf{v}_s) \cdot \mathbf{n}_s \tag{14.36}$$

where $\mathbf{v}_s = -\partial_t s \nabla s/|\nabla s|^2$ is the velocity and $\mathbf{n}_s = \nabla s/|\nabla s|$ the normal vector of the $s =$ constant surface. It is related to the Lagrangian derivative

$$\dot{s} := \frac{\mathrm{D}}{\mathrm{D}t}s \tag{14.37}$$

by

$$\tilde{w} = \frac{\dot{s}}{|\nabla s|} \tag{14.38}$$

Since ∇s is given in (x_1, x_2, s) coordinates by

$$\nabla s = \frac{1}{\partial z/\partial s}\left(-\frac{\partial z}{\partial x_1}, -\frac{\partial z}{\partial x_2}, 1\right) \tag{14.39}$$

one can solve (14.36) and (14.38) for the vertical velocity and obtain

$$w = (\partial_t + \mathbf{u}_\mathrm{h} \cdot \nabla_\mathrm{h} + \dot{s}\partial_s)\, z \tag{14.40}$$

Lagrangian derivative From the above formulae it follows that the Lagrangian derivative transforms as

$$(\partial_t + \mathbf{u}_\mathrm{h} \cdot \nabla_\mathrm{h} + w\partial_z)\, f = (\partial_t + \mathbf{u}_\mathrm{h} \cdot \nabla_\mathrm{h} + \dot{s}\partial_s)\, f \tag{14.41}$$

where f is again a scalar function of (x_1, x_2, z, t) on the left-hand side and a function of (x_1, x_2, s, t) on the right-hand side. The Lagrangian derivative has the same form in both coordinate systems. If s is a materially conserved tracer, then $\dot{s} = 0$ and the advection operator becomes two-dimensional, a major simplification. Note that $\mathbf{u}_\mathrm{h}$ is the horizontal velocity vector, not the velocity in a s = constant surface.

Tracer equation The tracer equation

$$\partial_t f + \nabla \cdot (f\mathbf{u}) = 0 \tag{14.42}$$

transforms into

$$\partial_t(hf) + \nabla_\mathrm{h} \cdot (\mathbf{u}_\mathrm{h} hf) + \partial_s(\dot{s}hf) = 0 \tag{14.43}$$

where

$$h = -\Delta s\ \partial z/\partial s \tag{14.44}$$

is the thickness of the layer between the surfaces s = constant and $s + \Delta s$ = constant. For $f = 1$ the tracer equation becomes the "incompressibility" condition $\nabla \cdot \mathbf{u} = 0$.

Pressure gradient and hydrostatic balance The horizontal pressure gradient is given by

$$\begin{aligned} \nabla_\mathrm{h}^z p &= \nabla_\mathrm{h} p - \frac{\partial p/\partial s}{\partial z/\partial s}\nabla_\mathrm{h} z \\ &= \nabla_\mathrm{h} p + \rho g \nabla_\mathrm{h} z \\ &= \nabla_\mathrm{h} M - g z \nabla_\mathrm{h} \rho \end{aligned} \tag{14.45}$$

if one uses the hydrostatic balance and introduces the *Montgomery* or *acceleration potential*

$$M := p + \rho g z \tag{14.46}$$

In terms of this potential the hydrostatic balance takes the form

$$\frac{\partial M}{\partial s} = \frac{\partial \rho}{\partial s} g z \tag{14.47}$$

14.3 Isopycnal coordinates

The most common choices of s in oceanography are $s = \rho$ and $s = \rho_{\text{pot}}$. These choices lead to isopycnal coordinate systems. They are motivated by the fact that some physical processes might align themselves more with isopycnal rather than with horizontal surfaces. Stirring by mesoscale eddies is often assumed to take place along isopycnal surfaces whereas mixing across isopycnal surfaces is assumed to be caused by nearly isotropic turbulence (generated for example by breaking internal waves). These same ideas were also behind the rotation of the eddy diffusion tensor so that its principal axes align with isopycnals, as discussed in Section 12.2.

The choice $s = \rho$ has the advantage that the horizontal pressure gradient is solely given by the gradient of the Montgomery potential ($\nabla_{\text{h}}\rho = 0$ for $s = \rho$) and that the hydrostatic balance takes the simple form $\partial M/\partial \rho = gz$. It has the disadvantage that (adiabatic) compressibility effects enter $\dot{\rho}$. These effects are eliminated by the choice $s = \rho_{\text{pot}}$ but then one needs to calculate the density since it enters the horizontal pressure gradient and hydrostatic balance. Often, one tries to combine the advantages of both choices by assuming that $\partial p/\partial z = -\rho_{\text{pot}} g$.

An overall complication of isopycnal coordinates is that the top-to-bottom (potential) density range varies with geographic location. Isopycnals thus outcrop at the surface and intersect the bottom. If one solves the equations in a fixed ρ-interval one has to allow for the vanishing or merging of isopycnals in z-space.

Equations Using the transformation formulae of the previous section it is straightforward to formulate the primitive equations in isopycnal coordinates. For an incompressible ocean ($s = \rho = \rho_{\text{pot}}$) they take the form

$$(\partial_t + \mathbf{u}_{\text{h}} \cdot \nabla_{\text{h}} + \dot{\rho}\partial_\rho)\mathbf{u}_{\text{h}} + f\hat{\mathbf{u}}_{\text{h}} = -\frac{1}{\rho_0}\nabla_{\text{h}} M + \frac{1}{\rho_0}\mathbf{F}_{\text{h}} \tag{14.48}$$

$$\frac{\partial M}{\partial \rho} = gz \tag{14.49}$$

$$\partial_t h + \nabla_{\text{h}} \cdot (\mathbf{u}_{\text{h}} h) + \frac{\partial}{\partial \rho}(\dot{\rho} h) = 0 \tag{14.50}$$

$$\frac{\partial z}{\partial \rho} = -\frac{h}{\Delta \rho} \tag{14.51}$$

$$\dot{\rho} = -\rho_0 \tilde{\alpha} D_\theta + \rho_0 \tilde{\beta} D_S \tag{14.52}$$

where

$$h := -\Delta\rho\, \partial z/\partial \rho \tag{14.53}$$

is the height between two isopycnal surfaces separated by $\Delta\rho$. The horizontal momentum balance can be rewritten by using the vector identity $\mathbf{u}_\mathrm{h} \cdot \nabla \mathbf{u}_\mathrm{h} = \nabla(\mathbf{u}_\mathrm{h} \cdot \mathbf{u}_\mathrm{h}/2) + \omega_\rho$.

The above equations constitute prognostic equations for $(\mathbf{u}_\mathrm{h}, h)$ and diagnostic equations for $(M, z, \dot{\rho})$. The pressure and vertical velocity do not enter but can be calculated from $p = M - \rho g z$ and $w = \mathrm{D}z/\mathrm{D}t$.

Eddy fluxes The eddy terms require two comments. First, the eddy terms $\mathbf{F}_\mathrm{h}$ and $D_{\theta,S}$ are generally not specified by transforming parametrization schemes that are formulated in height coordinates to isopycnal coordinates. The very reason for the introduction of isopycnal coordinates is to implement parametrization schemes that better represent eddy effects that might align themselves more with isopycnal rather than with height coordinates. The eddy terms are thus specified anew. These specifications have to be mathematically and physically consistent.

Second, the above equations can be obtained by transforming the primitive equations that were Reynolds averaged in height coordinates to isopycnal coordinates. The density ρ is then the mean density $\overline{\rho}(x_1, x_2, z, t)$ and $z = z(x_1, x_2, \overline{\rho}, t)$ the height of the mean density surface. Then $\mathbf{F}_\mathrm{h}$ and $D_{\theta,S}$ should have a superscript "eddy." Alternatively, one could have transformed the unaveraged primitive equations to isopycnal coordinates. The density ρ would then be the unaveraged density and z the height of the unaveraged density surface. There would be no eddy terms in the equations and $\mathbf{F}_\mathrm{h}$ and $D_{\theta,S}$ should have a superscript "mol." Reynolds averaging of theses equations then leads to equations that differ from the ones given above. First, the means are different. In the first case, they are taken as averages over z, in the second case they are averages over ρ. The mean height of a density surface is generally different from the height of the mean density surface. Second, there appear new types of eddy terms since, for example, the incompressibility condition is a nonlinear equation in isopycnal coordinates.

Potential vorticity If one chooses $\psi = \rho$ then Ertel's potential vorticity $q_\rho := \tilde{\omega}_\mathrm{h} \cdot \nabla\rho + (\omega_z + f)\partial_z \rho$ is given by

$$q_\rho = -\Delta\rho \frac{\omega_\rho + f}{h} \tag{14.54}$$

in isopycnal coordinates. It does not depend on the component $\tilde{\omega}_\mathrm{h}^\rho := \partial_\rho \hat{\mathbf{u}}_\mathrm{h}$ but only on the component $\omega_\rho := \hat{\nabla}_\mathrm{h}^\rho \cdot \mathbf{u}_\mathrm{h}$. It resembles but differs from q_σ for a homogeneous ocean (see (13.61)). Again, this simple form of the potential vorticity is exploited in many interpretations of oceanographic phenomena. The evolution equation for the potential vorticity can be inferred from the vorticity equation

$$\frac{\mathrm{D}}{\mathrm{D}t}(\omega_\rho + f) = -(\omega_\rho + f)\nabla_{\mathrm{h}} \cdot \mathbf{u}_{\mathrm{h}} + \tilde{\omega}_{\mathrm{h}}^{\rho} \cdot \nabla_{\mathrm{h}}\dot{\rho} + \frac{1}{\rho_0}\hat{\nabla}_{\mathrm{h}} \cdot \mathbf{F}_{\mathrm{h}} \quad (14.55)$$

and the incompressibility condition (14.50) and is given by

$$h\frac{\mathrm{D}}{\mathrm{D}t}\frac{\omega_\rho + f}{h} = -\nabla_{\mathrm{h}} \cdot \mathbf{I}_{\mathrm{h}}^{q} - \frac{\partial}{\partial \rho} I_{\rho}^{q} \quad (14.56)$$

where

$$\mathbf{I}_{\mathrm{h}}^{q} := \frac{1}{\rho_0}\hat{\mathbf{F}}_{\mathrm{h}} - \tilde{\omega}_{\mathrm{h}}^{\rho}\dot{\rho} \quad (14.57)$$

$$I_{\rho}^{q} := -(\omega_\rho + f)\dot{\rho} \quad (14.58)$$

14.4 Sigma-coordinates

The choice $s = \sigma$ with

$$\sigma = \frac{z - \eta(x_1, x_2, t)}{H(x_1, x_2) + \eta(x_1, x_2, t)} \quad (14.59)$$

leads to the sigma-coordinate system that might efficiently represent the effects of bathymetry and side wall geometry since the "vertical" domain becomes uniform, from $\sigma = 0$ to $\sigma = -1$. The variable $z = z(x_1, x_2, \sigma = 0, t)$ then determines the location of the surface and the variable $z = z(x_1, x_2, \sigma = -1, t)$ is the location of the bottom.

For surface elevations that are small compared to the vertical scale D one expands the kinematic boundary condition about $z = 0$ (as we did in Section 13.2; see (13.28)) and introduces

$$\sigma = \frac{z}{H} \quad (14.60)$$

which is linear in z and time independent. In this case the horizontal pressure gradient becomes

$$\nabla_{\mathrm{h}}^{z} p = \nabla_{\mathrm{h}} p + \rho g \sigma \nabla_{\mathrm{h}} H \quad (14.61)$$

and the hydrostatic balance

$$\frac{\partial p}{\partial \sigma} = -\rho g H \quad (14.62)$$

The location of the surface has then to be determined from $\partial_t \eta = w(x_1, x_2, \sigma = 0, t)$. The complete set of primitive equations in sigma-coordinates will not be given here but can easily be constructed from the transformation formulae.

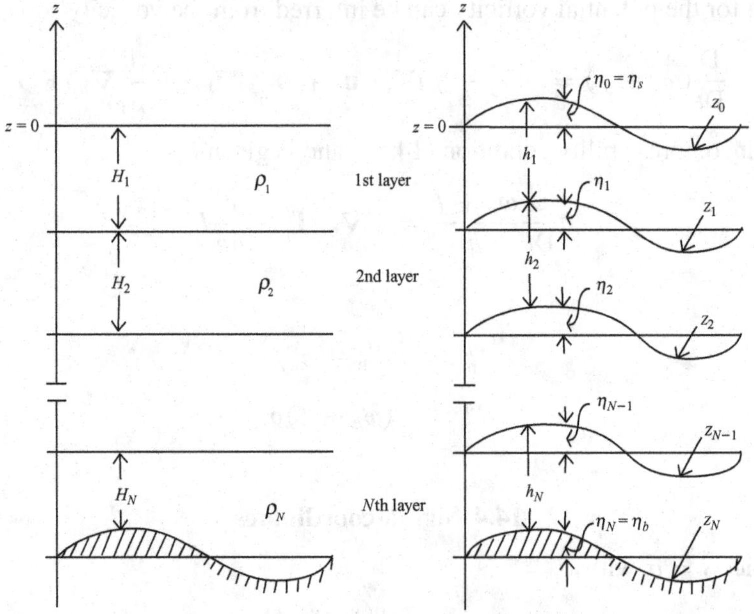

Figure 14.1. Basic geometry and definitions of N-layer model. The background state is sketched on the left and the perturbed state on the right.

14.5 Layer models

Layer models can either be viewed as a physical realizable system or as discrete versions of isopycnal models.

N-layer models

Background state. Consider a stack of N immiscible fluid layers (Figure 14.1) of densities and unperturbed thicknesses

$$\begin{array}{lll} \rho_i & i = 1, \ldots, N & \text{density of } i\text{th layer} \\ H_i & i = 1, \ldots, N & \text{unperturbed thickness of } i\text{th layer} \end{array} \qquad (14.63)$$

The densities are constant within each layer and are prescribed. We thus have to neglect density changes owing to compressibility and heat and salt diffusion. This excludes driving by the atmospheric and geothermal heat fluxes $Q^{\mathrm{a,b}}$ and salinity flux $(E - P)S$ but allows for driving by a mass flux $E - P$. The stratification is assumed to be stable, $\rho_1 < \rho_2 < \ldots < \rho_{N-1} < \rho_N$. The unperturbed thicknesses $H_1, \ldots, H_{N-1}$ are all constant. The bottom layer thickness is $H_N(x_1, x_2)$. The total depth is $H = \sum_{i=1}^{i=N} H_i$. There are $N + 1$ interfaces. The interface $i = 0$ is the surface. The interface $i = N$ is the bottom. The background state is motionless and in hydrostatic balance.

Variables. Perturbations about this background state will be described by the variables (see Figure 14.1):

$$
\begin{array}{lll}
z_i & i = 0, \ldots, N & \text{position of the interface} \\
 & & \text{between the } (i+1)\text{th and } i\text{th layers} \\
h_i & i = 1, \ldots, N & \text{thickness of } i\text{th layer} \\
\eta_i & i = 0, \ldots, N & \text{displacement of the } i\text{th interface} \\
p_i & i = 1, \ldots, N & \text{pressure in } i\text{th layer} \\
M_i & i = 1, \ldots, N & \text{Montgomery potential in } i\text{th layer} \\
\mathbf{u}_\mathrm{h}^i & i = 1, \ldots, N & \text{horizontal velocity in } i\text{th layer}
\end{array}
\tag{14.64}
$$

The surface is given by z_0 and the bottom by $z_N = -H$.

Equations. The governing equations consist of the primitive equations in each homogeneous layer. The equations are coupled since the pressure p, stress $\boldsymbol{\tau}_\mathrm{h}$, and normal velocity component must be continuous across the interfaces. The tangential velocities are allowed to be discontinuous. The layers can slide across each other. This coupling takes the form of simple recursion formulae.

Interface position and layer thickness are related by

$$z_{i-1} = z_i + h_i \tag{14.65}$$

or

$$z_{i-1} = -H + \sum_{j=N}^{i} h_j \tag{14.66}$$

This is an upward recursion formula for z_i if h_i is given starting at $z_N = -H$.

Integration of the hydrostatic balance from z_i to z in the ith and $(i+1)$th layer yields[2]

$$
\begin{aligned}
p_i(z) - p_i(z_i) &= -(z - z_i)\,\rho_i g \\
p_{i+1}(z_i) - p_{i+1}(z) &= -(z_i - z)\,\rho_{i+1} g
\end{aligned}
\tag{14.67}
$$

The Montgomery potentials in each layer are thus given by

$$
\begin{aligned}
M_i &= p_i(z) + \rho_i g z = p_i(z_i) + z_i \rho_i g \\
M_{i+1} &= p_{i+1}(z) + \rho_{i+1} g z = p_{i+1}(z_i) + z_i \rho_{i+1} g
\end{aligned}
\tag{14.68}
$$

and do not depend on z. If one implements the interface condition that the pressure is continuous across interfaces ($p_{i+1}(z_i) = p_i(z_i)$) then the Montgomery potentials

[2] In this and the following chapters we deviate from our standard notation and replace g_0 by g in order not to confuse the subscipt 0 with a layer index.

are related by the downward recursion formula

$$M_{i+1} = M_i + (\rho_{i+1} - \rho_i) g z_i \tag{14.69}$$

or

$$M_{i+1} = p_\mathrm{a} + \sum_{j=1}^{i} \rho_j g h_j + \rho_{i+1} g z_i \tag{14.70}$$

The recursion starts at $M_1 = p_\mathrm{a} + z_0 \rho_1 g$ where p_a is the atmospheric pressure.

The horizontal pressure gradient is given by the horizontal gradient of the Montgomery potential

$$\nabla_\mathrm{h} p = \nabla_\mathrm{h} M \tag{14.71}$$

and independent of depth within each layer. As has already been pointed out in Section 13.5 one requires that the eddy term $\mathbf{F}_\mathrm{h} = \mathbf{F}_\mathrm{h}^\mathrm{h} + \mathbf{F}_\mathrm{h}^\mathrm{v}$ with $\mathbf{F}_\mathrm{h}^\mathrm{v} = \partial_z \boldsymbol{\tau}_\mathrm{h}$ is independent of depth as well. Then the horizontal velocity field $\mathbf{u}_\mathrm{h}^i$ may be assumed to be independent of depth. This assumption implies that the stress $\boldsymbol{\tau}_\mathrm{h}$ must be a linear function of z such that

$$\partial_z \boldsymbol{\tau}_\mathrm{h}^i = \frac{\boldsymbol{\tau}_\mathrm{h}^i(z_{i-1}) - \boldsymbol{\tau}_\mathrm{h}^i(z_i)}{h_i} \tag{14.72}$$

The primitive equations in layer form then take the form

$$\partial_t \mathbf{u}_\mathrm{h}^i + \mathbf{u}_\mathrm{h}^i \cdot \nabla_\mathrm{h} \mathbf{u}_\mathrm{h}^i + f \hat{\mathbf{u}}_\mathrm{h}^i = -\frac{1}{\rho_i} \nabla_\mathrm{h} M_i + \frac{1}{\rho_i} \mathbf{F}_{\mathrm{h}i}^\mathrm{h} + \frac{\boldsymbol{\tau}_\mathrm{h}^i(z_{i-1}) - \boldsymbol{\tau}_\mathrm{h}^i(z_i)}{\rho_i h_i} \tag{14.73}$$

$$\partial_t h_i + \nabla_\mathrm{h} \cdot (h_i \mathbf{u}_\mathrm{h}^i) = -\frac{E - P}{\rho_0} \delta_{i1} \tag{14.74}$$

$$z_{i-1} = z_i + h_i, \quad z_N = -H \tag{14.75}$$

$$M_{i+1} = M_i + (\rho_{i+1} - \rho_i) g z_i, \quad M_1 = p_\mathrm{a} + z_0 \rho_1 g \tag{14.76}$$

In (14.74) we allowed for forcing by a freshwater flux $E - P$ in the first layer. The equations within each layer are the primitive equations (13.54) and (13.57) for a homogeneous ocean. These equations constitute prognostic equations for $(\mathbf{u}_\mathrm{h}^i, h_i)$ and recursion formulae for (z_i, M_i). The pressure and vertical velocity do not enter the equations but can be calculated from the Montgomery potential and incompressibility condition. The interface displacement can be calculated from the upward recursion formula $\eta_{i-1} = \eta_i + h_i - H_i$ starting at the bottom $\eta_N = \eta_\mathrm{b}$.

Stresses. The eddy term $\mathbf{F}^{\mathrm{h}}_{\mathrm{h}}$ and stress $\boldsymbol{\tau}_{\mathrm{h}}$ need to be specified or parametrized. The surface stress is usually the prescribed atmospheric stress, i.e., $\boldsymbol{\tau}_{\mathrm{h}}(z_0) = \boldsymbol{\tau}^{\mathrm{a}}_{\mathrm{h}}$. The layer equations are then forced by the atmospheric windstress $\boldsymbol{\tau}^{\mathrm{a}}_{\mathrm{h}}$, pressure p_{a}, and freshwater flux $E - P$. The bottom stress $\boldsymbol{\tau}_{\mathrm{h}}(z_N)$ is usually parametrized as linear or quadratic bottom friction. The interfacial stresses are often neglected.

Rigid lid approximation. In the rigid lid approximation M_1 is not given by (14.76) but has to be calculated from the constraint

$$\nabla_{\mathrm{h}} \cdot \mathbf{M}_{\mathrm{h}} = -\frac{E-P}{\rho_0} \tag{14.77}$$

where $\mathbf{M}_{\mathrm{h}} = \sum_{i=1}^{N} h_i \mathbf{u}^i_{\mathrm{h}}$ and $\sum_{i=1}^{N} h_i = H$. The constraint is usually incorporated by decomposing the flow field into its barotropic and baroclinic components.

Discretization. The layer equations can also be obtained by discretizing the primitive equations in isopycnal coordinates if $\dot{\rho} = 0$ is assumed and the following discretizations for the ρ derivatives are implemented

$$\frac{\partial M}{\partial \rho} = gz \to \frac{M_{i+1} - M_i}{\rho_{i+1} - \rho_i} = gz_i \to M_{i+1} = M_i + (\rho_{i+1} - \rho_i) g z_i \tag{14.78}$$

$$h = -\Delta\rho \frac{\partial z}{\partial \rho} \to h_i = -\Delta\rho \frac{z_i - z_{i-1}}{\Delta\rho} \to z_{i-1} = z_i + h_i \tag{14.79}$$

$$\frac{1}{\rho_0}\frac{\partial \boldsymbol{\tau}_{\mathrm{h}}/\partial\rho}{\partial z/\partial\rho} \to \frac{1}{\rho_0}\frac{\boldsymbol{\tau}^i_{\mathrm{h}} - \boldsymbol{\tau}^{i-1}_{\mathrm{h}}}{z_i - z_{i-1}} \to \frac{1}{\rho_0}\frac{\boldsymbol{\tau}^{i-1}_{\mathrm{h}} - \boldsymbol{\tau}^i_{\mathrm{h}}}{h_i} \tag{14.80}$$

Vorticity equations. The vorticity equation (13.58) and potential vorticity equation (13.65) hold for each layer separately. The relative height of a fluid particle in the ith layer is $\sigma_i = (z - z_{i-1})/h_i$. The freshwater flux $E - P$ only appears in the potential vorticity equation for the first layer. The vorticities and potential vorticities in the different layers are coupled through the curl of the interfacial stresses $\boldsymbol{\tau}_{\mathrm{h}}(z_i)$.

1- and 2-layer models For 1- and 2-layer models the recursion formulae for z and M are

$$\begin{aligned} z_0 &= -H + h_1 \quad M_1 = \rho_1 g z_0 + p_{\mathrm{a}} \\ z_1 &= -H \end{aligned} \tag{14.81}$$

and

$$\begin{aligned} z_0 &= -H + h_1 + h_2 \quad M_1 = \rho_1 g z_0 + p_{\mathrm{a}} \\ z_1 &= -H + h_2 \qquad\quad M_2 = \rho_1 g z_0 + (\rho_2 - \rho_1) g z_1 \\ z_2 &= -H \end{aligned} \tag{14.82}$$

Reduced gravity models Reduced gravity models eliminate the barotropic flow component by assuming an infinitely deep lowest layer without any horizontal motion. A reduced gravity model with N finite layers and one infinite layer is called a $(N+1/2)$ reduced gravity model. Since the lowest layer has no horizontal motions the gradient of its Montgomery potential must vanish

$$\nabla_{\mathrm{h}} M_{N+1} = 0 \tag{14.83}$$

which implies without loss of generality $M_{N+1} = 0$. The downward regression of the Montgomery potential thus implies

$$0 = \sum_{i=1}^{N} \rho_i g h_i + \rho_{N+1} g z_N \tag{14.84}$$

which can be used as the starting value for the upward regression of the interface displacement. For a $1\frac{1}{2}$ layer model one specifically finds

$$z_1 = -\frac{\rho_1}{\rho_2} h_1 \tag{14.85}$$

$$z_0 = \frac{\rho_2 - \rho_1}{\rho_2} h_1 \tag{14.86}$$

$$M_1 = \rho_1 \frac{\rho_2 - \rho_1}{\rho_2} g h_1 = \rho_1 g_1' h_1 \tag{14.87}$$

where

$$g_1' = \frac{\rho_2 - \rho_1}{\rho_2} g \tag{14.88}$$

is the reduced gravity.

14.6 Projection onto normal modes

The vertical structure of any variable $\psi(\mathbf{x}_{\mathrm{h}}, z, t)$ can be represented by expanding the variable with respect to a complete set of basis functions $\phi_n(z)$, $n = 0, \ldots, \infty$

$$\psi(\mathbf{x}_{\mathrm{h}}, z, t) = \sum_{n=0}^{\infty} a_n(\mathbf{x}_{\mathrm{h}}, t) \phi_n(z) \tag{14.89}$$

where $a_n(\mathbf{x}_{\mathrm{h}}, t)$ are the expansion coefficients or amplitudes. The equations of motion can then be projected onto these amplitudes. If the basis functions are orthogonal, $\int_{-H}^{0} \mathrm{d}z \phi_n(z) \phi_m(z) = \delta_{nm}$, then this projection is simply accomplished by substituting the above expansion into the equations of motion, multiplying by $\phi_n(z)$ and integrating over z. There is no restriction as to the choice of the basis functions. Of course, if anything is to be gained by such an expansion the basis functions should be representative of either the geometry or the dynamics. One set of basis functions often used are the eigenfunctions of the linearized equations of motion. They decouple the linear problem and facilitate the analysis of weakly nonlinear problems. Here, we discuss the eigenfunctions for the linearized primitive equations, in both their continuous and discrete forms.

Continuous vertical eigenvalue problem When the primitive equations are linearized about a motionless background state with stratification $N^2(z)$ one obtains

$$\partial_t \mathbf{u}_{\mathrm{h}} + f \hat{\mathbf{u}}_{\mathrm{h}} = -\frac{1}{\rho_0} \nabla_{\mathrm{h}} \delta p \tag{14.90}$$

$$0 = \frac{\partial \delta p}{\partial z} + \delta \rho g \tag{14.91}$$

$$\nabla_{\mathrm{h}} \cdot \mathbf{u}_{\mathrm{h}} + \frac{\partial w}{\partial z} + 0 \tag{14.92}$$

$$\partial \delta \rho - \frac{\rho_0}{g} N^2 w = 0 \tag{14.93}$$

These equations have separable eigensolutions. The vertical eigenfunctions $\phi(z)$ of $(\mathbf{u}_{\mathrm{h}}, \delta p)$ satisfy the vertical eigenvalue problem

$$\frac{\mathrm{d}}{\mathrm{d}z} \frac{1}{N^2(z)} \frac{\mathrm{d}}{\mathrm{d}z} \phi(z) = -\epsilon \phi(z) \tag{14.94}$$

$$N^2 \phi + g \frac{\mathrm{d}}{\mathrm{d}z} \phi = 0 \quad \text{at} \quad z = 0 \tag{14.95}$$

$$\frac{\mathrm{d}}{\mathrm{d}z} \phi = 0 \quad \text{at} \quad z = -H_0 \tag{14.96}$$

where ϵ is the eigenvalue. It is the limit $\omega^2 \ll N^2, \bar{c}^{-2} = 0$ of the eigenvalue problem (8.42) to (8.43). It does not contain the frequency ω anymore and constitutes a Sturm–Liouville problem for which powerful mathematical theorems are available. If $N^2(z) > 0$ then:

1. There exists a denumerable set of eigenfunctions $\phi_n(z)$ and eigenvalues $\epsilon_n, n = 0, 1, 2, \ldots$ The eigenmodes of the second kind $n = -1, -2, \ldots$ are not present any more.
2. The eigenvalues increase monotonically and without bound with increasing mode number.
3. The eigenfunction $\phi_n(z)$ has n zeroes in the interval $[-H, 0]$.
4. There exists between two adjacent zeroes of $\phi_n(z)$ a zero of $\phi_{n+1}(z)$.

5. The eigenfunctions are orthogonal

$$\frac{1}{H_0}\int_{-H_0}^{0} \mathrm{d}z\, \phi_n(z)\phi_m(z) = \delta_{nm} \tag{14.97}$$

6. The eigenfunctions are complete.

Instead of the eigenvalue ϵ_n one often uses the equivalent depth $(g\epsilon_n)^{-1}$. The lowest order mode $n = 0$ is called the barotropic mode. The modes $n = 1, 2, 3, \ldots$ are called the baroclinic modes. The integral form of the eigenvalue problem is

$$g\int_{-H}^{0} \mathrm{d}z'\, \phi(z') + \int_{z}^{0} \mathrm{d}z'\, N^2(z') \int_{-H}^{z'} \mathrm{d}z''\, \phi(z'') = \frac{1}{\epsilon}\phi(z) \tag{14.98}$$

The eigenfunction $\psi(z)$ of the vertical velocity w is given by

$$\psi(z) \sim \frac{1}{N^2(z)}\frac{\mathrm{d}}{\mathrm{d}z}\phi(z) \tag{14.99}$$

and satisfies the eigenvalue problem

$$\frac{\mathrm{d}}{\mathrm{d}z}\frac{\mathrm{d}}{\mathrm{d}z}\psi = -\epsilon N^2(z)\psi(z) \tag{14.100}$$

$$\frac{\mathrm{d}}{\mathrm{d}z}\psi - \epsilon g\psi = 0 \quad \text{at} \quad z = 0 \tag{14.101}$$

$$\psi = 0 \quad \text{at} \quad z = -H_0 \tag{14.102}$$

This is not a proper Sturm–Liouville problem since the eigenvalues appear in boundary conditions.

The eigenvalue problem contains the parameter $N^2H_0/g = H_0/\tilde{D}_{\mathrm{d}}$, generally assumed to be small. A perturbation expansion then yields to lowest order the rigid lid approximation $g \to \infty$

$$\frac{\mathrm{d}\phi}{\mathrm{d}z} = 0 \quad \text{or} \quad \psi = 0 \quad \text{at} \quad z = 0, -H_0 \tag{14.103}$$

The barotropic eigenfunction and eigenvalue are then given by $\phi_0^{(0)}(z) = 1$ and $\epsilon_0^{(0)} = 0$. The barotropic mode becomes depth independent. The baroclinic modes have zero vertical average

$$\frac{1}{H}\int_{-H}^{0} \mathrm{d}z\, \phi_n^{(0)}(z) = 0 \quad \text{for} \quad n = 1, 2, 3, \ldots \tag{14.104}$$

and can be determined most easily from the integral eigenvalue problem

$$\left(1 - \frac{1}{H_0}\int_{-H_0}^{0} \mathrm{d}z\right)\int_{z}^{0} \mathrm{d}z'\, N^2(z') \int_{-H}^{z'} \mathrm{d}z''\, \phi_n^{(0)}(z'') = \frac{1}{\epsilon_n^{(0)}}\phi_n^{(0)}(z) \tag{14.105}$$

Discrete vertical eigenvalue problem Linearization of the primitive N-layer equations about the basic state yields

$$\partial_t \mathbf{u}_\mathrm{h}^i + f\hat{\mathbf{u}}_\mathrm{h}^i = -\frac{1}{\rho_0}\nabla_\mathrm{h} M_i \tag{14.106}$$

$$\partial_t (z_{i-1} - z_i) + H_i \nabla \cdot \mathbf{u}_\mathrm{h}^i = 0 \tag{14.107}$$

$$M_{i+1} = M_i + (\rho_{i+1} - \rho_i)\, g z_i \tag{14.108}$$

where $z_{i-1} - z_i = h_i$ is the layer thickness. These equations have separable solutions of the form

$$\mathbf{u}_\mathrm{h}^i = \mathbf{u}_\mathrm{h}(\mathbf{x}_\mathrm{h}, t)\phi_i \tag{14.109}$$

$$M_i = M(\mathbf{x}_\mathrm{h}, t)\phi_i \tag{14.110}$$

$$(z_{i-1} - z_i) = \epsilon \rho_0^{-1} M(\mathbf{x}_\mathrm{h}, t)\phi_i H_i \tag{14.111}$$

where $\phi = (\phi_1, \ldots, \phi_N)$ satisfies the vertical eigenvalue problem

$$A_i(\phi_i - \phi_{i-1}) - B_i(\phi_{i+1} - \phi_i) = \epsilon \phi_i \quad i = 1, \ldots, N \tag{14.112}$$

with

$$A_1 := \frac{1}{g H_1} \tag{14.113}$$

$$A_i := \frac{1}{g'_{i-1} H_i} \quad i = 2, \ldots, N \tag{14.114}$$

$$B_i := \frac{1}{g'_i H_i} \quad i = 1, \ldots, N-1 \tag{14.115}$$

$$B_N = 0 \tag{14.116}$$

and reduced gravities

$$g'_i := g\frac{\rho_{i+1} - \rho_i}{\rho_0} \tag{14.117}$$

The vertical eigenvalue problem has N eigensolutions $n = 0, \ldots, N-1$ with eigenvalues $\epsilon^{(n)}$ and eigenvectors $\phi^{(n)}$. The mode $n = 0$ is called the barotropic mode and the modes $n = 1, \ldots, N-1$ are called the baroclinic modes. The modes are orthogonal

$$\frac{1}{H}\sum_{i=1}^{N} H_i \phi_i^{(n)} \phi_i^{(m)} = \delta_{nm} \tag{14.118}$$

Examples:

1. 1-layer model with free surface:

$$\phi_1^{(0)} = 1 \qquad \epsilon^{(0)} = \frac{1}{gH_0} \tag{14.119}$$

2. N-layer model with rigid surface. In this case $A_1 = 0$ and

$$\begin{pmatrix} \phi_1^{(0)} \\ \vdots \\ \phi_N^{(0)} \end{pmatrix} = \begin{pmatrix} 1 \\ \vdots \\ 1 \end{pmatrix} \qquad \epsilon^{(0)} = 0 \tag{14.120}$$

The baroclinic modes satisfy

$$\frac{1}{H}\sum_{i=1}^{N} H_i \phi_i^{(n)} = 0 \quad n = 1, \ldots, N-1 \tag{14.121}$$

In the case of a 2-layer model

$$\begin{pmatrix} \phi_1^{(1)} \\ \phi_2^{(1)} \end{pmatrix} = \frac{1}{\sqrt{H_1 H_2}} \begin{pmatrix} H_2 \\ -H_1 \end{pmatrix} \qquad \epsilon^{(1)} = \frac{1}{g_1'} \frac{H}{H_1 H_2} \tag{14.122}$$

3. Reduced gravity model. The $N + 1/2$ reduced gravity model has N modes. They are the baroclinic modes of a $N + 1$ layer model in the limit $H_{N+1} \to \infty$ or $A_{N+1} \to 0$. For the $1\frac{1}{2}$ layer model one obtains

$$\epsilon^{(1)} = \frac{1}{g_1' H_1} \tag{14.123}$$

15

Ekman layers

In the following chapters we consider motions for which vertical eddy friction is small. Formally, these motions are characterized by small vertical eddy Ekman numbers E_v. However, one cannot simply neglect the vertical eddy friction term altogether since it is the term with the highest z-derivative in the momentum equation. One would not be able to satisfy the dynamic boundary conditions at the surface and bottom. In the limit $E_v \ll 1$, the friction is negligible in the fluid interior but becomes prominent close to the boundary where it forms a frictional boundary layer. The Ekman boundary layer is such a boundary where the frictional force is balanced by the Coriolis force. Because of the boundary layer the interior flow will experience a modified kinematic boundary condition that includes an Ekman pumping (or suction) velocity. Ekman theory is not applicable at the equator where the Coriolis force vanishes.

15.1 Ekman number

The vertical eddy Ekman number is defined as the ratio of frictional to Coriolis force. In the diffusion parametrization, the vertical eddy friction in the horizontal momentum balance is given by

$$\frac{\partial}{\partial z}\boldsymbol{\tau}_h = \rho_0 \frac{\partial}{\partial z} A_v \frac{\partial}{\partial z} \mathbf{u}_h \tag{15.1}$$

and the Ekman number then given by

$$E_v := \frac{A_v}{|f| D^2} \tag{15.2}$$

If one introduces the frictional time scale

$$T_f := \frac{D^2}{A_v} \tag{15.3}$$

and the inertial time scale $T_\mathrm{i} := |f|^{-1}$ then

$$E_\mathrm{v} = \frac{T_\mathrm{i}}{T_\mathrm{f}} \tag{15.4}$$

Motions with small vertical eddy Ekman numbers are thus motions for which the frictional decay time is much larger than the inertial time scale.

15.2 Boundary layer theory

When the vertical eddy Ekman number is small, the vertical friction term is neglected in the interior but not in the Ekman boundary layer. There the frictional force is balanced by the Coriolis force. The other forces are neglected, which formally requires $\tilde{\mathrm{R}}\mathrm{o}$, Ro, $E_\mathrm{h} \ll 1$. It is a stationary boundary layer. Its depth d is given by

$$d^2 = \frac{A_\mathrm{v}}{|f|} \tag{15.5}$$

which implies

$$\frac{d^2}{D^2} = \frac{A_\mathrm{v}}{|f|D^2} = E_\mathrm{v}^{\mathrm{eddy}} \tag{15.6}$$

If the Ekman layer depth d is assumed to be $O(50)$m then A_v has to be $O(10^{-2}\,\mathrm{m}^2\,\mathrm{s}^{-1})$ in midlatitudes.

Formally, one analyzes boundary layers by decomposing all state variables ψ into an interior and a boundary component

$$\psi = \psi^\mathrm{I} + \psi^\mathrm{B} \tag{15.7}$$

and requires the boundary component to vanish far away from the boundary. Alternatively, one can decompose all state variables

$$\psi = \begin{cases} \tilde{\psi}^\mathrm{B} & \text{in boundary layer} \\ \tilde{\psi}^\mathrm{I} & \text{in interior} \end{cases} \tag{15.8}$$

One must then require that $\tilde{\psi}^\mathrm{B}$ matches $\tilde{\psi}^\mathrm{I}$ at the interface between the boundary layer and the interior.

Since $d \ll D$ one can exemplify the boundary layer approach by looking at the interface between two infinitely deep homogeneous layers. The interface is fixed at $z = 0$. If time-dependence and horizontal eddy friction are neglected the governing equations for layer "1" and "2" become

$$f\hat{\mathbf{u}}_\mathrm{h}^{1,2} = -\frac{1}{\rho_0}\nabla_\mathrm{h} p^{1,2} + \frac{1}{\rho_0}\frac{\partial}{\partial z}\boldsymbol{\tau}_\mathrm{h}^{1,2} \tag{15.9}$$

$$\partial_z p^{1,2} = -\rho_0^{1,2} g \tag{15.10}$$

$$\nabla_\mathrm{h} \cdot \mathbf{u}_\mathrm{h}^{1,2} + \partial_z w^{1,2} = 0 \tag{15.11}$$

with boundary conditions

$$\mathbf{u}_\mathrm{h}^1 = \mathbf{u}_\mathrm{h}^2, \quad w^1 = w^2 \quad p^1 = p^2 \quad \boldsymbol{\tau}_\mathrm{h}^1 = \boldsymbol{\tau}_\mathrm{h}^2 \text{ at } z = 0 \tag{15.12}$$

The hydrostatic balance implies that the horizontal pressure gradient and hence the dynamically active part of the pressure are depth independent. Decomposing each variable into an interior and boundary component we find in each layer that

$$f\hat{\mathbf{u}}_\mathrm{h}^\mathrm{I} = -\frac{1}{\rho_0}\boldsymbol{\nabla}_\mathrm{h} p^\mathrm{I} \tag{15.13}$$

$$\boldsymbol{\nabla}_\mathrm{h}\cdot\mathbf{u}_\mathrm{h}^\mathrm{I} + \partial_z w^\mathrm{I} = 0 \tag{15.14}$$

and

$$f\hat{\mathbf{u}}_\mathrm{h}^\mathrm{B} = \frac{1}{\rho_0}\frac{\partial}{\partial z}\boldsymbol{\tau}_\mathrm{h} \tag{15.15}$$

$$\boldsymbol{\nabla}_\mathrm{h}\cdot\mathbf{u}_\mathrm{h}^\mathrm{B} + \partial_z w^\mathrm{B} = 0 \tag{15.16}$$

There is no friction term in the momentum equation for the interior component because of the assumption $E_\mathrm{v} \ll 1$. The interior velocity $\mathbf{u}_\mathrm{h}^\mathrm{I}$ is hence depth independent. There is no pressure term in the momentum equation for the boundary layer component since $\lim_{z=\rightarrow\pm\infty} p^\mathrm{B} = 0$ can only be achieved by $p_\mathrm{B} = 0$ for a depth independent pressure field.

15.3 Ekman transport

The transports of the boundary layer component can be obtained by integration of the horizontal momentum balance

$$\mathbf{M}_\mathrm{Ek}^1 = \int_0^\infty \mathrm{d}z'\, \mathbf{u}_\mathrm{h}^{\mathrm{B},1} = \frac{1}{f\rho_0^1}\hat{\boldsymbol{\tau}}_\mathrm{h}(0) \tag{15.17}$$

$$\mathbf{M}_\mathrm{Ek}^2 = \int_{-\infty}^0 \mathrm{d}z'\, \mathbf{u}_\mathrm{h}^{\mathrm{B},2} = -\frac{1}{f\rho_0^2}\hat{\boldsymbol{\tau}}_\mathrm{h}(0) \tag{15.18}$$

where it has been assumed that $\boldsymbol{\tau}_\mathrm{h}$ vanishes for $z =\rightarrow \pm\infty$. At the air–sea interface where $\boldsymbol{\tau}_\mathrm{h}(0)$ is given by the atmospheric windstress $\boldsymbol{\tau}_\mathrm{h}^\mathrm{a}$ this formula determines the Ekman transport. At the bottom where the stress is not known it constitutes a relation between bottom stress and transport. The Ekman transports are perpendicular to the surface stress and cancel each other

$$\rho_0^1\mathbf{M}_\mathrm{Ek}^1 + \rho_0^2\mathbf{M}_\mathrm{Ek}^2 = 0 \tag{15.19}$$

15.4 Ekman pumping

Integration of the incompressibility condition $\boldsymbol{\nabla} \cdot \mathbf{u} = 0$ yields

$$w^{\mathrm{B},1} = \boldsymbol{\nabla}_{\mathrm{h}} \cdot \mathbf{M}_{\mathrm{Ek}}^{1} \tag{15.20}$$

$$w^{\mathrm{B},2} = -\boldsymbol{\nabla}_{\mathrm{h}} \cdot \mathbf{M}_{\mathrm{Ek}}^{2} \tag{15.21}$$

The kinematic boundary $w = w^{\mathrm{I}} + w^{\mathrm{B}} = 0$ thus becomes

$$w^{\mathrm{I},1} = w_{\mathrm{Ek}}^{1} \tag{15.22}$$

$$w^{\mathrm{I},2} = w_{\mathrm{Ek}}^{2} \tag{15.23}$$

where

$$w_{\mathrm{Ek}}^{1,2} = \mp \boldsymbol{\nabla}_{\mathrm{h}} \cdot \mathbf{M}_{\mathrm{h}}^{1,2} = \frac{1}{\rho_0^{1,2}} \hat{\boldsymbol{\nabla}}_{\mathrm{h}} \cdot \left(\frac{\boldsymbol{\tau}_{\mathrm{h}}}{f} \right) \tag{15.24}$$

is the Ekman pumping (or suction) velocity. The divergence of the Ekman transport thus acts like a vertical mass flux in the kinematic boundary condition. This vertical mass flux is continuous

$$\rho_0^1 w_{\mathrm{Ek}}^1 = \rho_0^2 w_{\mathrm{Ek}}^2 \tag{15.25}$$

The Ekman pumping velocity is determined by the atmospheric windstress at the surface and by the unknown bottom stress at the bottom.

15.5 Laminar Ekman layers

If the turbulent stress $\boldsymbol{\tau}_{\mathrm{h}}$ is given by the laminar diffusive form

$$\boldsymbol{\tau}_{\mathrm{h}} = \rho_0 A_{\mathrm{v}} \partial_z \mathbf{u}_{\mathrm{h}} \tag{15.26}$$

with constant friction coefficient A_{v}, then the boundary layer velocity and stress fields can be calculated explicitly. We introduce $\mu_{\mathrm{B}} = u_{\mathrm{B}} + \mathrm{i} v_{\mathrm{B}}$. The Ekman momentum balance then reads

$$f \mu_{\mathrm{B}}^{1,2} = -\mathrm{i} A_{\mathrm{v}} \partial_z \partial_z \mu_{\mathrm{B}} \tag{15.27}$$

It has the general solution

$$\mu_{\mathrm{B}}^{1,2} = B_{+}^{1,2} \exp^{\gamma_{+}^{1,2} z} + B_{-}^{1,2} \exp^{\gamma_{-}^{1,2} z} \tag{15.28}$$

where

$$\gamma_{\pm}^{1,2} = \pm \frac{1+\mathrm{i}}{\sqrt{2}} \frac{1}{d_{1,2}} \tag{15.29}$$

for $f > 0$ (northern hemisphere), which will be assumed throughout this section. The requirement that the solutions decay away from the interface implies $B_+^1 = 0$ and $B_-^2 = 0$. Hence

$$\mu_\text{B}^1 = B_-^1 \exp^{\gamma_-^1 z} \tag{15.30}$$

$$\mu_\text{B}^2 = B_+^2 \exp^{\gamma_+^2 z} \tag{15.31}$$

At the interface the stress must be continuous

$$\rho_0^1 A_\text{v}^1 \partial_z \mu_\text{B}^1 = \rho_0^2 A_\text{v}^2 \partial_z \mu_\text{B}^2 \quad \text{at} \quad z = 0 \tag{15.32}$$

Hence

$$\mu_\text{B}^1 = -\epsilon^{-1} B_+^2 \exp^{\gamma_-^1 z} \tag{15.33}$$

$$\mu_\text{B}^2 = B_+^2 \exp^{\gamma_+^2 z} \tag{15.34}$$

where

$$\epsilon := \frac{\rho_1 d_1}{\rho_2 d_2} \tag{15.35}$$

Continuity of the horizontal velocity

$$\mu_\text{I}^1 + \mu_\text{B}^1 = \mu_\text{I}^2 + \mu_\text{B}^2 \quad \text{at} \quad z = 0 \tag{15.36}$$

leads to

$$\mu_\text{B}^1 = -\frac{1}{1+\epsilon} \Delta\mu_\text{I} \exp^{\gamma_-^1 z} \tag{15.37}$$

$$\mu_\text{B}^2 = \frac{\epsilon}{1+\epsilon} \Delta\mu_\text{I} \exp^{\gamma_+^2 z} \tag{15.38}$$

where $\Delta\mu_\text{I} := \mu_\text{I}^1 - \mu_\text{I}^2$. This is the explicit solution for the boundary layer component. The solution describes a spiral decaying away from the interface.

At the interface, the velocity is given by

$$\mu(z = 0) = \mu_\text{I}^1 + \mu_\text{B}^1 = \mu_\text{I}^2 + \frac{\epsilon}{1+\epsilon} \Delta\mu_\text{I} = \mu_\text{I}^1 - \frac{1}{1+\epsilon} \Delta\mu_\text{I} \tag{15.39}$$

and the stress $\sigma = \tau_\varphi + \text{i}\tau_\theta$ by

$$\sigma(0) = \rho_1 A_\text{v}^1 \partial_z \mu_\text{B}^1|_{z=0} = \alpha \Delta\mu_\text{I} \tag{15.40}$$

where

$$\alpha := \frac{1+\text{i}}{\sqrt{2}} f \left(\frac{1}{\rho_1 d_1} + \frac{1}{\rho_2 d_2} \right)^{-1} \tag{15.41}$$

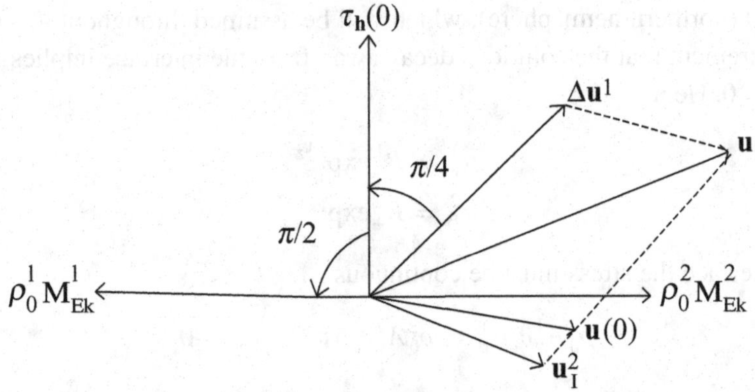

Figure 15.1. Geometric relations between stress $\tau_h(0)$, interior velocities $\mathbf{u}_I^{1,2}$, interface velocity $\mathbf{u}(0)$, and Ekman transports $\mathbf{M}_{Ek}^{1,2}$ for laminar Ekman layers above and below an interface.

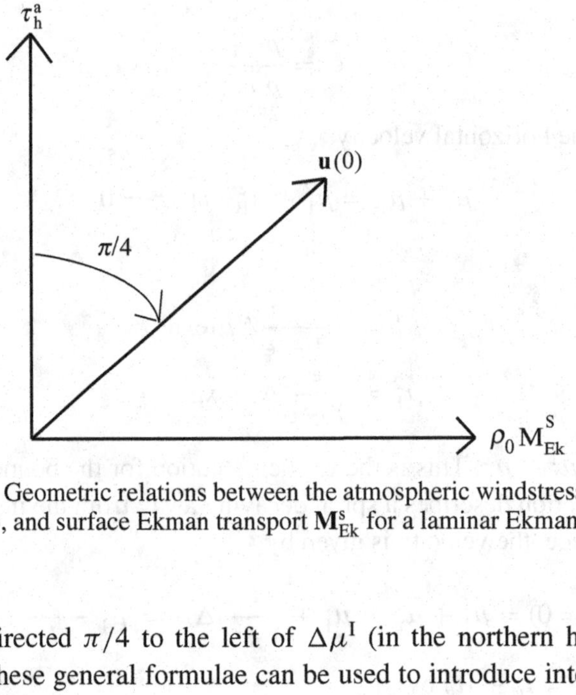

Figure 15.2. Geometric relations between the atmospheric windstress τ_h^a, surface velocity $\mathbf{u}(0)$, and surface Ekman transport $\mathbf{M}_{Ek}^s$ for a laminar Ekman layer below the surface.

The stress is directed $\pi/4$ to the left of $\Delta\mu^I$ (in the northern hemisphere, see Figure 15.1). These general formulae can be used to introduce interfacial friction into layer models.

At the ocean surface, layer 1 is the atmosphere and layer 2 the ocean. One assumes $|\mu_I^2| \ll |\mu_I^1|$ and $\epsilon \ll 1$. Then $\sigma(0) = \alpha\mu_I^1 =: \sigma^a$ is determined solely by the atmospheric interior velocity and assumed to be prescribed. The surface velocity is then $\pi/4$ to the right of the stress σ^a (in the northern hemisphere, see Figure 15.2). Explicit formulae are given in Table 15.1.

Table 15.1. *Characteristics of the surface and bottom Ekman layers*

	Surface Ekman layer	Bottom Ekman layer
Horizontal velocity profile	$\mu_B^2(z) = \frac{\sigma^a}{\rho_0^2 A_v^2 \gamma_+^2} \exp\{\gamma_+^2 z\}$	$\mu_B^1(z) = -\mu^o \exp\{\gamma_-^1 (z + H)\}$
Horizontal velocity at boundary	$\mu_B^2(0) = \frac{\sigma^a}{\rho_0^2 A_v^2 \gamma_+^2}$	$\mu_B^1(-H) = -\mu^o =$ prescribed
Stress profile	$\sigma^2(z) = \sigma^a \exp\{\gamma_+^2 z\}$	$\sigma^1(z) = -\rho_0^1 A_v^1 \gamma_-^1 \mu^o \exp\{\gamma_-^1 (z + H)\}$
Stress at boundary	$\boldsymbol{\tau}_h^2(0) = \boldsymbol{\tau}_h^a =$ prescribed	$\boldsymbol{\tau}_h^1(-H) = \boldsymbol{\tau}_h^b = \frac{1}{\sqrt{2}} \rho_0 d f (\mathbf{u}_h + \hat{\mathbf{u}}_h)$
Ekman transport	$\mathbf{M}_{Ek}^s = -\frac{1}{\rho_0 f} \hat{\boldsymbol{\tau}}_h^a$	$\mathbf{M}_{Ek}^b = -\frac{d}{\sqrt{2}} \mathbf{u}_h + \frac{d}{\sqrt{2}} \hat{\mathbf{u}}_h$
Ekman pumping velocity	$w_{Ek}^s = \frac{1}{\rho_0} \hat{\nabla}_h \cdot \left(\frac{\boldsymbol{\tau}_h^a}{f}\right)$	$w_{Ek}^b = \nabla_h \cdot \left(\frac{d}{\sqrt{2}} \mathbf{u}_h\right) + \hat{\nabla}_h \cdot \left(\frac{d}{\sqrt{2}} \mathbf{u}_h\right)$

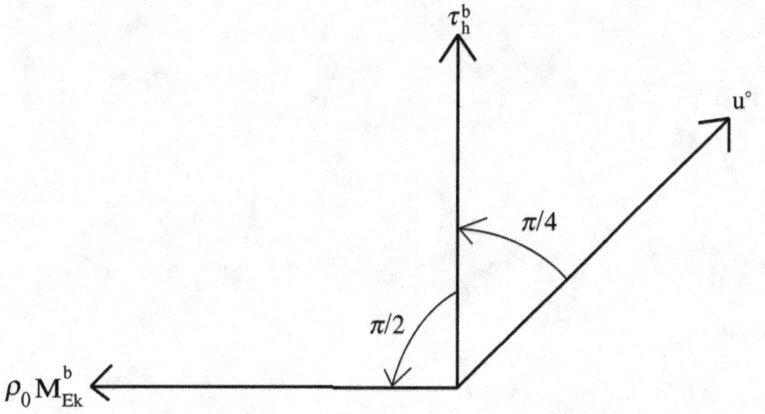

Figure 15.3. Geometric relations between the bottom stress $\boldsymbol{\tau}_h^b$, interior velocity $\mathbf{u}^o$, and bottom Ekman transport $\mathbf{M}_{Ek}^b$ for a laminar Ekman layer above the bottom.

At the ocean bottom, layer 1 is the ocean and layer 2 the underlying bottom. One assumes $\mu_I^2 = 0$ and $\epsilon = 0$. Then $\Delta\mu_I = \mu_I^1 =: \mu^o$ is determined by the oceanic interior velocity and assumed to be prescribed. The bottom stress is directed $\pi/4$ to the left of the velocity μ^o (in the northern hemisphere, see Figure 15.3). Explicit formulae are given in Table 15.1.

15.6 Modification of kinematic boundary condition

The main result of Ekman theory for the interior flow is a modification of the kinematic boundary condition. When vertical eddy friction is neglected the kinematic boundary conditions at the surface and bottom become

$$\partial_t \eta = w - w^{\mathrm{s}}_{\mathrm{Ek}} - \frac{E-P}{\rho_0} \quad \text{at } z = 0 \tag{15.42}$$

$$-\mathbf{u}_{\mathrm{h}} \cdot \nabla_{\mathrm{h}} H = w - w^{\mathrm{b}}_{\mathrm{Ek}} \quad \text{at } z = -H \tag{15.43}$$

For layer models the layer thickness equation becomes

$$\partial_t h_{\mathrm{i}} + \nabla_{\mathrm{h}} \cdot \left(h_{\mathrm{i}} \mathbf{u}^{\mathrm{i}}_{\mathrm{h}} \right) = w_{\mathrm{Ek}}(z_{\mathrm{i}}) - w_{\mathrm{Ek}}(z_{\mathrm{i}-1}) \tag{15.44}$$

Note that $h_{\mathrm{i}}\mathbf{u}^{\mathrm{i}}_{\mathrm{h}}$ is the *interior* transport, the total transport minus the Ekman transports. The Ekman pumping velocity is given in all cases by

$$w_{\mathrm{Ek}} = \frac{1}{\rho_0} \hat{\nabla}_{\mathrm{h}} \cdot \left(\frac{\tau_{\mathrm{h}}}{f} \right) \tag{15.45}$$

16

Planetary geostrophic flows

We now consider geostrophic flows. Formally, geostrophic flows require that the Rossby and Ekman numbers are much smaller than one. The horizontal momentum balance reduces to the geostrophic balance. The pressure force is balanced by the Coriolis force. The horizontal velocity field adjusts instantaneously. The gravity mode of motion is eliminated. Geostrophic flows are potential vorticity carrying motions. The geostrophic approximation has far-reaching consequences.

In this chapter we treat planetary or large-scale geostrophic flows. Small-scale or quasi-geostrophic flows are treated in Chapter 19. We treat the barotropic and baroclinic components separately. The equations for the barotropic part form the basis for the analysis of the wind-driven circulation; the equations for the baroclinic part form the basis for the analysis of the pycnocline.

16.1 The geostrophic approximation

If one assumes $\tilde{\text{Ro}}$, Ro, E_{v}, $E_{\text{h}} \ll 1$ the horizontal momentum balance reduces to the geostrophic balance

$$f\hat{\mathbf{u}}_{\text{h}}^{\text{g}} = -\frac{1}{\rho_0}\nabla_{\text{h}} p \tag{16.1}$$

The pressure force is balanced by the Coriolis force. The horizontal velocity is given by

$$\mathbf{u}_{\text{h}}^{\text{g}} = \frac{1}{\rho_0 f}\hat{\nabla}_{\text{h}} p \tag{16.2}$$

The geostrophic balance is assumed to hold only in the interior, not in boundary layers where friction becomes important.

The geostrophic approximation has far-reaching consequences:

1. In each horizontal plane, the streamlines of the horizontal velocity are isolines of pressure
$$\mathbf{u}_{\rm h}^{\rm g} \cdot \nabla_{\rm h} p = 0 \tag{16.3}$$
On the northern hemisphere ($f > 0$) the direction of the flow is such that the higher pressure is to the right of the flow. A "high-pressure center" hence represents clockwise, anticyclonic flow. A "low-pressure center" represents anticlockwise, cyclonic flow. In the southern hemisphere ($f < 0$) a "high" represents anticlockwise, anticyclonic flow and a "low" clockwise, cyclonic flow.
2. The vertical shear is related to the horizontal density gradient by the *thermal wind relation*
$$\partial_z \mathbf{u}_{\rm h}^{\rm g} = -\frac{g}{\rho_0 f} \hat{\nabla}_{\rm h} \rho \tag{16.4}$$
which is obtained by combining geostrophic and hydrostatic balance. If $\rho = \rho_0 =$ constant then $\partial \mathbf{u}_{\rm h}^{\rm g} / \partial z = 0$. The fluid moves in vertical columns. This is the Taylor–Proudman theorem.
3. The geostrophic relative vorticity is given by
$$\omega_z^{\rm g} := \hat{\nabla}_{\rm h} \cdot \mathbf{u}_{\rm h}^{\rm g} = \frac{\beta}{f} u_{\rm h}^{\rm g} + \frac{1}{\rho_0 f} \Delta_{\rm h} p = O\left(\gamma \frac{U}{L}\right) + O\left(\frac{U}{L}\right) \tag{16.5}$$
where $\gamma := L/r_0$. The ratio of relative to planetary vorticity is given by
$$\frac{\omega_z}{f} = O(\gamma \mathrm{Ro}) + O(\mathrm{Ro}) \tag{16.6}$$
and the ratio of the gradients by
$$\frac{|\nabla_{\rm h} \omega_z|}{|\nabla_{\rm h} f|} = O(\mathrm{Ro}) + O\left(\frac{\mathrm{Ro}}{\gamma}\right) \tag{16.7}$$
4. The vorticity balance takes the form
$$\hat{\nabla}_{\rm h} \cdot \left(f \hat{\mathbf{u}}_{\rm h}^{\rm g}\right) = f \nabla_{\rm h} \cdot \mathbf{u}_{\rm h}^{\rm g} + \mathbf{u}_{\rm h}^{\rm g} \cdot \nabla_{\rm h} f = 0 \tag{16.8}$$
or
$$f \partial_z w^{\rm g} = \mathbf{u}_{\rm h}^{\rm g} \cdot \nabla_{\rm h} f \tag{16.9}$$
when combined with the incompressibility condition. Vortex stretching is balanced by the advection of planetary vorticity. This vorticity balance is the backbone for many heuristic arguments about large-scale ocean dynamics.

One has to distinguish between planetary or large-scale geostrophic flows and quasi-geostrophic or small-scale geostrophic flows. Planetary geostrophic motions are defined as geostrophic motions for which $\gamma := L/r_0 = O(1)$. Quasi-geostrophic motions are defined as geostrophic motions for which $\gamma \ll 1$. For planetary flows the gradients of the relative vorticity are also small compared to the gradients of the planetary vorticity, whereas for quasi-geostrophic flows both gradients become comparable. In this chapter we discuss planetary geostrophic flows. Quasi-geostrophic flows are discussed in Chapter 19. For planetary geostrophic flows vertical or bottom friction is generally assumed to be more effective than lateral friction.

16.2 The barotropic problem

The barotropic planetary geostrophic equations are the basis for the theoretical analysis of the wind-induced horizontal transport.

Potential vorticity equation The derivation of the barotropic planetary geostrophic equations starts either from the primitive equations (13.54) to (13.56) for a homogeneous ocean or from the 1-layer equations (14.73) to (14.76). The momentum equation is approximated by the geostrophic balance. The vertical stresses are taken into account through Ekman pumping. Since $|\eta| \ll H$ is assumed we can replace the layer thickness h by H in all terms except the time derivative. The layer thickness equation then takes the form

$$\partial_t \eta + \nabla_{\mathrm{h}} \cdot \mathbf{M}_{\mathrm{h}} = -\frac{E-P}{\rho_0} \tag{16.10}$$

where $\mathbf{M}_{\mathrm{h}}$ is the total volume transport

$$\mathbf{M}_{\mathrm{h}} := \mathbf{M}_{\mathrm{g}} + \mathbf{M}_{\mathrm{Ek}}^{\mathrm{s}} + \mathbf{M}_{\mathrm{Ek}}^{\mathrm{b}} \tag{16.11}$$

consisting of the geostrophic transport

$$\mathbf{M}_{\mathrm{g}} := H\mathbf{u}_{\mathrm{h}}^{\mathrm{g}} \tag{16.12}$$

and the Ekman transports

$$\mathbf{M}_{\mathrm{Ek}}^{\mathrm{s}} := -\frac{1}{\rho_0 f}\hat{\boldsymbol{\tau}}_{\mathrm{h}}^{\mathrm{a}} \tag{16.13}$$

$$\mathbf{M}_{\mathrm{Ek}}^{\mathrm{b}} := \frac{1}{\rho_0 f}\hat{\boldsymbol{\tau}}_{\mathrm{h}}^{\mathrm{b}} \tag{16.14}$$

and where

$$\mathbf{u}_{\mathrm{h}}^{\mathrm{g}} = \frac{1}{\rho_0 f}\hat{\nabla}_{\mathrm{h}} M \tag{16.15}$$

$$M = \rho_0 g \eta + p_{\mathrm{a}} \tag{16.16}$$

An expression for the bottom stress needs to be added. These equations constitute a well-defined problem, with prognostic variable η and diagnostic variables M, $\mathbf{M}_{\mathrm{h}}$, and $\mathbf{u}_{\mathrm{h}}^{\mathrm{g}}$. The flow is forced by the atmospheric windstress $\boldsymbol{\tau}_{\mathrm{h}}^{\mathrm{a}}$, freshwater flux $E-P$ and pressure p_{a}.

Since geostrophic motions are vortical motions their dynamics are more succinctly expressed by potential vorticity. The potential vorticity for a 1-layer fluid is $q_\sigma = (f+\omega_z)/h$. For planetary geostrophic flows $|\omega_z| \ll f$ and $|\boldsymbol{\nabla}_{\mathrm{h}}\omega_z| \ll |\boldsymbol{\nabla}_{\mathrm{h}} f|$. Hence, the planetary potential vorticity reduces to

$$q_\sigma = \frac{f}{h} \tag{16.17}$$

and the potential vorticity equation to

$$\partial_t\left(\frac{f}{h}\right) = \frac{f}{H^2}\boldsymbol{\nabla}_{\mathrm{h}}\cdot\mathbf{M}_{\mathrm{h}} + \frac{f}{H^2}\frac{E-P}{\rho_0} \tag{16.18}$$

where again we have replaced h by H in all terms but the time derivative. The potential vorticity equation is equivalent to the layer thickness equation but shows explicitly that geostrophic motions are vorticity carrying motions.

The planetary geostrophic equations do not have any global free wave solutions. For $\eta = \hat{\eta}(\theta)\exp i(m\varphi - \omega t)$ the linear unforced equations reduce to

$$\left(\frac{\omega r_0^2 2\Omega\sin\theta^2}{gH_0} + m\right)\hat{\eta}(\theta) = 0 \tag{16.19}$$

or $(\hat{\omega}\hat{\epsilon}\mu^2 + m)\hat{P} = 0$ in the notation of Section 8.4 with only the trivial solution $\eta = 0$ and $\mathbf{u}_{\mathrm{h}}^{\mathrm{g}} = 0$. For non-trivial solutions to exist the geostrophic balance must be relaxed at the equator.

Steady-state problem Equation (16.18) describes the time-dependent adjustment to forcing. When one considers the steady-state or time-independent problem, then the time derivative can be dropped and the potential vorticity equation be rewritten as

$$\mathbf{M}_{\mathrm{h}}\cdot\boldsymbol{\nabla}_{\mathrm{h}}\frac{f}{H} = \hat{\boldsymbol{\nabla}}_{\mathrm{h}}\cdot\left(\frac{\boldsymbol{\tau}_{\mathrm{h}}^{\mathrm{a}}}{\rho_0 H}\right) - \hat{\boldsymbol{\nabla}}_{\mathrm{h}}\cdot\left(\frac{\boldsymbol{\tau}_{\mathrm{h}}^{\mathrm{b}}}{\rho_0 H}\right) + \frac{f}{H}\frac{E-P}{\rho_0} \tag{16.20}$$

It determines the total volume transport $\mathbf{M}_{\mathrm{h}}$ across given f/H-contours in response to forcing by the atmospheric windstress and freshwater flux and to bottom friction. It must be augmented by the boundary condition that there is no volume transport through any side wall

$$\mathbf{M}_{\mathrm{h}}\cdot\mathbf{n}_{\mathrm{h}} = 0 \tag{16.21}$$

If we do not consider a 1-layer model but the barotropic component of a depth dependent planetary geostrophic flow then the momentum balance is given by (see (14.12))

$$f\hat{\mathbf{u}}_{\mathrm{h}}^{\mathrm{g}} = -\frac{1}{\rho_0}\boldsymbol{\nabla}_{\mathrm{h}} M - \frac{\mathbf{T}}{H} \tag{16.22}$$

One would thus have to add a term

$$-\hat{\boldsymbol{\nabla}}_{\mathrm{h}} \cdot \left(\frac{\mathbf{T}}{H}\right) = \frac{1}{\rho_0 H^2}(\hat{\boldsymbol{\nabla}}_{\mathrm{h}} H) \cdot (\boldsymbol{\nabla}_{\mathrm{h}} E_{\mathrm{pot}}) \tag{16.23}$$

to the potential vorticity equation. The baroclinic pressure affects the barotropic potential vorticity equation only if there is bottom topography. The term $-\hat{\boldsymbol{\nabla}}_{\mathrm{h}} \cdot (\mathbf{T}/H)$ is called the JEBAR term, describing the **j**oint **e**ffect of **b**aroclinicity **a**nd **b**ottom **r**elief.

The potential vorticity equation has been formulated in terms of the total volume transport $\mathbf{M}_{\mathrm{h}}$, instead of the geostrophic velocity $\mathbf{u}_{\mathrm{h}}$. In this way, one can easily account for the lateral boundary condition. On the other hand, bottom friction generally requires the specification of the geostrophic velocity $\mathbf{u}_{\mathrm{h}}^{\mathrm{g}}$. For linear bottom friction $\boldsymbol{\tau}_{\mathrm{h}}^{\mathrm{b}} = \frac{1}{\sqrt{2}}\rho_0 d_0 f(\mathbf{u}_{\mathrm{h}}^{\mathrm{g}} + \hat{\mathbf{u}}_{\mathrm{h}}^{\mathrm{g}})$ (see Table 15.1) the geostrophic velocity can be obtained by inverting the expression for the total volume transport

$$\mathbf{M}_{\mathrm{h}} = H\mathbf{u}_{\mathrm{h}}^{\mathrm{g}} - \frac{1}{\rho_0 f}\hat{\boldsymbol{\tau}}_{\mathrm{h}}^{\mathrm{a}} - \frac{d_0}{\sqrt{2}}\mathbf{u}_{\mathrm{h}}^{\mathrm{g}} + \frac{d_0}{\sqrt{2}}\hat{\mathbf{u}}_{\mathrm{h}}^{\mathrm{g}} \tag{16.24}$$

which yields

$$f\mathbf{u}_{\mathrm{h}}^{\mathrm{g}} = \frac{f}{H}(\mathbf{M}_{\mathrm{h}} - \gamma\hat{\mathbf{M}}_{\mathrm{h}}) + \frac{1}{\rho_0 f}(\hat{\boldsymbol{\tau}}_{\mathrm{h}}^{\mathrm{a}} + \gamma\boldsymbol{\tau}_{\mathrm{h}}^{\mathrm{a}}) \tag{16.25}$$

since $\gamma := d_0/\sqrt{2}H = \sqrt{E_{\mathrm{v}}/2} \ll 1$. This geostrophic velocity would have to be substituted into the bottom friction term of the potential vorticity equation, leading to quite a complicated expression. This problem is often circumvented by assuming $\tilde{\mathrm{R}}\mathrm{o}$, Ro, $E_{\mathrm{h}} \ll 1$ in the momentum equation but not $E_{\mathrm{v}} \ll 1$. Then one does not introduce Ekman layers and geostrophically balanced interior flows but starts from the momentum balance

$$f\hat{\mathbf{u}}_{\mathrm{h}} = -\frac{1}{\rho_0}\boldsymbol{\nabla}_{\mathrm{h}} M + \frac{\boldsymbol{\tau}_{\mathrm{h}}^{\mathrm{a}} - \boldsymbol{\tau}_{\mathrm{h}}^{\mathrm{b}}}{\rho_0 H} \tag{16.26}$$

Taking the curl leads to the same potential vorticity equation where now $\mathbf{u}_{\mathrm{h}} = \mathbf{M}_{\mathrm{h}}/H$. Instead of, or in addition to, bottom friction one often introduces a Rayleigh friction term, $-\epsilon\mathbf{u}_{\mathrm{h}}$, into the momentum balance that gives rise to a term $-\epsilon\hat{\boldsymbol{\nabla}}_{\mathrm{h}} \cdot (\mathbf{M}_{\mathrm{h}}/H)$ in the potential vorticity balance. Adding Rayleigh friction, the potential vorticity equation takes the form

$$\mathbf{M}_{\mathrm{h}} \cdot \nabla_{\mathrm{h}} \frac{f}{H} = \hat{\nabla}_{\mathrm{h}} \cdot \left(\frac{\boldsymbol{\tau}_{\mathrm{h}}^{\mathrm{a}}}{\rho_0 H} \right) - \hat{\nabla}_{\mathrm{h}} \cdot \left(\frac{\boldsymbol{\tau}_{\mathrm{h}}^{\mathrm{b}}}{\rho_0 H} \right) + \frac{f}{H} \frac{E-P}{\rho_0} - \epsilon \hat{\nabla}_{\mathrm{h}} \cdot \left(\frac{\mathbf{M}_{\mathrm{h}}}{H} \right) \tag{16.27}$$

which we will use in the following.

If forcing by the freshwater flux is neglected, then $\nabla_{\mathrm{h}} \cdot \mathbf{M}_{\mathrm{h}} = 0$ and there exists a streamfunction Ψ such that $\mathbf{M}_{\mathrm{h}} = \hat{\nabla}_{\mathrm{h}} \Psi$. The lateral boundary condition $\mathbf{M}_{\mathrm{h}} \cdot \mathbf{n}_{\mathrm{h}} = 0$ then becomes

$$\Psi = \text{constant} \tag{16.28}$$

on all boundaries δS_j ($j = 1, \ldots, N$) (see Figure 19.1). For a simply connected domain the choice of the constant does not affect the transport. However, if the domain is multiply connected then the difference of the Ψ values on different boundaries matters since it determines the total volume transport between the boundaries. The values of Ψ on the boundaries can be inferred from integrability conditions.

Integrability conditions When using the potential vorticity equation to determine the flow field, the auxiliary relations have to be cast into a form so that they determine the flow variables in terms of the potential vorticity. This constitutes the invertibility problem. A condition that must be included to solve this invertibility problem is the integrability condition. It arises from the fact that the vorticity equation is derived by taking the curl of the momentum balance. This operation eliminates the Montgomery potential. If the momentum balance is written as

$$\nabla_{\mathrm{h}} M = \mathbf{B}_{\mathrm{h}} \tag{16.29}$$

where $\hat{\nabla}_{\mathrm{h}} \cdot \mathbf{B}_{\mathrm{h}} = 0$, the Montgomery potential (and hence the pressure) can be reconstructed from $\mathbf{B}_{\mathrm{h}}$ by

$$\int_{\mathbf{x}_0}^{\mathbf{x}} \mathrm{d}\mathbf{x}' \cdot \nabla_{\mathrm{h}} M = M(\mathbf{x}) - M(\mathbf{x}_0) = \int_{\mathbf{x}_0}^{\mathbf{x}} \mathrm{d}\mathbf{x}' \cdot \mathbf{B}_{\mathrm{h}} \tag{16.30}$$

In order that this integration leads to a unique single-valued expression for $M(\mathbf{x})$ one has to require that

$$\oint_{\delta S_j} \mathrm{d}\mathbf{x} \cdot \mathbf{B}_{\mathrm{h}} = 0 \tag{16.31}$$

for all "islands" δS_j ($j = 2, \ldots, N$) (see Figure 19.1). These integrability conditions constitute additional lateral boundary conditions. Since η is determined by M, the constant $M(\mathbf{x}_0)$ matters as well. It can be inferred from the volume conservation

$$\int\int_S \mathrm{d}^2x M = 0 \qquad (16.32)$$

which constitutes an additional integrability condition.

As an example take the momentum balance

$$f\hat{\mathbf{u}}_\mathrm{h} = -\frac{1}{\rho_0}\nabla_\mathrm{h} M + \frac{\boldsymbol{\tau}_\mathrm{h}^\mathrm{a} - \boldsymbol{\tau}_\mathrm{h}^\mathrm{b}}{\rho_0 H} - \epsilon\mathbf{u}_\mathrm{h} \qquad (16.33)$$

One then must require

$$\oint_{\delta S_j} \mathrm{d}\mathbf{x}\cdot\left(\frac{\boldsymbol{\tau}_\mathrm{h}^\mathrm{a} - \boldsymbol{\tau}_\mathrm{h}^\mathrm{b}}{\rho_0 H}\right) - \oint_{\delta S_j} \mathrm{d}\mathbf{x}\cdot\epsilon\mathbf{u}_\mathrm{h} = 0 \qquad (16.34)$$

for all islands. The integral over $f\hat{\mathbf{u}}_\mathrm{h} = f\hat{\mathbf{M}}_\mathrm{h}/H$ vanishes since $\mathbf{M}_\mathrm{h}\cdot\mathbf{n}_\mathrm{h}$ vanishes on the boundary.

16.3 The barotropic general circulation

The planetary geostrophic equations are used to study the oceanic general circulation. One must distinguish between channel geometry with closed f/H contours and basin geometry with blocked f/H contours.

Free inertial flows Free inertial flows are obtained when all forcing and dissipation processes are neglected. The barotropic planetary potential vorticity equation then reduces to

$$\mathbf{M}_\mathrm{h}\cdot\nabla_\mathrm{h}\left(\frac{f}{H}\right) = 0 \qquad (16.35)$$

or

$$\hat{\nabla}_\mathrm{h}\Psi\cdot\nabla_\mathrm{h}\left(\frac{f}{H}\right) = 0 \qquad (16.36)$$

with

$$\Psi = \text{constant} \qquad (16.37)$$

on all boundaries δS_j. Any function

$$\Psi = \Psi\left(\frac{f}{H}\right) \qquad (16.38)$$

is a solution of the potential vorticity equation.

For basin geometry with blocked f/H contours the boundary conditions imply Ψ = constant in the whole domain. There are no free inertial flows.

For channel geometry with closed f/H contours, any function $\Psi = \Psi(f/H)$ is a solution. There is an infinite number of free inertial flows.

The wind-driven circulation In calculating the wind-driven circulation one again has to distinguish between basin and channel geometry.

Basin geometry. In this case the f/H contours intersect the boundaries. The analysis starts from the Sverdrup balance

$$\mathbf{M}_{\mathrm{h}} \cdot \nabla_{\mathrm{h}} \left(\frac{f}{H} \right) = \hat{\nabla}_{\mathrm{h}} \cdot \left(\frac{\tau_{\mathrm{h}}^{\mathrm{a}}}{\rho_0 H} \right) \tag{16.39}$$

The vorticity input by the wind is balanced by a change in planetary vorticity. The Sverdrup balance determines the volume transport across f/H contours.

The Sverdrup balance is incomplete in two respects. First, since $\mathbf{M}_{\mathrm{h}} = \hat{\nabla}_{\mathrm{h}} \Psi$ the Sverdrup balance is a first order differential equation in Ψ and, hence, can only satisfy the kinematic boundary condition $\Psi =$ constant at one of the two end points of the f/H contour. Second, in general the total transport across an f/H contour is not zero. The streamlines are not closed. There then must be a return flow.

Both problems can be alleviated by adding friction to the vorticity balance. A closed circulation requires that a particle regains its original vorticity after completing one loop. The vorticity it acquires from the wind must then be dissipated by friction. Changes in planetary vorticity are inconsequential since a particle will regain its original planetary vorticity after one loop. For small Ekman numbers the return flow is in a narrow, frictionally controlled boundary layer where changes in planetary vorticity are balanced by friction, for example by Rayleigh friction for simplicity

$$\mathbf{M}_{\mathrm{h}} \cdot \nabla_{\mathrm{h}} \left(\frac{f}{H} \right) = -\epsilon \hat{\nabla}_{\mathrm{h}} \cdot \left(\frac{\mathbf{M}_{\mathrm{h}}}{H} \right) \tag{16.40}$$

The boundary layer has to be on the "western" side of the basin. The "western" side is where higher f/H values are to the left. This result can be inferred from an integration of the vorticity balance over an area surrounded by a closed streamline. Applying Stokes' theorem one finds, again in the case of Rayleigh friction

$$\oint \mathrm{d}\mathbf{x} \cdot \hat{\mathbf{M}}_{\mathrm{h}} \frac{f}{H} = \int_A \mathrm{d}^2 x \, \hat{\nabla}_{\mathrm{h}} \cdot \left(\frac{\tau_{\mathrm{h}}^{\mathrm{a}}}{\rho_0 H} \right) - \oint \mathrm{d}\mathbf{x} \cdot \mathbf{M}_{\mathrm{h}} \frac{\epsilon}{H} \tag{16.41}$$

where the integration along the streamline has to be taken in a counterclockwise sense (see Stokes' theorem (B.55)). The term on the left-hand side vanishes since $\hat{\mathbf{M}}_{\mathrm{h}} = -\nabla_{\mathrm{h}} \psi$. If the vorticity input by the wind is positive the interior Sverdrup flow is northward. This flow has to connect to a western boundary return flow to form a counterclockwise circulation such that the friction term can balance the vorticity input. If the vorticity input is negative the interior flow is southward. This flow then has to connect to a western boundary return flow as well such that a clockwise circulation is formed where friction balances the vorticity input.

The streamfunction in the interior is thus obtained by integrating the Sverdrup relation along an f/H contour starting at the eastern boundary

$$\Psi = \Psi_b - \int_{x_b}^{x} dx \frac{1}{\left|\nabla_h \frac{f}{H}\right|} \hat{\nabla}_h \cdot \left(\frac{\boldsymbol{\tau}_h^a}{\rho_0 H} \right) \tag{16.42}$$

where Ψ_b is the prescribed value at the eastern boundary. This is an integration of the incompressibility condition $\nabla_h \cdot \mathbf{M}_h = 0$, with the cross flow given by the Sverdrup relation.

Channel geometry. Channel geometry is characterized by closed f/H contours that do not intersect any boundary. The vorticity balance

$$\mathbf{M}_h \cdot \nabla_h \left(\frac{f}{H} \right) = \hat{\nabla}_h \cdot \left(\frac{\boldsymbol{\tau}_h^a}{\rho_0 H} \right) - \epsilon \hat{\nabla}_h \cdot \left(\frac{\mathbf{M}_h}{H} \right) \tag{16.43}$$

with boundary condition

$$\mathbf{M}_h \cdot \mathbf{n}_h = 0 \tag{16.44}$$

still determines the flow across f/H contours. The flow along f/H contours, however, remains undetermined. It cannot be determined by integrating the incompressibility condition $\nabla_h \cdot \mathbf{M}_h = 0$ from the coast, friction just determining that one should start this integration at the eastern coast. The flow along f/H contours is determined by the integrability conditions

$$\frac{1}{\rho_0} \oint_{\delta S_i} d\mathbf{x} \cdot \boldsymbol{\tau}_h^a - \epsilon \oint_{\delta S_i} d\mathbf{x} \cdot \mathbf{M}_h = 0 \tag{16.45}$$

$$\iint_S d^2x \, M = 0 \tag{16.46}$$

As an example, consider the "Antarctic Circumpolar Current" in a channel with coasts at θ_1 and θ_2. Assume $\tau_\varphi^a =$ constant, $\tau_\theta^a = 0$, and $H = H_0 =$ constant. The volume transport is then $M_\varphi = \tau_\varphi^a / \rho_0 \epsilon$ and $M_\theta = 0$ everywhere, by virtue of the integrability condition around the "island" Antarctica. The surface elevation can be calculated from the volume conservation condition.

16.4 The baroclinic problem

The baroclinic planetary geostrophic equations form the basis for the theoretical analysis of the pycnocline. As for the barotropic problem, we distinguish between strictly and viscous geostrophic equations.

Strictly geostrophic flows In the strictly geostrophic case the velocity and pressure fields are determined by the geostrophic balance, hydrostatic balance, and incompressibility condition

$$f\hat{\mathbf{u}}_{\mathrm{h}} = -\frac{1}{\rho_0}\nabla_{\mathrm{h}} p \tag{16.47}$$

$$\partial_z p = -\rho g \tag{16.48}$$

$$\nabla \cdot \mathbf{u} = 0 \tag{16.49}$$

subject to wind-induced Ekman pumping at the surface and friction-induced Ekman pumping at the bottom. These equations have to be augmented by the equation of state $\rho = \rho(z, \theta, S)$ and the evolution equations for θ and S

$$\frac{\mathrm{D}}{\mathrm{D}t}\theta = K_{\mathrm{h}}\Delta_{\mathrm{h}}\theta + K_{\mathrm{v}}\partial_z\partial_z\theta \tag{16.50}$$

$$\frac{\mathrm{D}}{\mathrm{D}t}S = K_{\mathrm{h}}\Delta_{\mathrm{h}}S + K_{\mathrm{v}}\partial_z\partial_z S \tag{16.51}$$

if the standard parametrization of eddy fluxes is used. These are the only prognostic equations of the problem. Usually, one considers these equations in the steady limit. In the presence of a barotropic circulation the advection operator must include the barotropic advection velocities. These equations generally cannot satisfy the lateral kinematic boundary conditions. They are valid only in the interior. The major challenge is to define suitable boundary conditions that are physically sensible and compatible with the mathematical structure of the equations.

Theoretical analyses often disregard the compressibility and two-component structure of sea water and replace the potential temperature and salinity equation by a single density equation of the same structure. The baroclinic planetary geostrophic potential vorticity

$$q_\rho = f\partial_z\rho \tag{16.52}$$

then satisfies

$$\frac{\mathrm{D}}{\mathrm{D}t}q_\rho = f\partial_z(K_{\mathrm{h}}\Delta_{\mathrm{h}}\rho + K_{\mathrm{v}}\partial_z\partial_z\rho) \tag{16.53}$$

It is again at the heart of the physics of the problem but is often subjugated to mathematical considerations as in the following two examples.

Example 1. The M equation. The strictly planetary geostrophic equations can be rewritten by introducing a function

$$M(\varphi, \theta, z) := \int_{z_0}^{z} \mathrm{d}z' \frac{p}{\rho_0} + M_0(\varphi, \theta) \tag{16.54}$$

where

$$M_0(\varphi, \theta) = M(\varphi, \theta, z = z_0) \tag{16.55}$$

Then

$$\frac{p}{\rho_0} = \partial_z M \tag{16.56}$$

$$\mathbf{u}_{\mathrm{h}} = \frac{1}{f} \hat{\boldsymbol{\nabla}}_{\mathrm{h}} \partial_z M \tag{16.57}$$

$$w = \frac{1}{f^2} \boldsymbol{\nabla}_{\mathrm{h}} f \cdot \hat{\boldsymbol{\nabla}}_{\mathrm{h}} M \tag{16.58}$$

$$\frac{\rho g}{\rho_0} = -\partial_z \partial_z M \tag{16.59}$$

if one chooses M_0 such that $\boldsymbol{\nabla}_{\mathrm{h}} f \cdot \hat{\boldsymbol{\nabla}}_{\mathrm{h}} M_0 = f^2 w_0$. Substitution into the density equation yields

$$\begin{aligned} &\partial_t \partial_z \partial_z M + f\left[-\partial_y \partial_z M \partial_x \partial_z \partial_z M + \partial_x \partial_z M \partial_y \partial_z \partial_z M\right] + \beta \partial_x M \partial_z \partial_z \partial_z M \\ &\quad = K_{\mathrm{h}} f^2 \Delta_{\mathrm{h}} \partial_z \partial_z M + K_{\mathrm{v}} f^2 \partial_z \partial_z \partial_z \partial_z M \end{aligned} \tag{16.60}$$

This is a fourth order differential equation in z. It is usually solved subject to the four vertical boundary conditions

1. Prescribed density distribution $\rho(\varphi, \theta, z_0)$ at the surface

$$M_{zz} = -\frac{g}{\rho_0} \rho(\varphi, \theta, z_0) \quad \text{at } z = 0 \tag{16.61}$$

2. Prescribed Ekman pumping velocity $w_{\mathrm{Ek}}^{\mathrm{s}}$ at the surface

$$\frac{1}{f^2} \boldsymbol{\nabla}_{\mathrm{h}} f \cdot \hat{\boldsymbol{\nabla}}_{\mathrm{h}} M = w_{\mathrm{Ek}}^{\mathrm{s}} \quad \text{at } z = 0 \tag{16.62}$$

3. No buoyancy flux at the bottom

$$M_{zzz} = 0 \quad \text{at } z = -H \tag{16.63}$$

4. No vertical velocity at the bottom

$$\frac{1}{f^2} \boldsymbol{\nabla}_{\mathrm{h}} f \cdot \hat{\boldsymbol{\nabla}}_{\mathrm{h}} M = 0 \quad \text{at } z = -H \tag{16.64}$$

Example 2. Ideal thermocline theory. If $K_{\mathrm{h}} = K_{\mathrm{v}} = 0$ then one obtains the ideal planetary geostrophic equations. Again the major challenge is to find suitable boundary conditions for these equations. The upstream–downstream concept is fuzzy. Weak solutions with discontinuities across internal boundaries exist. The equations are of a degenerate hyperbolic type with the vertical axis being a triple characteristic. For steady problems one has

$$\mathbf{u} \cdot \nabla_{\mathrm{h}} \rho + w \partial_z \rho = 0 \tag{16.65}$$

$$\mathbf{u} \cdot \nabla_{\mathrm{h}} q + w \partial_z q = 0 \tag{16.66}$$

$$\mathbf{u} \cdot \nabla_{\mathrm{h}} B + w \partial_z B = 0 \tag{16.67}$$

where

$$B = p + \rho g z \tag{16.68}$$

is the Bernoulli function. For each point where $\mathbf{u} \neq 0$ there thus exists a function

$$F(\rho, q, B) = 0 \tag{16.69}$$

where

$$\rho = -\frac{1}{g} \partial_z p \tag{16.70}$$

$$q = -\frac{f}{g} \partial_z \partial_z p \tag{16.71}$$

$$B = p - z \partial_z p \tag{16.72}$$

The pressure field must thus satisfy $F(-\partial_z p/g, -f\partial_z \partial_z p/g, p - z\partial_z p) = 0$. This relation may be used as a starting point for analysis.

In the ideal limit, one can also express the velocity field solely in terms of the density field by

$$\mathbf{u} = \frac{g}{\rho_0} \frac{\mathbf{e}_z \cdot (\nabla \rho \times \nabla q)}{(\nabla \rho \times \nabla q) \cdot \nabla \left(f \frac{\partial q}{\partial z} \right)} \nabla \rho \times \nabla q \tag{16.73}$$

One might use this formula for the determination of the oceanic velocity field from hydrographic data.

Viscous geostrophic flows To satisfy the lateral kinematic boundary conditions one often introduces Rayleigh friction into the horizontal and vertical momentum balance

$$f \hat{\mathbf{u}}_{\mathrm{h}} = -\frac{1}{\rho_0} \nabla_{\mathrm{h}} p - \epsilon \mathbf{u}_{\mathrm{h}} \tag{16.74}$$

$$0 = -\partial_z p - \rho g - \rho_0 \tilde{\epsilon} w \tag{16.75}$$

with friction coefficients ϵ and $\tilde{\epsilon}$. One can then recast the problem in the following way. From the momentum balances one finds the velocity field

$$\mathbf{u}_\mathrm{h} = \frac{1}{\rho_0}\frac{f}{f^2+\epsilon^2}\hat{\boldsymbol{\nabla}}_\mathrm{h} p - \frac{1}{\rho_0}\frac{\epsilon}{f^2+\epsilon^2}\boldsymbol{\nabla}_\mathrm{h} p \tag{16.76}$$

$$w = -\frac{\partial_z p + \rho g}{\rho_0 \tilde{\epsilon}} \tag{16.77}$$

Substitution into the incompressibility condition yields the elliptic equation

$$\tilde{\epsilon}\left[\hat{\boldsymbol{\nabla}}_\mathrm{h}\left(\frac{f}{f^2+\epsilon^2}\right)\cdot\boldsymbol{\nabla} p + \boldsymbol{\nabla}_\mathrm{h}\left(\frac{\epsilon}{f^2+\epsilon^2}\boldsymbol{\nabla}_\mathrm{h} p\right)\right] + \partial_z\partial_z p = -\partial_z \rho g \tag{16.78}$$

It can be regarded as an "advective–diffusion" equation for the pressure with a source term on the right-hand side. The kinematic boundary conditions become conditions for the pressure through the relations (16.76) and (16.77). Therefore one has to solve a "prognostic" equation for the density ρ, a diagnostic elliptic equation for the pressure p. The velocity is then determined by (16.76) and (16.77). If $\tilde{\epsilon}$ were set to zero one could not satisfy the kinematic boundary conditions for w.

17
Tidal equations

Tidal motions are the motions directly forced by the gravitational potential of the Moon and Sun. As discussed in Chapter 5, the tidal force is a volume force and approximately constant throughout the water column. Tidal currents are hence depth-independent. The tidal force does not generate any vorticity within the fluid. Tidal motions thus represent the gravity mode of motion. This fact is, however, not exploited in the theory and models of the tides. Rather, these start from Laplace tidal equations, which are the primitive equations for a homogeneous fluid layer. The tidal potential is extremely well known. Nevertheless, uncertainties enter when parametrizing the subgridscale fluxes and when incorporating the Earth and load tides and gravitational self-attraction.

17.1 Laplace tidal equations

Laplace tidal equations are obtained from the primitive equations (13.54) and (13.55) (or from the 1-layer equations (14.73) and (14.74)) by adding the tidal potential ϕ_{T} and neglecting the atmospheric forcing terms. Thus

$$\begin{aligned} &\left(\frac{\partial}{\partial t} + \mathbf{u}_{\mathrm{h}} \cdot \nabla_{\mathrm{h}}\right) \mathbf{u}_{\mathrm{h}} + f \hat{\mathbf{u}}_{\mathrm{h}} = -g \nabla_{\mathrm{h}} \eta - \nabla_{\mathrm{h}} \phi_{\mathrm{T}} - \frac{\boldsymbol{\tau}_{\mathrm{h}}^{\mathrm{b}}}{\rho_0 (H + \eta)} + \frac{1}{\rho_0} \mathbf{F}_{\mathrm{h}}^{\mathrm{h}} \\ &\partial_t \eta + \nabla_{\mathrm{h}} \cdot [(H + \eta) \mathbf{u}_{\mathrm{h}}] = 0 \end{aligned} \tag{17.1}$$

In order that they also apply to shallow seas they do not assume that the surface elevation η is much smaller than the ocean depth H. The bottom stress is given either by linear ($\boldsymbol{\tau}_{\mathrm{h}}^{\mathrm{b}} = \rho_0 (H + \eta) r \mathbf{u}_{\mathrm{h}}$) or "quadratic" ($\boldsymbol{\tau}_{\mathrm{h}}^{\mathrm{b}} = \rho_0 c_{\mathrm{d}} |\mathbf{u}_{\mathrm{h}}| \mathbf{u}_{\mathrm{h}}$) bottom friction with friction coefficient r or drag coefficient c_{d}. The term $H + \eta$ in the denominator introduces an additional nonlinearity in the case of quadratic friction. The horizontal friction term $\mathbf{F}_{\mathrm{h}}^{\mathrm{h}}$ is usually parametrized in terms of a lateral eddy viscosity coefficient A_h. Often the tidal potential is written as

$$\phi_\mathrm{T} = -g\eta_\mathrm{equ} \tag{17.2}$$

where η_equ is the equilibrium tidal displacement. The lateral boundary conditions are

$$\boldsymbol{u}_\mathrm{h} \cdot \boldsymbol{n}_\mathrm{h} = 0 \ \text{ at coastlines} \tag{17.3}$$

$$\eta(t) = \eta_0(t) \ \text{ at open boundaries} \tag{17.4}$$

where $\boldsymbol{n}_\mathrm{h}$ is the normal vector of the coastline and $\eta_0(t)$ a prescribed water level.

In order to solve Laplace tidal equations one has to specify the parameters r or c_d, A_h, and prescribe the tidal potential ϕ_T or equilibrium tidal displacement η_equ, and the surface elevation $\eta_0(t)$ at open boundaries. Laplace tidal equations contain nonlinearities in the advection and bottom friction terms and in the volume balance. If one introduces the depth integrated volume transport

$$\mathbf{M}_\mathrm{h} := \int_{-H}^{\eta} \mathrm{d}z\, \boldsymbol{u}_\mathrm{h} \tag{17.5}$$

instead of the velocity $\boldsymbol{u}_\mathrm{h}$, then the volume balance becomes strictly linear

$$\partial_t \eta + \boldsymbol{\nabla}_\mathrm{h} \cdot \mathbf{M}_\mathrm{h} = 0 \tag{17.6}$$

The momentum equation is then usually approximated by

$$\partial_t \mathbf{M}_\mathrm{h} + f\hat{\mathbf{M}}_\mathrm{h} = -gH\boldsymbol{\nabla}_\mathrm{h}(\eta - \eta_\mathrm{equ}) - \frac{\boldsymbol{\tau}_\mathrm{h}^\mathrm{b}}{\rho_0} + \frac{H}{\rho_0}\mathbf{F}_\mathrm{h}^\mathrm{h} \tag{17.7}$$

All nonlinearities are eliminated, except for a possibly nonlinear bottom friction term.

When a strictly linear version of the tidal equations is assumed then one can solve the equations by expanding the tidal potential $\eta_\mathrm{equ}(t)$, the tidal elevation $\eta(t)$, and the tidal transport $\mathbf{M}_\mathrm{h}(t)$ into sums of tidal harmonics $\exp(-i\omega_j t)$, $j = 1, \ldots, J$. The complex tidal amplitudes $\eta(\omega_j)$ and $\mathbf{M}_h(\omega_j)$ can then be calculated separately for each tidal component j. The time derivative $\partial/\partial t$ is replaced by $-i\omega_j$ and the problem becomes purely algebraic.

17.2 Tidal loading and self-gravitation

Because of the relative simplicity of Laplace tidal equations and the absence of strong nonlinearities, the first numerical tidal models already provided satisfying results. Later the high accuracy demands of regional water level forecasts for navigation and coastal protection, and the need for correcting geodetical data, demonstrated some deficiencies in global tidal models. A closer inspection of the problem led to the finding that the conventional Laplace tidal equations needed to be modified to account for three more processes, namely the Earth tides, tidal loading, and ocean self-attraction. The tidal potential causes tides of the solid but elastic Earth, called the Earth tides, which deform the ocean bottom. The moving tidal water bulge also causes an elastic deformation of the ocean bottom, the load tides. At the same time the water mass exerts variable gravitational self-attraction, depending on its own distribution. Technically, these effects can be included by modifying Laplace tidal equations as follows.

The bottom elevation of the earth tide is denoted by η_{e} and the bottom elevation of the load tide by η_{l}. If these bottom elevations are included then the displacement in the pressure term of Laplace tidal equations is not the elevation η of the water column but the geocentric elevation

$$\eta_{\mathrm{g}} = \eta + \eta_{\mathrm{e}} + \eta_{\mathrm{l}} \tag{17.8}$$

In addition, the earth tides, the load tides, and the ocean tides redistribute mass and modify the gravitational potential by amounts $\delta\eta^{\mathrm{e}}_{\mathrm{equ}}$, $\delta\eta^{\mathrm{l}}_{\mathrm{equ}}$, and $\delta\eta^{\circ}_{\mathrm{equ}}$. This modification is the "gravitational self-attraction." The full tidal forcing is thus given by

$$\phi_{\mathrm{T}} = -g\left(\eta_{\mathrm{equ}} + \delta\eta^{\mathrm{e}}_{\mathrm{equ}} + \delta\eta^{\mathrm{l}}_{\mathrm{equ}} + \delta\eta^{\circ}_{\mathrm{equ}}\right) \tag{17.9}$$

The calculation of the Earth and load tides must take into account the elastic properties of the solid Earth. The calculation of the gravitational self-attraction involves a spatial convolution with a known Green's function (see (5.3)). Under simplifying assumptions these calculations result in factors α and β modifying the elevations and equilibrium tidal displacement in the momentum balance (17.7)

$$\partial_t \mathbf{M}_{\mathrm{h}} + f\hat{\mathbf{M}}_{\mathrm{h}} = -gH\boldsymbol{\nabla}_{\mathrm{h}}(\beta\eta - \alpha\eta_{\mathrm{equ}}) - \frac{\tau^{\mathrm{b}}_{\mathrm{h}}}{\rho_0} + \frac{H}{\rho_0}\mathbf{F}^{\mathrm{h}}_{\mathrm{h}} \tag{17.10}$$

Equations (17.10) and (17.6) represent the zeroth order physics: volume conservation, momentum changes due to the Coriolis, pressure, astronomical, and frictional forces. Solid Earth tides, the load tides, and tidal self-attraction are included in an approximate manner. The frictional force may account for the generation of baroclinic tides at topography in a stratified ocean.

18

Medium-scale motions

Medium-scale motions are defined here as motions whose horizontal space scales L are much smaller than the Earth's radius r_0 but still large enough that they are not directly affected by molecular diffusion. They are the middle component of our triple decomposition into large-, medium-, and small-scale motions that arises from two Reynolds decompositions. The first decomposition separates large- from medium- and small-scale motions. The second decomposition separates large- and medium-scale motions from small-scale motions. Large-scale motions must thus parametrize the eddy fluxes caused by medium- and small-scale motions. Medium-scale motions must prescribe the large-scale fields and parametrize the eddy fluxes caused by small-scale motions. Small-scale motions must prescribe the large- and medium-scale motions. Diffusion is molecular. Medium-scale motions thus face two closure problems, one with respect to larger scale motions and one with respect to smaller scale motions. The only simplification is one of geometry. The smallness of the parameter $\gamma := L/r_0$ allows the spherical geometry to be approximated by a variety of "planar" geometries. These approximations are only valid locally and include:

- midlatitude beta-plane approximation;
- equatorial beta-plane approximation;
- f-plane approximation; and
- polar plane approximation.

These are geometric approximations that are similar to the spherical approximation that relied on the smallness of Earth's eccentricity d_0^2/r_0^2 and of the parameter H/r_0, and led to the introduction of pseudo-spherical coordinates.

Medium-scale motions have to prescribe a background state. Here we only consider a background stratification.

18.1 Geometric approximations

Field variables can be represented in any kind of coordinate system. The Mercator, polar stereographic, and Lambert conformal projections are often used to map the sphere onto a plane. The equations of motion can also be represented in any kind of coordinate system. The metric coefficients in the equations then change. This is not an approximation but a change of representation. A geometric *approximation* is made when the metric coefficients are approximated, e.g., by expanding them with respect to $\gamma = L/r_0$ and truncating the expansion. Approximating the metric coefficients rather than the equations has the advantage that it does not affect the general properties of the equations. The pressure force in the momentum equation remains a gradient. Its integral along a closed circuit vanishes exactly; its curl vanishes exactly. The incompressibility condition is still a divergence. Volume is conserved.

Boussinesq equations in planar coordinates Instead of longitude and latitude we introduce the arclengths

$$\begin{aligned} x &= \varphi r_0 \cos\theta_0 \\ y &= r_0(\theta - \theta_0) \end{aligned} \tag{18.1}$$

where θ_0 is a reference latitude. The scale factors of the metric then become

$$\begin{aligned} h_x &= \frac{\cos\theta}{\cos\theta_0} = \frac{\cos(\theta_0 + y/r_0)}{\cos\theta_0} \\ h_y &= 1 \\ h_z &= 1 \end{aligned} \tag{18.2}$$

This is still the metric of the non-Euclidean pseudo-spherical coordinate system, but expressed in a new set of coordinates (x, y).

For $|y| \ll r_0$ the metric may be approximated by

$$\begin{aligned} h_x &= 1 - \frac{y}{r_0}\tan\theta_0 \\ h_y &= h_z = 1 \end{aligned} \tag{18.3}$$

The Coriolis parameter may be similarly approximated by

$$f = f_0 + \beta_0 y \tag{18.4}$$

where

$$f_0 = 2\Omega \sin\theta_0 \tag{18.5}$$

and the beta-parameter

$$\beta_0 = \frac{2\Omega}{r_0}\cos\theta_0 \tag{18.6}$$

The introduction of the coordinates (x, y) with these approximations constitutes the beta-plane approximation. If the reference latitude is at midlatitudes, $\theta_0 \neq 0$, then it is called the midlatitude beta-plane approximation. If the reference latitude is the equator, $\theta_0 = 0$, it is called the equatorial beta-plane approximation. The coordinates (x, y) are not Cartesian coordinates but arclengths, though the beta-plane metric is Euclidean.

If one further approximates

$$h_x = h_y = h_z = 1 \tag{18.7}$$

and

$$f = f_0 \tag{18.8}$$

then one arrives at the midlatitude *f*-plane approximation.

For polar problems one often introduces the coordinates (φ, r) with

$$r = r_0 \left(\theta - \frac{\pi}{2}\right) \tag{18.9}$$

Then

$$h_\varphi = r_0 \cos\left(\frac{\pi}{2} + \frac{r}{r_0}\right) \tag{18.10}$$

$$h_r = 1 \tag{18.11}$$

The approximation

$$h_\varphi = -r \tag{18.12}$$

and

$$f = 2\Omega\left(1 - \frac{1}{2}\frac{r^2}{r_0^2}\right) \tag{18.13}$$

then constitute the polar plane approximation.

For any metric $(h_x, h_y = 1, h_z = 1)$ the Boussinesq equations in the traditional approximation take the general form

$$\partial_t u + \frac{1}{h_x} u\, \partial_x u + v\, \partial_y u + \frac{1}{h_x} uv \frac{\partial h_x}{\partial y} + w\, \partial_z u - f v = -\frac{1}{\rho_0}\frac{1}{h_x} \partial_x p + \frac{1}{\rho_0} F_x \tag{18.14}$$

$$\partial_t v + \frac{1}{h_x} u\, \partial_x v + v\, \partial_y v - \frac{1}{h_x} u^2 \frac{\partial h_x}{\partial y} + w\, \partial_z v + f u = -\frac{1}{\rho_0} \partial_y p + \frac{1}{\rho_0} F_y \tag{18.15}$$

$$\partial_t w + \frac{1}{h_x} u\,\partial_x w + v\,\partial_y w + w\,\partial_z w = -\frac{1}{\rho_0}\partial_z p - \frac{\rho g}{\rho_0} + \frac{1}{\rho_0}F_z \qquad (18.16)$$

$$\partial_x u + \partial_y(h_x v) + h_x \partial_z w = 0 \qquad (18.17)$$

$$\partial_t \psi + \frac{1}{h_x} u\partial_x \psi + v\partial_y \psi + w\partial_z \psi = D_\psi \qquad (18.18)$$

where ψ is any scalar tracer.

Midlatitude beta-plane approximation If one substitutes $h_x = 1 - \frac{y}{r_0}\tan\theta_0$ and $f = f_0 + \beta_0 y$ into the above Boussinesq equations then one obtains the midlatitude beta-plane equations. In their general form, they are not widely used. They are mostly combined with other approximations such as the geostrophic approximation, leading to the quasi-geostrophic equations discussed in Chapter 19. In this case one additionally approximates

$$\frac{1}{h_x} = \frac{1}{1 - \frac{y}{r_0}\tan\theta_0} = 1 + \frac{y}{r_0}\tan\theta_0 + O(\epsilon^2) \qquad (18.19)$$

The beta-plane equations then take the explicit form

$$\partial_t u + \left(1 + \frac{y}{r_0}\tan\theta_0\right) u\,\partial_x u + v\,\partial_y u - \frac{uv}{r_0}\tan\theta_0 + w\,\partial_z u - (f_0 + \beta_0 y)v = -\frac{1}{\rho_0}\left(1 + \frac{y}{r_0}\tan\theta_0\right)\partial_x p + \frac{1}{\rho_0}F_x \qquad (18.20)$$

$$\partial_t v + \left(1 + \frac{y}{r_0}\tan\theta_0\right) u\,\partial_x v + v\,\partial_y v + \frac{u^2}{r_0}\tan\theta_0 + w\,\partial_z v + (f_0 + \beta_0 y)u = -\frac{1}{\rho_0}\partial_y p + \frac{1}{\rho_0}F_y \qquad (18.21)$$

$$\partial_t w + \left(1 + \frac{y}{r_0}\tan\theta_0\right) u\,\partial_x w + v\,\partial_y w + w\,\partial_z w = -\frac{1}{\rho_0}\partial_z p - \frac{\rho g}{\rho_0} + \frac{1}{\rho_0}F_z \qquad (18.22)$$

$$\partial_x u + \left(1 - \frac{y}{r_0}\tan\theta_0\right)\partial_y v - \frac{v}{r_0}\tan\theta_0 + \left(1 - \frac{y}{r_0}\tan\theta_0\right)\partial_z w = 0 \qquad (18.23)$$

$$\partial_t \psi + \left(1 + \frac{y}{r_0}\tan\theta_0\right) u\partial_x \psi + v\partial_y \psi + w\partial_z \psi = D_\psi \qquad (18.24)$$

This is now an asymptotic expansion, correct to order γ. The resulting equations do not correspond to a realizable physical system since they do not represent physical laws in a well-defined geometric space. For example, the curl of the pressure force is not exactly zero, but only zero up to order γ.

Equatorial beta-plane approximation If the reference latitude is the equator, $\theta_0 = 0$, then $h_x = 1$ and $f = \beta_0 y$ and one obtains the equatorial beta-plane equations

$$\partial_t u + u\,\partial_x u + v\,\partial_y u + w\,\partial_z u - \beta_0 y v = -\frac{1}{\rho_0}\partial_x p + \frac{1}{\rho_0}F_x \tag{18.25}$$

$$\partial_t v + u\,\partial_x v + v\,\partial_y v + w\,\partial_z v + \beta_0 y u = -\frac{1}{\rho_0}\partial_y p + \frac{1}{\rho_0}F_y \tag{18.26}$$

$$\partial_t w + u\,\partial_x w + v\,\partial_y w + w\,\partial_z w = -\frac{1}{\rho_0}\partial_z p - \frac{\rho g}{\rho_0} + \frac{1}{\rho_0}F_z \tag{18.27}$$

$$\partial_x u + \partial_y v + \partial_z w = 0 \tag{18.28}$$

$$\partial_t \psi + u\partial_x \psi + v\partial_y \psi + w\partial_z \psi = D_\psi \tag{18.29}$$

These equations are the starting point for the analysis of equatorial phenomena. The linearized equations about a motionless stratified background state are given by

$$\partial_t u + \beta_0 y v = -\frac{1}{\rho_0}\partial_x \delta p \tag{18.30}$$

$$\partial_t v + \beta_0 y u = -\frac{1}{\rho_0}\partial_y \delta p \tag{18.31}$$

$$\partial_t w = -\frac{1}{\rho_0}\partial_z \delta p - \frac{g}{\rho_0}\delta\rho \tag{18.32}$$

$$\partial_x u + \partial_y v + \partial_z w = 0 \tag{18.33}$$

$$\partial_t \delta\rho - w\frac{\rho_0}{g}N^2 = 0 \tag{18.34}$$

Separation of variables leads to the horizontal eigenvalue problem (in the notation of Section 8.4)

$$\hat{\omega}\hat{U} - \mu\hat{V} - m\hat{P} = 0 \tag{18.35}$$

$$\hat{\omega}\hat{V} - \mu\hat{U} - D\hat{P} = 0 \tag{18.36}$$

$$-\hat{\omega}\hat{\epsilon}\hat{P} - D\hat{V} + m\hat{U} = 0 \tag{18.37}$$

where $\mu = y/r_0$. This leads to

$$\left[\frac{\mathrm{d}^2}{\mathrm{d}\mu^2} - m^2 - \frac{m}{\hat{\omega}} + \hat{\epsilon}(\hat{\omega}^2 - \mu^2)\right]\hat{V} = 0 \tag{18.38}$$

and is the limit $\hat{\epsilon} \to +\infty$, $\hat{\omega}^2\hat{\epsilon}$ bounded, $\mu^2 \ll 1$ discussed in Section 8.4.

f-plane approximation The limit $h_x = 1$ and $f = f_0$ constitutes the midlatitude f-plane approximation. The primitive equations take the form

$$\partial_t u + u\,\partial_x u + v\,\partial_y u + w\,\partial_z u - f_0 v = -\frac{1}{\rho_0}\partial_x p + \frac{1}{\rho_0}F_x \tag{18.39}$$

$$\partial_t v + u\,\partial_x v + v\,\partial_y v + w\,\partial_z v + (f_0 + \beta_0 y)u = -\frac{1}{\rho_0}\partial_y p + \frac{1}{\rho_0}F_y \tag{18.40}$$

$$\partial_t w + u\,\partial_x w + v\,\partial_y w + w\,\partial_z w = -\frac{1}{\rho_0}\partial_z p - \frac{\rho g}{\rho_0} + \frac{1}{\rho_0}F_z \tag{18.41}$$

$$\partial_x u + \partial_y v + \partial_z w = 0 \tag{18.42}$$

$$\partial_t \psi + u\partial_x \psi + v\partial_y \psi + w\partial_z \psi = D_\psi \tag{18.43}$$

These equations differ from the equations for a rotating plane in that they do not contain the centrifugal force. The centrifugal force enters the definition of the geoid to which the f-plane is a local planar approximation. The equilibrium shape of a fluid in a rotating cylinder is a parabola.

18.2 Background stratification

For medium-scale flows, one also has to specify the large-scale background state. We demonstrate some of the intricacies of such specifications by considering the tracer equation

$$(\partial_t + u\partial_x + v\partial_y + w\partial_z)\psi = \partial_z K_{\mathrm{v}}^{\mathrm{s}} \partial_z \psi \tag{18.44}$$

where the superscript "s" on the vertical eddy diffusion coefficient K_{v} is to indicate that the diffusion is caused by the small-scale motions of a triple decomposition. If one decomposes the tracer field into a horizontally homogeneous, time-independent background and deviations therefrom, $\psi = \tilde{\psi}(z) + \delta\psi$, the tracer equation becomes

$$(\partial_t + u\partial_x + v\partial_y + w\partial_z)\delta\psi + w\frac{\mathrm{d}\tilde{\psi}}{\mathrm{d}z} = \partial_z K_{\mathrm{v}}^{\mathrm{s}} \partial_z \delta\psi + \frac{\mathrm{d}}{\mathrm{d}z} K_{\mathrm{v}}^{\mathrm{s}} \frac{\mathrm{d}\tilde{\psi}}{\mathrm{d}z} \tag{18.45}$$

The deviation field $\delta\psi$ changes because of vertical advection of the background profile, the last term on the left-hand side. This term is an important and genuine part of the interaction between background and deviations. The deviations have also to balance the diffusion of the background, the last term on the right-hand side. This is usually not the physical understanding. Rather, one presumes that this diffusion is balanced by some other large-scale process, such as advection by a vertical background velocity $\tilde{w}$

$$\tilde{w}\frac{\mathrm{d}\tilde{\psi}}{\mathrm{d}z} = \frac{\mathrm{d}}{\mathrm{d}z}\left(K_{\mathrm{v}}^{\mathrm{m}} + K_{\mathrm{v}}^{\mathrm{s}}\right)\frac{\mathrm{d}\tilde{\psi}}{\mathrm{d}z} \tag{18.46}$$

Here we have introduced two vertical diffusion coefficients to indicate that the diffusion of the large scale flow is caused by both medium- and small-scale motion. The tracer equation then becomes

$$(\partial_t + u\partial_x + v\partial_y + \delta w\partial_z)\delta\psi + \tilde{w}\partial_z\delta\psi + \delta w\frac{\mathrm{d}\tilde{\psi}}{\mathrm{d}z} = \partial_z K_{\mathrm{v}}^{\mathrm{s}}\partial_z\delta\psi - \frac{\mathrm{d}}{\mathrm{d}z}K_{\mathrm{v}}^{\mathrm{m}}\frac{\mathrm{d}\tilde{\psi}}{\mathrm{d}z} \tag{18.47}$$

with two new terms: advection by the background vertical velocity $\tilde{w}$, the second term on the left-hand side, and diffusion of the background with diffusion coefficient $K_{\mathrm{v}}^{\mathrm{m}}$, the last term on the right-hand side. These two terms might be neglected arguing that the background vertical velocity is small and that vertical (diapycnal) diffusion is caused by small-scale processes not by medium-scale processes so that $K_{\mathrm{v}}^{\mathrm{m}}$ can be neglected. Under these assumptions, the tracer equation reduces to

$$(\partial_t + u\partial_x + v\partial_y + w\partial_z)\delta\psi + w\frac{\mathrm{d}\tilde{\psi}}{\mathrm{d}t} = \partial_z K_{\mathrm{v}}^{\mathrm{s}}\partial_z\delta\psi \tag{18.48}$$

This is the view we take in the following but we stress that this form of the tracer equation follows from specific assumptions about the dynamics of the background flow. As discussed in Chapter 10 on Reynolds decomposition this specification represents a closure if the background state is regarded as the mean state.

Consider now a horizontally homogeneous, time-independent hydrostatically balanced background stratification ($\tilde{\theta}(z)$,$\tilde{S}(z)$) maintained by an advective–diffusive balance. Under the above assumptions fluctuations ($\delta\theta$, δS) about this background stratification are then governed by

$$\frac{\mathrm{D}}{\mathrm{D}t}\delta\theta + w\frac{\mathrm{d}\tilde{\theta}}{\mathrm{d}z} = (\partial_x K_{\mathrm{h}}\partial_x + \partial_y K_{\mathrm{h}}\partial_y + \partial_z K_{\mathrm{v}}\partial_z)\delta\theta \tag{18.49}$$

$$\frac{\mathrm{D}}{\mathrm{D}t}\delta S + w\frac{\mathrm{d}\tilde{S}}{\mathrm{d}z} = (\partial_x K_{\mathrm{h}}\partial_x + \partial_y K_{\mathrm{h}}\partial_y + \partial_z K_{\mathrm{v}}\partial_z)\delta S \tag{18.50}$$

when the standard parametrization (12.21) for the eddy fluxes and identical eddy diffusion coefficients for potential temperature and salinity (no differential mixing) are assumed. Density fluctuations $\delta\rho = \rho - \tilde{\rho}$ where $\tilde{\rho} = \rho(z, \tilde{\theta}, \tilde{S})$ are generally calculated from the linearized form

$$\delta\rho = -\rho\tilde{\alpha}\delta\theta + \rho\tilde{\beta}\delta S \tag{18.51}$$

where the thermodynamic coefficients are evaluated for the background state.

When temperature–salinity effects are disregarded and a one-component description with density as a prognostic variable is sought, one starts from the density equation (12.9) in the Boussinesq approximation

$$\frac{\mathrm{D}}{\mathrm{D}t}\rho = -c^{-2}(z, \theta, S) w \rho_0 g + D_\rho \tag{18.52}$$

or

$$\frac{\mathrm{D}}{\mathrm{D}t}\delta\rho = -w \left[\frac{\mathrm{d}\tilde{\rho}}{\mathrm{d}z} + \rho_0 g c^{-2}(z, \theta, S) \right] + D_\rho \tag{18.53}$$

where D_ρ represents the eddy effects. One then assumes that the speed of sound on the right-hand side is given by $\tilde{c} = c(z, \tilde{\theta}, \tilde{S})$. This assumption constitutes a linearization about the background state. Then

$$\frac{\mathrm{D}}{\mathrm{D}t}\delta\rho - \frac{\rho_0}{g} N^2(z) w = D_\rho \tag{18.54}$$

where

$$N^2(z) = -\frac{g}{\rho_0}\frac{\mathrm{d}\tilde{\rho}}{\mathrm{d}z} - \frac{g^2}{\tilde{c}^2} \tag{18.55}$$

is the buoyancy frequency.

For the eddy term one also adopts the form

$$D_\rho = (\partial_x K_\mathrm{h} \partial_x + \partial_y K_\mathrm{h} \partial_y + \partial_z K_\mathrm{v} \partial_z)\delta\rho \tag{18.56}$$

which can be "derived" from equations (18.49) to (18.51) under the assumption that the coefficients in (18.51) are constant (linear equation of state).

Instead of $\delta\rho$ one often introduces the buoyancy

$$b := -\frac{g}{\rho_0}\delta\rho \tag{18.57}$$

or the vertical displacement η. For an incompressible ideal fluid $\mathrm{D}\rho/\mathrm{D}t = 0$ and the vertical displacement η is related to the density by

$$\rho = \tilde{\rho}(z - \eta) \tag{18.58}$$

The density equation is then replaced by

$$\frac{\mathrm{D}}{\mathrm{D}t}(z - \eta) = 0 \tag{18.59}$$

and the buoyancy term in the vertical momentum balance is given by the expansion

$$b = -\frac{g}{\rho_0}(\rho - \tilde{\rho}) = -\frac{g}{\rho_0}(\tilde{\rho}(z - \eta) - \tilde{\rho}(z)) = -\eta N^2 + \ldots \tag{18.60}$$

19

Quasi-geostrophic flows

Geostrophic flows require that the Rossby and Ekman numbers are small, $\tilde{\text{Ro}}, \text{Ro}, E_{\text{h}}, E_{\text{v}} \ll 1$, so that the momentum balance reduces to the geostrophic balance. Quasi-geostrophic motions (or small-scale geostrophic motions) are geostrophic motions for which additionally:

- the horizontal length scale is much smaller than the radius of the Earth, $\gamma := L/r_0 \ll 1$; and
- the vertical displacement Z of isopycnals is much smaller than the vertical scale D, $\sigma := Z/D \ll 1$.

The second condition is equivalent to requiring the vertical strain to be smaller than one. The quasi-geostrophic theory also disregards temperature–salinity effects and assumes a one-component fluid. The smallness of γ is exploited by applying the midlatitude beta-plane approximation. We will carry out the perturbation expansion with respect to all of the small parameters explicitly since the zeroth order is degenerate and one has to go to the first order. The evolution of quasi-geostrophic flows is again governed by the potential vorticity equation. The quasi-geostrophic vorticity consists of the relative vorticity, planetary vorticity, and a vertical strain contribution. The boundary conditions also include zeroth and first order contributions. Because of the inherent approximations, the quasi-geostrophic equations cannot include all dissipation and forcing processes. Linearization of the quasi-geostrophic potential vorticity equation about a motionless stratified background state yields the Rossby wave solutions discussed in Chapter 8.

19.1 Scaling of the density equation

The scaled momentum balance and incompressibility condition are given in Section 13.1. Here we scale the density equation (18.54) in the midlatitude beta-plane approximation. We assume that $\delta\rho$ scales as $\eta\rho_0 N^2/g$ where η is the isopycnal displacement (see (18.60)) and that temporal changes are comparable to nonlinear advective changes. Then

$$\begin{array}{ccccc} \partial_t \delta\rho + \mathbf{u}\cdot\nabla\delta\rho + & \frac{y}{r_0}\tan\theta_0\, u\partial_x\delta\rho & - & \frac{\rho_0}{g}N^2 w & \\ \sigma \quad\quad \sigma & \sigma\epsilon & & 1 & \\ = & K_\text{h}\Delta_\text{h}\delta\rho & + & K_\text{v}\partial_z\partial_z\delta\rho & \\ & \sigma\frac{1}{P_\text{h}} & & \sigma\frac{1}{P_\text{v}} & \end{array} \tag{19.1}$$

where $\sigma := Z/D$, Z is the scale of η, and

$$P_\text{h} = \frac{UL}{K_\text{h}} \tag{19.2}$$

$$P_\text{v} = \frac{WH}{K_\text{v}} \tag{19.3}$$

are the Peclet numbers, the ratios of the diffusive to the advective time scales.

Quasi-geostrophic motions assume $\sigma \ll 1$. The time rate of change and the nonlinear advection term are then much smaller than the linear advection term. In addition, one assumes that the diffusive time scales are equal or larger than the advective time scales, requiring Peclet numbers comparable to or larger than one.

19.2 Perturbation expansion

Small-scale geostrophic motions are characterized by $\tilde{\text{Ro}}$, Ro, γ, E_h, E_v, $\sigma \ll 1$. One therefore expands all variables with respect to a common expansion parameter $\epsilon = O(\tilde{\text{Ro}}, \text{Ro}, \gamma, E_\text{h}, E_\text{v}, \sigma) \ll 1$.

$$(u, v, w, \delta p, \delta\rho) = \sum_{n=0}^{\infty} \epsilon^n \left(u^{(n)}, v^{(n)}, w^{(n)}, \delta p^{(n)}, \delta\rho^{(n)}\right) \tag{19.4}$$

To zeroth order in ϵ the beta-plane primitive equations (18.20) to (18.24) then become

$$-f_0 v^{(0)} = -\frac{1}{\rho_0}\partial_x \delta p^{(0)} \tag{19.5}$$

$$f_0 u^{(0)} = -\frac{1}{\rho_0}\partial_y \delta p^{(0)} \tag{19.6}$$

$$0 = \partial_z \delta p^{(0)} + \delta\rho^{(0)} g \tag{19.7}$$

$$-\frac{\rho_0}{g} N^2 w^{(0)} = 0 \tag{19.8}$$

The zeroth order vertical velocity vanishes, i.e., the vertical velocity is much smaller than δU. The geostrophic balance implies that the zeroth order flow is horizontally non-divergent. It can thus be described by a streamfunction ψ. The zeroth order variables are then given by

$$u^{(0)} = -\partial_y \psi \tag{19.9}$$

$$v^{(0)} = \partial_x \psi \tag{19.10}$$

$$w^{(0)} = 0 \tag{19.11}$$

$$\delta p^{(0)} = \rho_0 f_0 \psi \tag{19.12}$$

$$\delta \rho^{(0)} = -\frac{\rho_0 f_0}{g} \partial_z \psi \tag{19.13}$$

The zeroth order equations are degenerate in the sense that they determine all field variables in terms of the streamfunction ψ but do not provide an equation for ψ. This degeneracy is due to the fact that the incompressibility condition is satisfied by the zeroth order flow and does not provide an additional constraint.

19.3 Quasi-geostrophic potential vorticity equation

The dynamic evolution of the streamfunction is determined by higher order ageostrophic terms. To account systematically for these one takes the curl of the momentum balance. This eliminates the zeroth order geostrophic balance and yields an equation for the vertical component of the vorticity. We first neglect eddy diffusion and forcing processes in both the momentum and the density equation. They will be incorporated later. To the lowest order in ϵ the vorticity equation then takes the form

$$\left(\partial_t + u^{(0)}\partial_x + v^{(0)}\partial_y\right)\left(\omega_z^{(0)} + f_0 + \beta_0 y\right) = f_0 \partial_z w^{(1)} \tag{19.14}$$

where

$$\omega_z^{(0)} = \partial_x v^{(0)} - \partial_y u^{(0)} = (\partial_x \partial_x + \partial_y \partial_y)\psi \tag{19.15}$$

is the vertical component of the relative vorticity. Changes in the absolute vorticity are caused by vortex stretching. In deriving the vorticity equation use has been made of the first order incompressibility condition. Note that the relative vorticity is much smaller than the planetary vorticity, $\omega_z/f \sim \mathrm{Ro}$, but that changes in relative vorticity are comparable to changes in planetary vorticity, $\frac{D\omega_z}{Dt} / \frac{Df}{Dt} = O\left(\frac{\mathrm{Ro}}{\epsilon}\right)$.

The first order vertical velocity is obtained from the first order density equation

$$w^{(1)} = -\frac{f_0}{N^2}\left(\partial_t + u^{(0)}\partial_x + v^{(0)}\partial_y\right)\partial_z \psi \tag{19.16}$$

When substituted into the vorticity equation one finds

$$\frac{\mathrm{D}^{(0)}}{\mathrm{D}t} q = 0 \tag{19.17}$$

where

$$\frac{\mathrm{D}^{(0)}}{\mathrm{D}t} = \partial_t + u^{(0)}\partial_x + v^{(0)}\partial_y \tag{19.18}$$

and

$$\begin{array}{cccc} q = (\partial_x\partial_x + \partial_y\partial_y)\psi & + f_0 + & \beta_0 y & + \partial_z \frac{f_0^2}{N^2}\partial_z\psi \\ \mathrm{Ro} & 1 & \gamma & \sigma \end{array} \tag{19.19}$$

is the quasi-geostrophic potential vorticity. It is the sum of the relative vorticity, planetary vorticity, and a vertical strain contribution

$$\partial_z \frac{f_0^2}{N^2} \partial_z \psi = f_0 \partial_z \eta \tag{19.20}$$

The different contributions scale as indicated. The relative vorticity and the vertical strain contribution are comparable when the horizontal length scale L is comparable to the internal Rossby radius of deformation $R_\mathrm{i} := ND/f_0$ since $\sigma = \mathrm{Ro}L^2/R_\mathrm{i}^2$. Instead of the parameter σ one often introduces the Froude number Fr by $Fr^2 := \sigma\mathrm{Ro}$. The quasi-geostrophic potential vorticity (19.19) is the lowest order of an expansion with respect to ϵ of Ertel's potential vorticity $q_\rho = (f + \omega_z)\partial_z\rho + \boldsymbol{\omega}_\mathrm{h} \cdot \boldsymbol{\nabla}_\mathrm{h}\rho$ (see (13.43)) normalized by $d\tilde{\rho}/dz$.

Equation (19.17) is called the quasi-geostrophic potential vorticity equation. It states that quasi-geostrophic potential vorticity q is conserved along horizontal particle trajectories for ideal fluid motion. Flows that are governed by (19.9) to (19.13) and (19.17) are called quasi-geostrophic flows. The attribute "quasi" is used because the dynamic evolution of these geostrophically balanced flows is governed by ageostrophic effects. The potential vorticity equation may also be written as

$$\partial_t q + J(\psi, q) = 0 \tag{19.21}$$

where $J(a, b) = \partial_x a \partial_y b - \partial_y a \partial_x b$ is the Jacobian.

To determine the advection velocities in (19.17) one has to invert (19.19), i.e., one has to determine the streamfunction ψ from q. This requires solving a three-dimensional Poisson equation.

19.4 Boundary conditions

The quasi-geostrophic potential vorticity equation has to be augmented by vertical and lateral boundary conditions. Expanding the kinematic and dynamic boundary conditions for a free surface $z = \eta_s(x, y, t)$ one finds, to leading order in ϵ,

$$\frac{D^{(0)}}{Dt}(N^2\psi + g\partial_z\psi) = 0 \quad \text{at} \quad z = 0 \tag{19.22}$$

with surface elevation

$$\eta_s^{(0)} = g^{-1} f_0 \psi(z = 0) \tag{19.23}$$

In the rigid lid approximation ($g \to \infty$) the surface boundary condition reduces to

$$\frac{D^{(0)}}{Dt}\partial_z\psi = 0 \quad \text{at} \quad z = 0 \tag{19.24}$$

At a flat rigid bottom $z = -H_0$ the boundary condition is

$$\frac{D^{(0)}}{Dt}\partial_z\psi = 0 \quad \text{at} \quad z = -H_0 \tag{19.25}$$

For variable bottom topography $z = -H_0 + \eta_b(x, y)$ (see Figure 14.1) the quasi-geostrophic approximation remains valid as long as $|\eta_b|/D \leq \text{Ro}$. In this case the bottom boundary condition becomes

$$\frac{D^{(0)}}{Dt}\left(\frac{\partial\psi}{\partial z} + \frac{N^2}{f_0}\eta_b\right) = 0 \quad \text{at } z = -H_0 \tag{19.26}$$

At lateral rigid boundaries δS_j (see Figure 19.1) the kinematic boundary condition becomes, to zeroth order in ϵ

$$\frac{\partial\psi}{\partial s} = 0 \quad \text{on all } \delta S_j \tag{19.27}$$

This condition leaves the streamfunction ψ an unspecified function of z and t on each side wall. As auxiliary conditions one imposes the integrability conditions (see Section 16.2)

$$\oint_{\delta S_j} ds\, \partial_t \partial_n \psi = 0 \quad \text{for} \quad j = 2, \ldots, N \tag{19.28}$$

and

$$\iint_S d^2x \frac{1}{N^2}\frac{D^{(0)}}{Dt}\partial_z\psi = 0 \tag{19.29}$$

which are the line integrals of the first order momentum equation around each island and the first order of volume conservation. In (19.27) and (19.28) n denotes

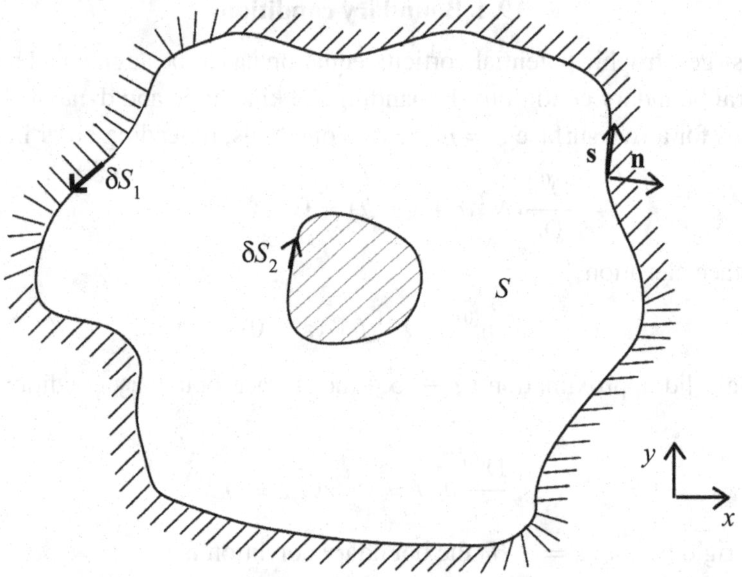

Figure 19.1. Schematic representation of lateral domain S with boundaries $\delta S_{1,2}$. The vectors $\mathbf{n}$, $\mathbf{s}$, and $\mathbf{e}_z$ form a right-handed orthonormal basis along the boundaries.

the normal and s the tangential coordinate of a right-handed orthonormal coordinate system ($\mathbf{n}$, $\mathbf{s}$, $\mathbf{e}_z$) along the lateral boundaries (see Figure 19.1). For a detailed derivation see McWilliams (1977).

In a laterally infinite ocean the linearized potential vorticity equation and boundary conditions

$$\partial_t \left[(\partial_x \partial_x + \partial_y \partial_y)\psi + \partial_z \frac{f_0^2}{N^2} \partial_z \psi \right] + \beta_0 \partial_x \psi = 0 \tag{19.30}$$

$$\partial_t (N^2 \psi + g \partial_z \psi) = 0 \quad \text{at} \quad z = 0 \tag{19.31}$$

$$\partial_t \partial_z \psi = 0 \quad \text{at} \quad z = -H_0 \tag{19.32}$$

have solutions $\psi \sim \phi_n(z) \exp i(kx + ly - \omega t)$ with

$$\omega = -\frac{\beta_0 k}{k^2 + l^2 + \epsilon_n f_0^2} \tag{19.33}$$

where $\phi_n(z)$ and ϵ_n are the eigenfunctions and eigenvalues of the vertical eigenvalue problem (14.94) to (14.96). These are the midlatitude Rossby waves discussed in Section 8.5.

19.5 Conservation laws

Inviscid, adiabatic, and unforced quasi-geostrophic flows conserve both energy and enstrophy. To lowest order in ϵ the energy equation takes the form

$$\frac{\mathrm{D}^{(0)}}{\mathrm{D}t}\tilde{e} + \boldsymbol{\nabla}\cdot\tilde{\mathbf{I}}_e = 0 \tag{19.34}$$

where

$$\tilde{e} := \frac{1}{2}\left(u^{(0)\,2} + v^{(0)\,2}\right) + \frac{1}{2}\frac{g^2}{\rho_0^2 N^2}\rho^{(0)\,2} \tag{19.35}$$

is the specific mechanical energy in the quasi-geostrophic approximation, consisting of (horizontal) kinetic and available potential energy, and

$$\tilde{\mathbf{I}}_e := \rho_0^{-1}\mathbf{u}^{(0)}p^{(1)} + \rho_0^{-1}\mathbf{u}^{(1)}p^{(0)} \tag{19.36}$$

the specific energy flux. The energy of an enclosed volume only changes if the pressure does work on the surroundings. For a fluid volume enclosed by free surfaces and rigid boundaries, energy is conserved.

Energy conservation can also be expressed in terms of the streamfunction. When the potential vorticity equation is multiplied by ψ one obtains

$$\frac{\mathrm{D}^{(0)}}{\mathrm{D}t}e + \boldsymbol{\nabla}\cdot\mathbf{I}_e = 0 \tag{19.37}$$

with

$$e := \frac{1}{2}\left((\partial_x\psi)^2 + (\partial_y\psi)^2 + \frac{f_0^2}{N^2}(\partial_z\psi)^2\right) \tag{19.38}$$

and

$$\mathbf{I}_e := \begin{pmatrix} -\psi\frac{\mathrm{D}^{(0)}}{\mathrm{D}t}\partial_x\psi - \frac{1}{2}\beta_0\psi^2 \\ -\psi\frac{\mathrm{D}^{(0)}}{\mathrm{D}t}\partial_y\psi \\ -\psi\frac{f_0^2}{N^2}\frac{\mathrm{D}^{(0)}}{\mathrm{D}t}\partial_z\psi \end{pmatrix} \tag{19.39}$$

One finds $\tilde{e} = e$ and $\boldsymbol{\nabla}\cdot\tilde{\mathbf{I}}_e = \boldsymbol{\nabla}\cdot\mathbf{I}_e$ but generally $\tilde{\mathbf{I}}_e \neq \mathbf{I}_e$. The proof of the conservation of energy for a physically enclosed volume now requires the boundary conditions (19.27), (19.28), and (19.29).

The enstrophy density is defined by

$$\zeta := \frac{1}{2}\left(\frac{\partial^2\psi}{\partial x^2} + \frac{\partial^2\psi}{\partial y^2} + \frac{\partial}{\partial z}\frac{f_0^2}{N^2}\frac{\partial\psi}{\partial z}\right)^2 \tag{19.40}$$

and its time evolution is governed by

$$\frac{\mathrm{D}^{(0)}}{\mathrm{D}t}\zeta + \nabla \cdot \mathbf{I}_\zeta = 0 \tag{19.41}$$

where

$$\mathbf{I}_\zeta := \beta_0 \begin{pmatrix} \frac{1}{2}\left(\frac{\partial\psi}{\partial x}\right)^2 - \frac{1}{2}\left(\frac{\partial\psi}{\partial y}\right)^2 - \frac{1}{2}\frac{f_0^2}{N^2}\left(\frac{\partial\psi}{\partial z}\right)^2 \\ \frac{\partial\psi}{\partial x}\frac{\partial\psi}{\partial y} \\ \frac{\partial\psi}{\partial x}\frac{f_0^2}{N^2}\frac{\partial\psi}{\partial z} \end{pmatrix} \tag{19.42}$$

The enstrophy of a fluid volume is conserved, except for fluxes through the boundaries. These fluxes do not generally vanish for a physically enclosed volume. Enstrophy can, e.g., be transferred through rigid zonal boundaries and through the top and bottom unless these are isopycnals, $\partial_z\psi = 0$. These fluxes can break the powerful constraints implied by the conservation of energy and enstrophy.

19.6 Diffusion and forcing

Eddy diffusion and forcing can be included by adding appropriate source terms to the primitive equations and repeating the expansion procedure. These source terms must not upset the lowest order geostrophic balance. The resulting source terms in the quasi-geostrophic potential vorticity equation and in its boundary conditions are given below for a few specific diffusion and forcing mechanisms. For simplicity of notation a rigid lid and a flat bottom are assumed. A systematic derivation is given in McWilliams (1977) and Frankignoul and Müller (1979).

Interior diffusion First include momentum and density diffusion with constant eddy diffusion coefficients A_h, A_v, K_h, and K_v in the primitive equations. In order that these diffusion terms do not upset the zeroth order balances we must require that the Ekman numbers satisfy $E_\mathrm{h}, E_\mathrm{v} \ll 1$ and the Peclet numbers $P_\mathrm{h}, P_\mathrm{v} \geq 1$. The incorporation of diffusion terms also requires additional dynamic boundary conditions. For the unforced case, these are that there are no fluxes of tangential momentum and density through any of the boundaries:

$$\left.\begin{aligned} A_\mathrm{v}\partial_z\mathbf{u}_\mathrm{h} &= 0 \\ K_\mathrm{v}\partial_z\delta\rho &= 0 \end{aligned}\right\} \text{ at } z = 0, -H_0 \tag{19.43}$$

$$\left.\begin{aligned} A_\mathrm{h}\partial_n(\mathbf{u}_\mathrm{h}\cdot\hat{\mathbf{n}}_\mathrm{h}) &= 0 \\ K_\mathrm{h}\partial_n\delta\rho &= 0 \end{aligned}\right\} \text{ on all } \delta S_j \tag{19.44}$$

Out of the four different diffusion processes only the horizontal diffusion of momentum through A_h and the vertical diffusion of density through K_v can be incorporated directly into the quasi-geostrophic potential vorticity equation (McWilliams (1977)). Otherwise the boundary conditions (19.43) and (19.44) over-specify the problem. We thus have to set $A_\mathrm{v} = 0$ and $K_\mathrm{h} = 0$. Expansion of the primitive equations and boundary conditions then leads to the quasi-geostrophic potential vorticity equation

$$\frac{\mathrm{D}^{(0)}}{\mathrm{D}t} q = A_\mathrm{h}(\partial_x\partial_x + \partial_y\partial_y)(\partial_x\partial_x + \partial_y\partial_y)\psi + K_\mathrm{v}\partial_z\frac{f_0^2}{N^2}\partial_z\partial_z\partial_z\psi \tag{19.45}$$

with surface and bottom boundary conditions

$$\left.\begin{array}{l}\frac{\mathrm{D}^{(0)}}{\mathrm{D}t}\partial_z\psi = K_\mathrm{v}\partial_z\partial_z\partial_z\psi \\ K_\mathrm{v}\partial_z\partial_z\psi = 0\end{array}\right\} \text{ at } z = 0, -H_0 \tag{19.46}$$

and lateral boundary conditions

$$\partial_s\psi = 0 \quad \text{on all } \delta S_j \tag{19.47}$$

$$A_\mathrm{h}\partial_n\partial_n\psi = 0 \quad \text{on all } \delta S_j \tag{19.48}$$

$$\oint_{\delta S_j} \mathrm{d}s\ \partial_t\partial_n\psi = A_h \oint_{\delta S_j} \mathrm{d}s\ \partial_n(\partial_x\partial_x + \partial_y\partial_y)\psi \quad \text{for } j = 2, \ldots, N \tag{19.49}$$

$$\int\!\!\int_S \mathrm{d}^2x \frac{1}{N^2}\left(\frac{\mathrm{D}^{(0)}}{\mathrm{D}t}\partial_z\psi - K_\mathrm{v}\partial_z\partial_z\partial_z\psi\right) = 0 \tag{19.50}$$

Bottom friction The effect of the vertical viscosity A_v and the horizontal diffusivity K_h has to be resolved in ageostrophic boundary layers. These boundary layers have thicknesses that are much smaller than the scales of the interior quasi-geostrophic flow. The most important boundary layers are the Ekman layers at the top and bottom of the ocean, which transmit the effects of boundary stresses to the quasi-geostrophic interior. At the bottom, the fluid-induced bottom stress leads to bottom friction.

Linear bottom friction is obtained when the Ekman pumping velocity (see Table 15.1)

$$w_\mathrm{Ek}^\mathrm{b} = \boldsymbol{\nabla}_\mathrm{h}\cdot\left(\frac{d}{\sqrt{2}}\mathbf{u}_\mathrm{h}\right) + \hat{\boldsymbol{\nabla}}_\mathrm{h}\cdot\left(\frac{d}{\sqrt{2}}\mathbf{u}_\mathrm{h}\right) \approx r\frac{H_0}{f_0}(\partial_x\partial_x + \partial_y\partial_y)\psi \tag{19.51}$$

with bottom friction coefficient

$$r = \left(\frac{A_\mathrm{v} f_0}{2H_0^2}\right)^{1/2} = \left(\frac{1}{2}E_\mathrm{v}\right)^{1/2} f_0 \tag{19.52}$$

is used in the kinematic boundary condition (15.44). One then obtains to leading order

$$\frac{\mathrm{D}^{(0)}}{\mathrm{D}t}\partial_z\psi = -\frac{N^2}{f_0}w^{\mathrm{b}}_{\mathrm{Ek}} = -r\frac{N^2}{f_0^2}H(\partial_x\partial_x + \partial_y\partial_y)\psi \quad \text{at } z = -H_0 \tag{19.53}$$

Windstress forcing A prescribed surface windstress $\boldsymbol{\tau}^{\mathrm{a}}_{\mathrm{h}}$ similarly causes an Ekman pumping velocity

$$w^{\mathrm{s}}_{\mathrm{Ek}} = \frac{1}{\rho_0 f_0}\hat{\boldsymbol{\nabla}}_{\mathrm{h}}\cdot\boldsymbol{\tau}^{\mathrm{a}}_{\mathrm{h}} \approx \frac{1}{\rho_0 f_0}\left(\partial_x\tau^{\mathrm{a}}_y - \partial_y\tau^{\mathrm{a}}_x\right) \tag{19.54}$$

which drives the quasi-geostrophic flow through the kinematic boundary condition

$$\frac{\mathrm{D}^{(0)}}{\mathrm{D}t}\partial_z\psi = -\frac{N^2}{f_0}w^{\mathrm{s}}_{\mathrm{Ek}} \quad \text{at } z = 0 \tag{19.55}$$

Surface buoyancy flux A surface buoyancy flux

$$B = \frac{g\tilde{\alpha}}{\rho_0 c_{\mathrm{p}}}Q^{\mathrm{a}} + \frac{g\tilde{\beta}}{\rho_0}(E-P)S \tag{19.56}$$

can be taken into account by solving

$$\frac{\mathrm{D}^{(0)}}{\mathrm{D}t}q = K_{\mathrm{v}}\partial_z\frac{f_0^2}{N^2}\partial_z\partial_z\partial_z\psi \tag{19.57}$$

subject to the boundary condition

$$K_{\mathrm{v}}\partial_z\partial_z\psi = \frac{1}{f_0}B \quad \text{at } z = 0 \tag{19.58}$$

Care must be taken that the vertical diffusion and surface buoyancy flux are consistent with the processes that maintain the background stratification.

19.7 Layer representation

Consider an N-layer model (see Figure 14.1) with constant layer densities $\rho_1 < \rho_2 < \ldots < \rho_N$ and constant layer thicknesses $H_1, H_2, \ldots, H_N$. In the presence of motion and bottom elevations, the layers have thicknesses $h_i = H_i + \eta_{i-1} - \eta_i$ $(i = 1, \ldots, N)$, where $\eta_0 = \eta_{\mathrm{s}}$ is the surface elevation, $\eta_2, \ldots, \eta_{N-1}$ are deviations of the interfaces from their mean positions, and $\eta_N = \eta_b$ is the given bottom elevation.

Quasi-geostrophic motions in an N-layer system are completely described by the N streamfunctions $\psi_i(x, y, t)$ $(i = 1, \ldots, N)$ for each layer. The equations of motion for the streamfunctions ψ_i can be obtained formally by setting

$$N^2(z) = \sum_{i=1}^{N-1} g'_i \, \delta \left(z + \sum_{j=1}^{i} H_j \right) \tag{19.59}$$

in (19.17) with reduced gravity $g'_i := g(\rho_{i+1} - \rho_i)/\rho_0$, $(i = 1, \ldots, N-1)$. The result is the layer form of the quasi-geostrophic potential vorticity equation

$$(\partial_t + u_i \partial_x + v_i \partial_y) q_i = \frac{f_0}{H_1} w^{\mathrm{s}}_{\mathrm{Ek}} \delta_{i1} - \frac{f_0}{H_N} \left(w^{\mathrm{b}}_{\mathrm{Ek}} - \partial_x \eta_i \partial_y \psi_i + \partial_y \eta_i \partial_x \psi_i \right) \delta_{iN} \tag{19.60}$$

with potential vorticity

$$q_i = (\partial_x \partial_x + \partial_y \partial_y) \psi_i + f_0 + \beta_0 y + \frac{f_0^2}{g'_i H_i} (\psi_{i+1} - \psi_i) - \frac{f_0^2}{g'_{i-1} H_i} (\psi_i - \psi_{i-1}) \tag{19.61}$$

where $g'_0 = g$ and $g'_N = \infty$ for convenience of notation. Equation (19.60) includes the Ekman suction velocities $w^{\mathrm{s}}_{\mathrm{Ek}}$ and $w^{\mathrm{b}}_{\mathrm{Ek}}$, and bottom topography. All other variables can be expressed in terms of the streamfunction as

$$\begin{aligned} u_i &= -\partial_y \psi_i \quad i = 1, \ldots, N \\ v_i &= \partial_x \psi_i \\ \eta_i &= \tfrac{f_0}{g'_i} (\psi_{i+1} - \psi_i) \\ M_i &= \rho_0 f_0 \psi_i \end{aligned} \tag{19.62}$$

The vertical velocities at the surface, mean interface, and bottom are given by

$$w_i = (\partial_t + u_i \partial_x + v_i \partial_y) \eta_i + w^{\mathrm{s}}_{\mathrm{Ek}} \delta_{i0} + w^{\mathrm{b}}_{\mathrm{Ek}} \delta_{iN} \quad i = 0, \ldots, N \tag{19.63}$$

and vary linearly within the layers.

For a one-layer model (homogeneous ocean) one specifically finds

$$\begin{aligned} (\partial_t + u \partial_x + v \partial_y) q = {} & \frac{1}{\rho_0 H_0} \left(\partial_x \tau^{\mathrm{a}}_y - \partial_y \tau^{\mathrm{a}}_x \right) - r (\partial_x \partial_x + \partial_y \partial_y) \psi \\ & + A_{\mathrm{h}} (\partial_x \partial_x + \partial_y \partial_y)(\partial_x \partial_x + \partial_y \partial_y) \psi \end{aligned} \tag{19.64}$$

and

$$\begin{array}{ccccccccc} q = & (\partial_x \partial_x + \partial_y \partial_y) \psi & + & f_0 & + & \beta_0 y & - \frac{f_0}{H_0} \eta_{\mathrm{s}} & + & \frac{f_0}{H_0} \eta_{\mathrm{b}} \\ & \mathrm{Ro} & & 1 & & \gamma & \sigma_{\mathrm{s}} & & \sigma_{\mathrm{h}} \end{array} \tag{19.65}$$

when the Ekman pumping velocities (19.52) and (19.54) are substituted and horizontal friction is included. The potential vorticity consists of the relative vorticity, planetary vorticity, and the vertical strain contribution. The quasi-geostrophic potential vorticity consists of the leading order terms when expanding Ertel's potential vorticity

$$\frac{f+\omega_z}{H_0+\eta_s-\eta_b}=\frac{1}{H_0}\left[\omega_z+f_0+\beta_0 y-\frac{f_0}{H_0}\eta_s+\frac{f_0}{H_0}\eta_b+\dots\right] \quad (19.66)$$

The different contributions to the potential vorticity scale as indicated with $\sigma_{s,b} := Z_{s,b}/H_0$ where $Z_{s,b}$ are the scales of the surface and bottom elevations. The scaling of the surface term can also be written as $\sigma_s = \mathrm{Ro}L^2/R_e^2$ where $R_e^2 = \frac{gH_0}{f_0^2}$ is the external Rossby radius of deformation. If atmospheric pressure forcing is included the surface elevation η_s is given by

$$\eta_s = \frac{f_0}{g}\psi - \frac{p_a}{\rho_0 g} \quad (19.67)$$

If $L^2 \ll R_e^2$ the surface elevation term is smaller than the relative vorticity term and we could have taken the rigid lid limit ($g \to \infty$, $R_e \to \infty$). Bottom friction dominates lateral friction for scales $L^2 \gg A_h/r$.

20

Motions on the f-plane

As the space and time scales L, D, and T become smaller the parameters $\delta := D/L$, $\tilde{\mathrm{Ro}} := 1/fT$, $\mathrm{Ro} := f/UL$, and $\sigma := Z/H$ become larger. The constraints exerted by rotation and stratification weaken. The motions become less geostrophically and hydrostatically balanced and more three dimensional, allowing for vertical motions $W \approx U$. The major simplification is one of geometry: one can apply the f-plane approximation. These f-plane motions represent medium-scale motions. They are still large enough that Reynolds averaging must be applied. Molecular diffusion is still replaced by eddy diffusion. Alternatively, they are small enough that larger-scale environmental fields must be prescribed as background states. The f-plane equations generally disregard differential mixing and thermobaricity. They will therefore be formulated here in the one-component approximation, but can be extended trivially to a two-component flow should boundary conditions or other circumstances so require.

Motions on the f-plane exhibit a rich variety of phenomena. Very few approximations can be imposed.

20.1 Equations of motion

The equations on the f-plane are given by

$$\rho_0 \frac{\mathrm{D}}{\mathrm{D}t}\mathbf{u}_\mathrm{h} + \rho_0 f_0 \hat{\mathbf{u}}_\mathrm{h} = -\boldsymbol{\nabla}_\mathrm{h}\delta p + \frac{1}{\rho_0}\mathbf{F}_\mathrm{h} \tag{20.1}$$

$$\frac{\mathrm{D}}{\mathrm{D}t} w = -\frac{1}{\rho_0}\partial_z \delta p + b + F_z \tag{20.2}$$

$$\boldsymbol{\nabla} \cdot \mathbf{u} = 0 \tag{20.3}$$

$$\frac{\mathrm{D}}{\mathrm{D}t} b + N^2 w = D_b \tag{20.4}$$

where

$$\frac{\mathrm{D}}{\mathrm{D}t} = \partial_t + u\partial_x + v\partial_y + w\partial_z \tag{20.5}$$

is the Lagrangian derivative with coordinates x eastward, y northward, and z upward and velocity components u eastward, v northward, and w upward,

$$b := -\delta\rho g/\rho_0 \tag{20.6}$$

the buoyancy, N the buoyancy frequency, and $f_0 = 2\Omega\sin\theta_0$ the Coriolis frequency.

The f-plane approximation is the lowest order of an expansion of the metric with respect to $\gamma := L/r_0$ resulting in scale factors $h_x = h_y = h_z = 1$. The physics becomes equivalent to that of a rotating plane, except that there are no centrifugal forces acting in the horizontal plane. The equations neglect without detailed justification the meridional component of the planetary vorticity, i.e., make the traditional approximation. The smallness of the aspect ratio cannot be used as a justification. Closing the dynamical equations through the buoyancy equation eliminates the temperature–salinity mode. The equations retain the gravity and potential vorticity carrying modes, often called the vortical mode in this context. When linearized about a motionless stratified background state one obtains the gravity wave solutions of Table 8.3. The dispersion relation of the vortical mode degenerates to $\omega = 0$ and represents steady geostrophically balanced currents. Nonlinear versions of these modes are considered in Sections 20.3 to 20.5.

There are some obvious limits that one can take: the hydrostatic limit $\delta \to 0$, the non-rotating limit $f_0 \to 0$, the non-stratified limit $N^2 \to 0$, and the ideal fluid limit $\mathbf{F}$, $D_b \to 0$.

Often these equations are applied to an interior part of the ocean that is not limited by the physical boundaries of the ocean. In this case boundary conditions may be applied that differ from those at free or rigid surfaces. Periodic boundary conditions are a typical example.

20.2 Vorticity equations

The vorticity and potential vorticity equations become

$$\frac{\mathrm{D}}{\mathrm{D}t}\boldsymbol{\omega} = \boldsymbol{\omega}\cdot\nabla\mathbf{u} + f_0\partial_z\mathbf{u} + \nabla b \times \mathbf{e}_z + \frac{1}{\rho_0}\nabla\times\mathbf{F} \tag{20.7}$$

$$\frac{\mathrm{D}}{\mathrm{D}t}q_\psi = -\nabla\cdot\mathbf{I}_q \tag{20.8}$$

where

$$q_\psi := \boldsymbol{\omega}\cdot\nabla\psi + f_0\partial_z\psi \tag{20.9}$$

and

$$\mathbf{I}_q := -b\mathbf{e}_z \times \nabla\psi - \frac{1}{\rho_0}\mathbf{F} \times \nabla\psi - (\boldsymbol{\omega} + f_0\mathbf{e}_z)\dot{\psi} \tag{20.10}$$

If one chooses $\psi = z - \eta$ in the definition of Ertel's potential vorticity, then it takes the form

$$q_{z-\eta} = f_0 + \omega_z - f_0\frac{\partial\eta}{\partial z} - \boldsymbol{\omega} \cdot \nabla\eta \tag{20.11}$$

20.3 Nonlinear internal waves

For simplicity, we assume an incompressible, ideal, non-rotating, and hydrostatically balanced flow, governed by the primitive equations of Chapter 13. The potential vorticity then reduces to $q_\rho = \boldsymbol{\omega} \cdot \nabla\rho$. The gravity mode of motion is characterized by $q_\rho \equiv 0$. This condition is most easily implemented in isopycnal coordinates, described in Section 14.3. With $\dot{\rho} = 0$ and a background stratification

$$\tilde{h} = \frac{\Delta\rho g}{\rho_0}\frac{1}{N^2} \tag{20.12}$$

equations (14.48) to (14.51) take the form

$$\partial_t\mathbf{u}_\mathrm{h} + \nabla_\mathrm{h}\left(\frac{1}{2}\mathbf{u}_\mathrm{h} \cdot \mathbf{u}_\mathrm{h}\right) + \omega_\rho\hat{\mathbf{u}}_\mathrm{h} = -\frac{1}{\rho_0}\nabla_\mathrm{h}\delta M \tag{20.13}$$

$$\frac{\partial\delta M}{\partial\rho} = g\delta z \tag{20.14}$$

$$\partial_t\delta h + \tilde{h}\nabla_\mathrm{h} \cdot \mathbf{u}_\mathrm{h} + \nabla \cdot (\mathbf{u}_\mathrm{h}\delta h) = 0 \tag{20.15}$$

$$\frac{\partial\delta z}{\partial\rho} = -\frac{\delta h}{\Delta\rho} \tag{20.16}$$

where we made use of the vector identity $\mathbf{u}_\mathrm{h} \cdot \nabla_\mathrm{h}\mathbf{u}_\mathrm{h} = \nabla_\mathrm{h}(\mathbf{u}_\mathrm{h} \cdot \mathbf{u}_\mathrm{h}/2) + \omega_\rho$ with $\omega_\rho := \hat{\nabla}_\mathrm{h} \cdot \mathbf{u}_\mathrm{h}$. All horizontal gradients are taken at constant ρ. In isopycnal coordinates $q_\rho = -\Delta\rho\omega_\rho/h$ and hence $q_\rho \equiv 0$ implies $\omega_\rho \equiv 0$, and one can introduce a velocity potential φ such that $\mathbf{u}_\mathrm{h} = \nabla_\mathrm{h}\varphi$. The equations then reduce to

$$\partial_t\varphi = -\frac{1}{\rho_0}\delta M - \frac{1}{2}\nabla_\mathrm{h}\varphi \cdot \nabla_\mathrm{h}\varphi \tag{20.17}$$

$$\partial_t\delta h + \tilde{h}\nabla_\mathrm{h} \cdot \nabla_\mathrm{h}\varphi + \nabla_\mathrm{h} \cdot (\nabla_\mathrm{h}\varphi\delta h) = 0 \tag{20.18}$$

$$\frac{\partial\delta M}{\partial\rho} = g\delta z \tag{20.19}$$

$$\frac{\partial\delta z}{\partial\rho} = -\frac{\delta h}{\Delta\rho} \tag{20.20}$$

These equations describe nonlinear internal waves in the non-rotating hydrostatic case. If the nonlinear terms are neglected one obtains

$$\partial_t \partial_t \partial_\rho \partial_\rho \varphi = -\frac{g^2}{\rho_0^2} \frac{1}{N^2} (\partial_x \partial_x + \partial_y \partial_y) \varphi \tag{20.21}$$

and recovers the internal wave dispersion relation in the non-rotating hydrostatic case

$$\omega^2 = N^2 \frac{k_{\mathrm{h}}^2}{k_z^2} \tag{20.22}$$

if one substitutes $k_\rho = -g k_z / \rho_0 N^2$.

20.4 Two-dimensional flows in a vertical plane

The gravity mode can also be isolated by considering non-rotating two-dimensional flow in a vertical (x, z)-plane. The conditions $\partial/\partial y \equiv 0$ and $v \equiv 0$ imply

$$q = \boldsymbol{\omega} \cdot \boldsymbol{\nabla}(z - \eta) \equiv 0 \tag{20.23}$$

Two-dimensional motions in a non-rotating vertical plane have zero potential vorticity and represent the gravity mode of motion.

20.5 Two-dimensional flows in a horizontal plane

The vortical mode is isolated by considering a strictly two-dimensional ideal fluid flow in a horizontal (x, y)-plane. The conditions $\partial/\partial z \equiv 0$ and $w \equiv 0$ then imply

$$\boldsymbol{\nabla}_{\mathrm{h}} \cdot \mathbf{u}_{\mathrm{h}} = 0 \tag{20.24}$$

$$(\partial_t + \mathbf{u}_{\mathrm{h}} \cdot \boldsymbol{\nabla}_{\mathrm{h}})\, \mathbf{u}_{\mathrm{h}} + f_0 \hat{\mathbf{u}}_{\mathrm{h}} = -\boldsymbol{\nabla}_{\mathrm{h}} \frac{p}{\rho_0} \tag{20.25}$$

The first equation implies that the velocity can be described by a streamfunction ψ. The vorticity equation is given by

$$(\partial_t + \mathbf{u}_{\mathrm{h}} \cdot \boldsymbol{\nabla}_{\mathrm{h}})\, \omega_z = 0 \tag{20.26}$$

with diagnostic relations

$$\omega_z = \Delta_{\mathrm{h}} \psi \tag{20.27}$$

$$\mathbf{u}_{\mathrm{h}} = \hat{\boldsymbol{\nabla}}_{\mathrm{h}} \psi \tag{20.28}$$

The vorticity is materially conserved. The generation term $(\omega_z + f_0)\partial w/\partial z$ vanishes. The only physical process is the advection of vorticity. Such flows are called two-dimensional turbulence. They are usually studied in wavenumber or spectral space, obtained by Fourier transformation of the equations. Nonlinear advection then becomes nonlinear triad interaction.

Two-dimensional turbulence satisfies the following theorems

$$\frac{\mathrm{d}}{\mathrm{d}t}\Gamma = 0 \tag{20.29}$$

$$\frac{\mathrm{d}}{\mathrm{d}t}\mathrm{E}_{\mathrm{kin}} = \frac{\mathrm{d}}{\mathrm{d}t}\iint_S \mathrm{d}^2x \, \frac{1}{2}\mathbf{u}_{\mathrm{h}} \cdot \mathbf{u}_{\mathrm{h}} = 0 \tag{20.30}$$

$$\frac{\mathrm{d}}{\mathrm{d}t}\iint_S \mathrm{d}^2x \, F(\omega_z) = 0 \tag{20.31}$$

where F is an arbitrary function of the vorticity. The last two theorems require the kinematic boundary condition $\mathbf{u}_{\mathrm{h}} \cdot \mathbf{n}_{\mathrm{h}} = 0$ on δS. The special case $F(\omega_z) = \omega_z^2/2$ is of particular importance. It constitutes the conservation of the enstrophy

$$Z := \iint \mathrm{d}^2x \, \frac{1}{2}\omega_z^2 \tag{20.32}$$

The simultaneous conservation of two quadratic quantities, E_{kin} and Z, has important consequences. If appropriate dissipation mechanisms are included, the analysis suggests the existence of steady solutions where enstrophy cascades from a source region at medium wavenumbers through an inertial range to a high-wavenumber dissipation region whereas energy cascades through a different inertial range to a low-wavenumber dissipation region. Nonlinear interactions cause a "blue" or forward cascade of enstrophy and a "red" or inverse cascade of energy.

The above equations for two-dimensional turbulence can be obtained from the quasi-geostrophic potential vorticity equation (19.64) for a homogeneous fluid if one takes the limit $\gamma, \sigma_{\mathrm{s}}, \sigma_{\mathrm{b}} \ll \mathrm{Ro}$ and if one neglects all forcing and diffusion processes. It can also be obtained from the quasi-geostrophic potential equation (19.17) in the limit $\gamma, \sigma \ll \mathrm{Ro}$. An essential difference is that ψ can then depend on the vertical coordinate z. The flow consists of two-dimensional turbulence evolving independently in each horizontal layer. The pressure is given by $\delta p = \rho_0 f_0 \psi$.

The two-dimensional turbulence equations can also be obtained from (20.1) to (20.4) in the limit of no rotation and strong stratification, $\sigma := Z/D \ll 1$. The density equation then implies $w \to 0$. Again, the flow evolves independently as two-dimensional turbulence in each horizontal layer. In contrast to the quasi-geostrophic case, the pressure is now given by the cyclostrophic balance $\Delta_{\mathrm{h}}\delta p = -\rho_0 \nabla_{\mathrm{h}} \cdot (\mathbf{u}_{\mathrm{h}} \cdot \nabla_{\mathrm{h}}\mathbf{u}_{\mathrm{h}})$.

21

Small-scale motions

Within our triple Reynolds decomposition, small-scale motions are defined as those motions that are directly affected by molecular diffusion. Their "subgridscale" motions are the molecular motions whose effect is "parametrized" by the molecular flux laws. No eddy fluxes need to be parametrized but the large- and medium-scale background fields need to be prescribed. Since the molecular diffusion coefficients for salt and heat differ, the temperature–salinity mode emerges again, in the form of double diffusive processes. The time and space scales of these motions are so small that rotation and the sphericity of the Earth can be neglected. One leaves the realm of geophysical fluid dynamics and enters the realm of regular fluid dynamics.

For small-scale motions, variations in temperature and salinity are usually assumed to be so small that all thermodynamic and molecular coefficients can either be linearized or be regarded as constant. If the density is assumed to be constant the equations for small-scale motions reduce to the Navier–Stokes equations. As for medium-scale motions, boundary conditions become less well defined when an open interior volume of the ocean is considered.

21.1 Equations

The starting point is the equations in the Boussinesq approximation for a non-rotating fluid:

$$\rho_0 \frac{\mathrm{D}}{\mathrm{D}t}\mathbf{u} = -\boldsymbol{\nabla} p - \rho \boldsymbol{\nabla}\phi + \boldsymbol{\nabla}\cdot\boldsymbol{\sigma}^{\mathrm{mol}} \tag{21.1}$$

$$\boldsymbol{\nabla}\cdot\mathbf{u} = 0 \tag{21.2}$$

$$\frac{\mathrm{D}}{\mathrm{D}t}T = \Gamma \frac{\mathrm{D}}{\mathrm{D}t} p_{\mathrm{r}} + \frac{1}{\rho_0 c_{\mathrm{p}}} D_{\mathrm{h}}^{\mathrm{mol}} \tag{21.3}$$

$$\frac{\mathrm{D}}{\mathrm{D}t}S = -\frac{1}{\rho_0}\boldsymbol{\nabla}\cdot\mathbf{I}_{\mathrm{s}}^{\mathrm{mol}} \tag{21.4}$$

and

$$\rho = \rho(p_r, T, S) \tag{21.5}$$

where

$$D_h^{mol} := -\nabla \cdot \mathbf{q}^{mol} + \boldsymbol{\sigma}^{mol} : \mathbf{D} - \mathbf{I}_s^{mol} \cdot \nabla \Delta h \tag{21.6}$$

$$\mathbf{q}^{mol} = -\rho c_p \lambda \nabla T + \frac{D'}{D} T S(1-S) \frac{\partial \Delta \mu}{\partial S} \mathbf{I}_s^{mol} \tag{21.7}$$

$$\mathbf{I}_s^{mol} = -\rho \left[D(\nabla S - \gamma \nabla p) + S(1-S) D' \nabla T \right] \tag{21.8}$$

$$\boldsymbol{\sigma}^{mol} = 2\rho\nu \mathbf{S} + 3\rho\nu' \mathbf{N} \tag{21.9}$$

In addition to double diffusion these equations describe a rich (and interesting) variety of phenomena that depend on the peculiarities of the molecular flux laws and thermodynamic properties. These include cross diffusion, heat of mixing, temperature-dependent friction, and many more. Most of these effects are eliminated by a set of thermodynamic approximations that one usually employs and that can be justified in most circumstances by simple scale analysis, but that can also be relaxed easily if circumstances so demand. These approximations are:

1. The "adiabatic" temperature change $\Gamma \mathrm{D} p_r / \mathrm{D}t$ in the temperature equation is neglected. Note that assuming $\Gamma = 0$ strictly would imply $\alpha = 0$ because of the thermodynamic relation (2.13).
2. The heating $\boldsymbol{\sigma}^{mol} : \mathbf{D}$ owing to the dissipation of kinetic energy and the heating $\mathbf{I}_S^{mol} \cdot \nabla \Delta h$ owing to mixing are neglected in the "diabatic" heating term D_h^{mol}.
3. Cross-diffusion is neglected, $D' = 0$.
4. Expansion viscosity is neglected, $\nu' = 0$.
5. All the phenomenological coefficients are assumed to be constant.
6. The divergence of $\rho D \nabla S_{equ} = \rho D \gamma \nabla p$ is neglected.
7. In the equation of state, one replaces p_r by a constant reference value p_0 and then linearizes in T and S about reference values T_0 and S_0.

The results of these approximations are the equations for small-scale motions:

$$\rho_0 \frac{\mathrm{D}}{\mathrm{D}t} \mathbf{u} = -\nabla p - \rho \nabla \phi + \mu_0 \Delta \mathbf{u} \tag{21.10}$$

$$\nabla \cdot \mathbf{u} = 0 \tag{21.11}$$

$$\frac{\mathrm{D}}{\mathrm{D}t} T = \lambda_0 \Delta T \tag{21.12}$$

$$\frac{\mathrm{D}}{\mathrm{D}t} S = D_0 \Delta S \tag{21.13}$$

$$\rho = \rho_0 \left[1 - \alpha_0 (T - T_0) + \beta_0 (S - S_0) \right] \tag{21.14}$$

The equations contain the temperature–salinity, vortical, and gravity modes of motion. Typical applications include the evolution of the flow on prescribed temperature, salinity, and velocity gradients.

21.2 The temperature–salinity mode

Consider a motionless hydrostatically balanced background state $(\tilde{T}(z), \tilde{S}(z))$. Temperature and salinity fluctuations about this state are governed by

$$\frac{\mathrm{D}}{\mathrm{D}t}\delta T + w\frac{\mathrm{d}\tilde{T}}{\mathrm{d}z} = \lambda_0 \Delta \delta T \tag{21.15}$$

$$\frac{\mathrm{D}}{\mathrm{D}t}\delta S + w\frac{\mathrm{d}\tilde{S}}{\mathrm{d}z} = D_0 \Delta \delta S \tag{21.16}$$

where the background state has been "subtracted" out (see discussion in Section 18.2). Instead of δT and δS introduce density $\delta\rho$ and spiciness $\delta\chi$ (see (2.31))

$$\delta\rho = \rho_0\left[-\alpha_0\delta T + \beta_0\delta S\right] \tag{21.17}$$

$$\delta\chi = \rho_0\left[\alpha_0\delta T + \beta_0\delta S\right] \tag{21.18}$$

The prognostic equations for density and spiciness then become

$$\frac{\mathrm{D}}{\mathrm{D}t}\delta\rho = w\frac{\rho_0}{g}N^2 + \frac{1}{2}(\lambda_0 + D_0)\delta\rho + \frac{1}{2}(D_0 - \lambda_0)\delta\chi \tag{21.19}$$

$$\frac{\mathrm{D}}{\mathrm{D}t}\delta\chi = w\frac{\rho_0}{g}M^2 + \frac{1}{2}(\lambda_0 + D_0)\delta\chi + \frac{1}{2}(D_0 - \lambda_0)\delta\rho \tag{21.20}$$

where

$$N^2 := g\alpha_0\frac{\mathrm{d}\tilde{T}}{\mathrm{d}z} - g\beta_0\frac{\mathrm{d}\tilde{S}}{\mathrm{d}z} \tag{21.21}$$

$$M^2 := -g\alpha_0\frac{\mathrm{d}\tilde{T}}{\mathrm{d}z} - g\beta_0\frac{\mathrm{d}\tilde{S}}{\mathrm{d}z} \tag{21.22}$$

are the buoyancy and "spiciness" frequencies. If the diffusivities for heat and salt are different then spiciness is an active tracer, influencing the evolution of the density and hence velocity. If the two diffusivities are assumed equal then spiciness becomes a passive tracer and the temperature–salinity mode is suppressed. The above equations form the starting point for the theoretical analysis of double diffusive phenomena.

21.3 Navier–Stokes equations

For a well mixed homogeneous fluid with constant density $\rho = \rho_0 =$ constant the equations for small-scale motions reduce to the Navier–Stokes equations

$$\rho_0 \frac{\mathrm{D}}{\mathrm{D}t}\mathbf{u} = -\boldsymbol{\nabla} p - \rho_0 \boldsymbol{\nabla}\phi + \mu_0 \Delta \mathbf{u} \tag{21.23}$$

$$\boldsymbol{\nabla} \cdot \mathbf{u} = 0 \tag{21.24}$$

with $\phi = gz$. The boundary conditions are $\mathbf{u} = 0$ at rigid boundaries and $(\partial_t + \mathbf{u}_\mathrm{h} \cdot \boldsymbol{\nabla}_\mathrm{h})\eta_\mathrm{s} = w$, and $p = 0$ at $z = \eta_\mathrm{s}$ for a free unforced surface. The gravitational term can be eliminated from the momentum equation by subtracting out a hydrostatically balanced pressure, i.e., by introducing the Montgomery potential $M = p + \rho_0 g z$. Gravitation will, however, re-enter the problem when a free surface is included since the dynamic boundary condition $p = 0$ at $z = \eta_\mathrm{s}$ becomes $M = \rho_0 g \eta_\mathrm{s}$ at $z = \eta_\mathrm{s}$. Temperature and salinity are passive tracers advected by the velocity field. The momentum equation can be written in the equivalent form

$$\partial_t \mathbf{u} + \boldsymbol{\omega} \times \mathbf{u} = -\boldsymbol{\nabla} B + \nu_0 \Delta \mathbf{u} \tag{21.25}$$

with the Bernoulli function

$$B = \frac{p}{\rho_0} + \phi + \frac{1}{2}\mathbf{u} \cdot \mathbf{u} \tag{21.26}$$

The diagnostic pressure equation becomes the three-dimensional Poisson equation

$$\partial_i \partial_i p = -\rho_0 G_{ij} G_{ji} \tag{21.27}$$

At rigid boundaries, the boundary condition $\mathbf{u} = 0$ implies

$$\frac{\partial p}{\partial x_n} = \mu_0 \frac{\partial^2 u_n}{\partial x_n \partial x_n} \tag{21.28}$$

The kinetic and potential energy equations become

$$\partial_t \left(\frac{1}{2} u_i u_i \right) + \partial_j \left(u_j \frac{1}{2} u_i u_i + p u_j - 2\nu_0 u_i D_{ij} \right) = -2\nu_0 D_{ij} D_{ij} - \rho_0 g w \tag{21.29}$$

$$\partial_t \phi + \partial_j (u_j \phi) = \rho_0 g w \tag{21.30}$$

There is an exchange of potential and kinetic energy. Kinetic energy is systematically dissipated by molecular viscosity.

The vorticity equation takes the form

$$\frac{\mathrm{D}}{\mathrm{D}t}\boldsymbol{\omega} = (\boldsymbol{\omega} \cdot \boldsymbol{\nabla})\mathbf{u} + \nu_0 \Delta \boldsymbol{\omega} \tag{21.31}$$

The Navier–Stokes equations contain the gravity and vortical mode. A major distinction is made between turbulent flows and laminar flows, depending on whether nonlinear advection or friction dominates the dynamic evolution. The ratio of advection to friction is measured by the Reynolds number $\mathrm{Re} := UL/\nu$. Friction cannot be neglected altogether but might be confined to boundary layers. In that case, the vorticity equation implies that one can have irrotational motion, $\boldsymbol{\omega} \equiv 0$, in the interior.

Three-dimensional turbulence Consider an unbounded (or periodic) domain. Gravity can then be eliminated and one obtains

$$\rho_0 \frac{\mathrm{D}}{\mathrm{D}t}\mathbf{u} = -\boldsymbol{\nabla} M + \mu_0 \Delta \mathbf{u} \tag{21.32}$$

$$\boldsymbol{\nabla} \cdot \mathbf{u} = 0 \tag{21.33}$$

These equations form the basis for the theoretical analysis of homogeneous three-dimensional isotropic turbulence. Since all coefficients are constant this analysis is usually carried out in wavenumber space, obtained by Fourier transformation of the equations. Nonlinear advection then becomes nonlinear triad interaction. There is no exchange between kinetic and potential energy. Kinetic energy is conserved, except for viscous dissipation. Energy is the only quadratic invariant. This fact distinguishes three-dimensional turbulence from two-dimensional turbulence, which has two quadratic invariants, energy and enstrophy. In the limit $\mu_0 \to 0$, the analysis suggests the existence of steady solutions where nonlinear interactions cascade energy from a source region at low wavenumbers through an inertial range to a viscous dissipation region at high wavenumbers.

Nonlinear surface gravity waves The vorticity equation (21.31) implies that in the inviscid[1] case there exist irrotational motions, $\boldsymbol{\omega} \equiv 0$. They are governed by

$$\partial_t \mathbf{u} = -\boldsymbol{\nabla} B \tag{21.34}$$

$$\boldsymbol{\nabla} \cdot \mathbf{u} = 0 \tag{21.35}$$

with Bernoulli function

$$B = \frac{p}{\rho_0} + \phi + \frac{1}{2}\mathbf{u} \cdot \mathbf{u} \tag{21.36}$$

[1] Inviscid surface gravity waves are discussed in this chapter on small-scale motions although their dissipation mechanisms may be either molecular friction or turbulent friction in bottom boundary layers and wave breaking, depending on whether they are short or long waves.

For unforced waves, these equations must be solved subject to the boundary conditions

$$(\partial_t + \mathbf{u}_h \cdot \nabla_h)\eta_s = w \quad \text{at} \quad z = \eta_s \tag{21.37}$$
$$p = 0 \quad \text{at} \quad z = \eta_s \tag{21.38}$$
$$\mathbf{u}_h \cdot \nabla_h \eta_b = w \quad \text{at} \quad z = -H_0 + \eta_b \tag{21.39}$$

Since $\omega \equiv 0$ one can introduce a velocity potential φ such that

$$\mathbf{u} = \nabla\varphi \tag{21.40}$$

The incompressibility condition then becomes the Laplace equation $\Delta\varphi = 0$, and the momentum equation becomes $\nabla(\partial_t\varphi + B) = 0$, which implies without loss of generality $\partial_t\varphi + B = 0$ or

$$p = -\rho_0 \left(\partial_t\varphi + \phi + \frac{1}{2}\nabla\varphi \cdot \nabla\varphi \right) \tag{21.41}$$

which is a diagnostic equation for the pressure in terms of the velocity potential φ. When these diagnostic relations for the velocity vector and pressure are substituted one obtains a coupled system of equations for η_s and φ. The surface displacement η_s has to be determined from the prognostic equation

$$(\partial_t + \nabla_h\varphi \cdot \nabla_h)\eta_s = \partial_z\varphi \quad \text{at} \quad z = \eta_s \tag{21.42}$$

The velocity potential φ has to be determined from the Laplace equation

$$\Delta\varphi = 0 \tag{21.43}$$

subject to the boundary condition

$$\partial_t\varphi + g\eta_s + \frac{1}{2}\nabla\varphi \cdot \nabla\varphi = 0 \quad \text{at} \quad z = \eta_s \tag{21.44}$$
$$(\nabla_h\varphi \cdot \nabla_h)\eta_b = \partial_z\varphi \quad \text{at} \quad z = -H_0 + \eta_b \tag{21.45}$$

The solution of these coupled equations is complicated by the fact that the domain in which the Laplace equation has to be solved is time-dependent and by the fact that the surface boundary condition for the Laplace equation contains the tendency term $\partial_t\varphi$. The equations are characterized by the three dimensionless parameters $\sigma_s := Z_s/H_0$, $\sigma_b := Z_b/H_0$, and $\delta = H_0/L$ (or any other three combinations of them) where $Z_{s,b}$ are the scales of the surface and bottom displacements $\eta_{s,b}$ and L the horizontal scale. If any of these parameters is small solutions can be constructed by asymptotic expansions.

For linear waves ($\sigma_s \ll 1$) and a flat bottom ($\sigma_b = 0$) the equations reduce to

$$\partial_t \eta_s = \partial_z \varphi \quad \text{at} \quad z = 0 \tag{21.46}$$

$$\Delta \varphi = 0 \tag{21.47}$$

$$\partial_t \varphi + g\eta_s = 0 \quad \text{at} \quad z = 0 \tag{21.48}$$

$$\partial_z \varphi = 0 \quad \text{at} \quad z = -H_0 \tag{21.49}$$

and have solutions

$$\varphi = A(z) e^{i(\mathbf{k}_h \cdot \mathbf{x}_h - \omega t)} \tag{21.50}$$

with vertical structure

$$A(z) = \cosh k(z + H_0) \tag{21.51}$$

and dispersion relation

$$\omega^2 = g k_h \tanh k_h H_0 \tag{21.52}$$

In the limit $k_h H_0 \ll 1$ and $k_h H_0 \gg 1$ one recovers the dispersion relations for long and short surface gravity waves of Table 8.3.

The shallow water limit (or Airy approximation) $\delta = H_0/L \ll 1$ of the nonlinear equations can also be obtained from the primitive equations (13.54) and (13.55) for a homogeneous ocean. In the inviscid unforced case they allow irrotational motions $\omega_z = 0$ or $\mathbf{u}_h = \nabla_h \varphi$ where φ does not depend on z. The layer thickness equation and the Bernoulli expression become the prognostic equations for φ and η_s

$$\partial_t \eta_s + \nabla_h \cdot [\nabla_h \varphi (H_0 - \eta_b + \eta_s)] = 0 \tag{21.53}$$

$$\partial_t \varphi + g\eta_s + \frac{1}{2} \nabla_h \varphi \cdot \nabla_h \varphi = 0 \tag{21.54}$$

In the linear limit we obtain the dispersion relation $\omega^2 = g H_0 k_h^2$.

If terms of up to order $O(\sigma_s)$ and order $O(\delta^2)$ are retained one obtains the Boussinesq theory of water waves.

22
Sound waves

In Chapter 7 sound (or acoustic) waves were identified as irrotational motions in a homentropic ideal fluid that owe their existence to the compressibility of sea water. In Chapter 8 solutions were constructed for a horizontally homogeneous and motionless background state. Here we derive the equation for acoustic waves propagating in a more general background state. To do so we assume that the (Lagrangian) time scale T_a of acoustic motions is much shorter than the (Lagrangian) time scale $\tilde{T}$ of the background and perform a two-time scale expansion with respect to the parameter $\epsilon_a := T_a/\tilde{T} \ll 1$. This expansion requires the sound speed c to be much larger than the phase speeds and advection velocities of the background flow. The same two-time scale expansion was used in Chapter 11 to derive the Boussinesq equations. There we considered the equations for the slow background flow. Here we consider the equations for the fast acoustic motions. As in the case of the Boussinesq approximation, the resulting acoustic wave equation can be simplified by assuming additionally that the vertical scales D of sound waves are much smaller than the adiabatic and diabatic depth scales $\tilde{D}_a$ and $\tilde{D}_d$ of the density field. The wave equation then contains the background sound speed $\tilde{c}$ and velocity field $\tilde{\mathbf{u}}$ as the only relevant background fields.

Time variations of the background sound speed field $\tilde{c}$ are neglected during a single acoustic pulse. This leads to the Helmholtz equation, which is the basis for most acoustic studies.

22.1 Sound speed

The sound speed c is the most fundamental quantity that determines acoustic propagation. It is defined by the thermodynamic relationship

$$c^{-2} := \frac{\partial \rho(p, \theta, S)}{\partial p} \tag{22.1}$$

where $\rho(p, \theta, S)$ is the equation of state. The sound speed in sea water varies roughly between 1450 m s^{-1} and 1550 m s^{-1}. There exist numerical formulae (e.g., by Del Grosso (1974) and MacKenzie (1981)) that give the sound speed as a function of pressure (or depth), potential temperature (or temperature), and salinity. These formulae are discussed in Dushaw *et al.* (1993). The sound speed increases with pressure, temperature, and salinity. The effect of salinity is relatively small.

In the ocean, pressure, potential temperature, and salinity are functions of position $\mathbf{x}$ and time t. So is the sound speed, $c(\mathbf{x}, t) = c(p(\mathbf{x}, t), \theta(\mathbf{x}, t), S(\mathbf{x}, t))$. The sound speed at a specific horizontal position and at a specific time as a function of the vertical coordinate z is called a sound speed profile $c(z)$. Sound speed profiles in the deep ocean often show a minimum at about 1 km depth, the sound channel. The increase above the minimum is the result of temperature increasing towards the surface. The increase towards the ocean bottom is due to the increasing pressure in the nearly isothermal deep waters. There is also a surface duct corresponding to the surface mixed layer.

The gradient of sound speed is given by

$$\frac{\mathrm{d}c}{\mathrm{d}z} = \left(\frac{\partial c}{\partial \theta}\right)_{p,S} \frac{\mathrm{d}\theta}{\mathrm{d}z} + \left(\frac{\partial c}{\partial S}\right)_{p,\theta} \frac{\mathrm{d}S}{\mathrm{d}z} + \left(\frac{\partial c}{\partial p}\right)_{\theta,S} \frac{\mathrm{d}p}{\mathrm{d}z} \tag{22.2}$$

The first two terms constitute the diabatic gradient

$$\left(\frac{\mathrm{d}c}{\mathrm{d}z}\right)_{\mathrm{d}} := \left(\frac{\partial c}{\partial \theta}\right)_{p,S} \frac{\mathrm{d}\theta}{\mathrm{d}z} + \left(\frac{\partial c}{\partial S}\right)_{p,\theta} \frac{\mathrm{d}S}{\mathrm{d}z} \tag{22.3}$$

The third term is the adiabatic gradient. It can be written

$$\left(\frac{\mathrm{d}c}{\mathrm{d}z}\right)_{\theta,S} = -\gamma c \tag{22.4}$$

where the rate of change is given by

$$\gamma := -\frac{1}{c}\left(\frac{\partial c}{\partial p}\right)_{\theta,S} \frac{\mathrm{d}p}{\mathrm{d}z} = \frac{\rho g}{c}\left(\frac{\partial c}{\partial p}\right)_{\theta,S} \tag{22.5}$$

and where the second equality sign assumes the hydrostatic approximation. The adiabatic rate varies little about the value $\gamma = 0.0113\ \mathrm{km}^{-1}$.

22.2 The acoustic wave equation

The starting point for the derivation of the acoustic wave equation is equations (7.1) to (7.5) for a non-rotating ideal fluid.[1] Loss processes can be easily included but are

[1] This derivation is based on comments from Frank Henyey (University of Washington).

disregarded here since losses for sound waves can often not be formulated in terms of molecular diffusion coefficients. If one takes the time derivative of the pressure equation (7.4)

$$\frac{1}{c^2}\frac{\mathrm{D}}{\mathrm{D}t}p = -\rho\boldsymbol{\nabla}\cdot\mathbf{u} \tag{22.6}$$

one obtains

$$\begin{aligned}\frac{\mathrm{D}}{\mathrm{D}t}\frac{1}{c^2}\frac{\mathrm{D}}{\mathrm{D}t}p &= -\frac{\mathrm{D}}{\mathrm{D}t}\left(\rho\boldsymbol{\nabla}\cdot\mathbf{u}\right)\\ &= \rho\boldsymbol{\nabla}\cdot\left(\frac{1}{\rho}\boldsymbol{\nabla}p\right) + \rho\mathbf{G}:\mathbf{G} + \rho\left(\boldsymbol{\nabla}\cdot\mathbf{u}\right)^2\end{aligned} \tag{22.7}$$

where $\mathbf{G} = \boldsymbol{\nabla}\mathbf{u}$ is the velocity gradient tensor. The first term on the right-hand side arises from $\mathrm{D}\mathbf{u}/\mathrm{D}t$, the second term from the commutator of $\mathrm{D}/\mathrm{D}t$ and $\boldsymbol{\nabla}$, and the third term from $\mathrm{D}\rho/\mathrm{D}t$. The gradient of the gravitational acceleration has been neglected. The pressure and all other variables are now decomposed into a background state $\tilde{\psi}$ and acoustic fluctuations $\delta\psi$. In order to obtain the acoustic wave equation one then makes the following two essential assumptions:

1. The (Lagrangian) time scale T_{a} of the acoustic fluctuations is assumed to be much shorter than the (Lagrangian) time scale $\tilde{T}$ of the background. In the language of Section 9.2 the acoustic fluctuations are fast variables and the background is a slow variable. This assumption requires that the sound speed c is much larger than the advection and phase speeds of the background flow, $c^2 \gg \tilde{u}^2, gH_0, \ldots$ As a consequence the term $\mathrm{D}\tilde{p}/\mathrm{D}t$ in the pressure equation is neglected with respect to $\mathrm{D}\delta p/\mathrm{D}t$. As shown in Section 9.2 this result can formally be obtained by a two-time scale expansion with respect to $\epsilon_{\mathrm{a}} := T_{\mathrm{a}}/\tilde{T}$.
2. The amplitudes of the acoustic fluctuations are assumed to be so small that the equations can be linearized in $\delta\psi$. All terms of $O((\delta\psi)^2)$ and higher are neglected. One of the formal non-dimensional expansion parameters is the Mach number $\delta u/c \ll 1$. As a consequence, the velocity in the Lagrangian derivative is given by the background velocity $\tilde{\mathbf{u}}$ and the sound speed given by the background sound speed $\tilde{c}$.

These two assumptions imply that the operator on the left-hand side of (22.7) can be approximated by

$$\frac{\mathrm{D}}{\mathrm{D}t}\frac{1}{c^2}\frac{\mathrm{D}}{\mathrm{D}t}p \to \frac{\tilde{\mathrm{D}}}{\mathrm{D}t}\frac{1}{\tilde{c}^2}\frac{\tilde{\mathrm{D}}}{\mathrm{D}t}\delta p \tag{22.8}$$

where

$$\frac{\tilde{\mathrm{D}}}{\mathrm{D}t} = \partial_t + \tilde{\mathbf{u}}\cdot\boldsymbol{\nabla} \tag{22.9}$$

When the first operator $\tilde{\mathrm{D}}/\mathrm{D}t$ acts on the $\tilde{\mathbf{u}}$-term in the second operator it generates terms $O(\epsilon)$ that can be neglected.

The right-hand side of (22.7) is simplified by the following considerations:

1. The density ρ can be replaced by a constant ρ_0 wherever it appears as a factor since the adiabatic and diabatic depth scales $\tilde{D}_{\mathrm{a,d}}$ of the density are much larger then the vertical scale D of the acoustic fluctuations.
2. In the first term, the pressure can be regarded as the acoustic pressure since $\nabla \tilde{p}/\nabla \delta p = O(\epsilon)$. The density inside the parentheses is then the background density. When the gradient operator is acting on this density it generates a term $-\tilde{\rho}^{-2}\nabla\tilde{\rho}\cdot\nabla\delta p$, which is $D/\tilde{D}_{\mathrm{a,d}}$ times smaller than the term $\tilde{\rho}^{-1}\nabla\cdot\nabla\delta p$. The first term thus reduces to $\nabla\cdot\nabla\delta p$.
3. The second term consists of three contributions. The contribution $\tilde{\mathbf{G}}:\tilde{\mathbf{G}}$ represents a (turbulent) sound source (Lighthill, 1952). It is disregarded here. The contribution $2\tilde{\mathbf{G}}:\delta\mathbf{G}$ is ϵ times smaller than $\nabla\cdot\nabla\delta p$ and can be neglected. The third contribution is quadratic in the acoustic fluctuations.
4. The third term in (22.7) can be neglected as well. It is quadratic in the acoustic fluctuations since the background satisfies $\nabla\cdot\tilde{\mathbf{u}} = 0$.

One thus arrives at the acoustic wave equation

$$\frac{\tilde{\mathrm{D}}}{\mathrm{D}t}\frac{1}{\tilde{c}^2}\frac{\tilde{\mathrm{D}}}{\mathrm{D}t}\delta p = \nabla\cdot\nabla\delta p \tag{22.10}$$

It governs the propagation of acoustic waves through a background field of sound speed $\tilde{c}$ and velocity $\tilde{\mathbf{u}}$. The equation neglects nonlinear acoustics and sources and losses. At interfaces, one usually requires that δp and the normal component of $\tilde{\rho}^{-1}\nabla\delta p$ are continuous. The forward or direct acoustic problem is to calculate the acoustic pressure for a given background field. The inverse problem consists of inferences about the background field from measurements of the acoustic pressure.

The potential temperature, salinity, density, and momentum signals associated with acoustic fluctuations can be inferred from the equations

$$\frac{\tilde{\mathrm{D}}}{\mathrm{D}t}\delta\theta = 0 \tag{22.11}$$

$$\frac{\tilde{\mathrm{D}}}{\mathrm{D}t}\delta S = 0 \tag{22.12}$$

$$\begin{array}{ccccc} \tilde{\rho}\frac{\tilde{\mathrm{D}}}{\mathrm{D}t}\delta\mathbf{u} = & -\nabla\tilde{p} & -\nabla\delta p & -\tilde{\rho}g- & \delta\rho g \\ & \epsilon & 1 & \epsilon & D/\tilde{D}_{\mathrm{a}} \end{array} \tag{22.13}$$

where again we have neglected the time derivative of the slowly varying background fields. From the first two equations one infers $\delta\theta = 0$ and $\delta S = 0$. No potential temperature and salinity signal is associated with acoustic waves and hence

$$\delta\rho = \tilde{c}^{-2}\delta p \tag{22.14}$$

The terms on the right-hand side of the momentum equation scale as indicated. The momentum equation thus reduces to $\rho_0 \tilde{D}\delta\mathbf{u}/Dt = -\boldsymbol{\nabla}\delta p$ and is not affected by gravity.

If the background sound speed and velocity field are constant the acoustic wave equation has plane wave solutions $\delta p \sim \exp\{i(\mathbf{k}\cdot\mathbf{x} - \omega t)\}$ where wavenumber vector $\mathbf{k}$ and frequency ω are related by the dispersion relation

$$\omega = \tilde{c}_0 k + \mathbf{k}\cdot\tilde{\mathbf{u}}_0 \tag{22.15}$$

The Doppler shift $\mathbf{k}\cdot\tilde{\mathbf{u}}_0$ depends on the direction of the wavenumber vector whereas the intrinsic frequency $\tilde{c}_0 k$ depends only on its magnitude. When the Doppler shift is neglected, $\omega = ck$ and all waves propagate with the same phase speed. Acoustic waves are then non-dispersive.

The acoustic wave equation also reproduces the mode solutions of Chapter 8. For a motionless background state with sound speed profile $\tilde{c}(z)$ the wave equation has separable solutions of the form $\delta p(\varphi, \theta, z, t) = P(\varphi, \theta)\,\phi(z)\exp\{-i\omega t\}$ where the vertical eigenfunction $\phi(z)$ satisfies

$$\left[\frac{d^2}{dz^2} + \omega^2\left(\tilde{c}^{-2}(z) - \epsilon\right)\right]\phi(z) = 0 \tag{22.16}$$

$$\phi = 0 \quad \text{at} \quad z = 0 \tag{22.17}$$

$$\frac{d\phi}{dz} = 0 \quad \text{at} \quad z = -H_0 \tag{22.18}$$

with separation constant ϵ. This vertical eigenvalue problem is the $N = 0$ and $g = 0$ limit of the eigenvalue problem (8.36) to (8.38) of Chapter 8. It is a standard Sturm–Liouville eigenvalue problem. The solutions are a countable set of eigen or normal modes $\phi_n(z)$ with eigenvalues ϵ_n $(n = -1, -2, \ldots)$. The normal mode index is chosen to be negative to distinguish the acoustic normal modes from the external $(n = 0)$ and internal modes $n = 1, 2, \ldots$ The nth mode has $|n|$ zero crossings in the interval $(-H_0, 0]$. From Table 8.1 we find for $\tilde{c} = \tilde{c}_0 =$ constant, $\phi_n(z) = \sin k_n z$ with vertical wavenumber $k_n = -\pi(n + 1/2)/H_0$, and separation constant $\epsilon_n = \tilde{c}_0^{-2} - k_n^2/\omega^2$. Other bottom boundary conditions might also be applied.

The horizontal eigenfunction $P(\varphi, \theta)$ must satisfy

$$\left(\nabla_h^2 + \omega^2\epsilon\right)\, P(\varphi, \theta) = 0 \tag{22.19}$$

which is the horizontal eigenvalue problem (8.70) for a non-rotating sphere, with eigensolutions and eigenvalues given by (8.71) and (8.72).

If $\tilde{c} = \tilde{c}(x, y, z, t)$ one can still use the vertical normal modes to represent the vertical structure of the acoustic field. However, the normal modes then do not propagate independently but become coupled.

22.3 Ray equations

Plane waves are also a good local solution when the wavelengths and periods of the acoustic waves are much smaller than the scales of the background fields $\tilde{c}(\mathbf{x}, t)$ and $\tilde{\mathbf{u}}(\mathbf{x}, t)$. A global solution of the form

$$\delta p \sim \exp\{i\theta(\mathbf{x}, t)\} \tag{22.20}$$

is then assumed with local wavenumber $\mathbf{k} = \boldsymbol{\nabla}\theta$, local frequency $\omega = -\partial_t\theta$ and local dispersion relation

$$\omega = \tilde{c}(\mathbf{x}, t)k + \mathbf{k} \cdot \tilde{\mathbf{u}}(\mathbf{x}, t) \tag{22.21}$$

The envelope of wave packets (or wave trains or wave groups) then moves along ray paths with group velocity (see Section E.3)

$$\frac{\mathrm{d}\mathbf{x}}{\mathrm{d}t} = \frac{\tilde{c}}{k}\mathbf{k} + \tilde{\mathbf{u}} \tag{22.22}$$

Changes of the wavenumber $\mathbf{k}(t)$ along these ray paths are governed by the rate of refraction

$$\frac{\mathrm{d}\mathbf{k}}{\mathrm{d}t} = -k\boldsymbol{\nabla}\tilde{c} - \boldsymbol{\nabla}(\mathbf{k} \cdot \tilde{\mathbf{u}}) \tag{22.23}$$

The Doppler advection speed $\tilde{u}$ is of course much smaller than the intrinsic phase speed c. It is even small compared to changes Δc in the intrinsic phase speed and matters only in reciprocal transmissions that eliminate Δc. If the Doppler shift is neglected the group velocity also becomes the velocity with which a point on a surface of constant phase moves in normal direction. The propagation of phases is then also along ray paths.

22.4 Helmholtz equation

When the time dependence of the background fields is neglected during the duration of a single acoustic pulse, each acoustic frequency component propagates independently. For the case $\tilde{\mathbf{u}} = 0$, Fourier transformation of the wave equation (22.10) in time then leads to the Helmholtz equation

$$\left(\nabla^2 + \omega^2\tilde{c}^{-2}(\mathbf{x})\right)\ \hat{p}(\mathbf{x}) = 0 \tag{22.24}$$

where ω is the acoustic frequency and $\hat{p}(\mathbf{x})$ the Fourier amplitude of the acoustic pressure. Often, one specifies a reference sound speed $\tilde{c}_0$ and introduces the index of refraction $n := \tilde{c}_0/\tilde{c}$. Global wave solutions are again of the form $\hat{p} \sim \exp\{i\theta(\mathbf{x})\}$ with local wavenumber $\mathbf{k} = \boldsymbol{\nabla}\theta$. The phase $\theta(\mathbf{x})$ then satisfies the Eikonal equation

$$\nabla\theta \cdot \nabla\theta = \omega^2 \tilde{c}^{-2} \tag{22.25}$$

When the position along a ray path is measured by the path length l then the ray path $\mathbf{x}(l)$ is determined by the ray equation

$$\frac{\mathrm{d}}{\mathrm{d}l}\left(\tilde{c}^{-1}(\mathbf{x})\frac{\mathrm{d}\mathbf{x}}{\mathrm{d}l}\right) = \nabla\tilde{c}^{-1}(\mathbf{x}) \tag{22.26}$$

This ray equation is a second order ordinary differential equation. It is the same for all acoustic frequencies owing to the non-dispersive nature of acoustic waves.

22.5 Parabolic approximation

The parabolic approximation assumes acoustic propagation to proceed in only one direction along some selected coordinate. This direction is usually the direction from the source to the receiver. If x is the selected coordinate one writes the Helmholtz equation in the form

$$\frac{\partial^2 \hat{p}}{\partial x^2} + k_0^2(1+Y)\hat{p} = 0 \tag{22.27}$$

where

$$Y := \frac{1}{k_0^2}\left(\frac{\partial^2}{\partial y^2} + \frac{\partial^2}{\partial z^2} + \omega^2\tilde{c}^{-2} - k_0^2\right) \tag{22.28}$$

Here k_0 is the reference wavenumber $k_0 := \omega/\tilde{c}_0$ and $\tilde{c}_0$ a reference sound speed. The wave equation can then be factored into incoming and outgoing propagation

$$\left[\frac{\partial}{\partial x} + \mathrm{i}k_0(1+Y)^{1/2}\right]\left[\frac{\partial}{\partial x} - \mathrm{i}k_0(1+Y)^{1/2}\right]\hat{p} = -\mathrm{i}\hat{p}k_0\frac{\partial}{\partial x}(1+Y)^{1/2} \tag{22.29}$$

Neglecting the commutator on the right-hand side and assuming that the outgoing propagation dominates one obtains the outgoing wave equation or parabolic approximation

$$\frac{\partial}{\partial x}\hat{p} = \mathrm{i}k_0(1+Y)^{1/2}\hat{p} \tag{22.30}$$

The major simplification achieved by the parabolic approximation is that (22.30) can (in principle) be solved by integrating along x outwardly away from the source.

The narrow angle parabolic approximation is obtained by assuming $Y \ll 1$ so that $(1+Y)^{1/2} \approx 1 + Y/2$. Equation (22.30) then reduces to the Schrödinger equation

$$i\frac{\partial}{\partial x}\hat{p} = -\frac{1}{2k_0}\left(\frac{\partial^2}{\partial y^2} + \frac{\partial^2}{\partial z^2}\right)\hat{p} - \frac{k_0^2 + \omega^2\tilde{c}^{-2}}{2k_0}\hat{p} \tag{22.31}$$

Often one also neglects the operator $\partial^2/\partial y^2$ and accounts for the energy loss owing to geometric spreading by a factor $1/x$ in the intensity.

Appendix A
Equilibrium thermodynamics

Thermodynamics is a macroscopic theory of matter that ignores its discrete molecular structure and makes the continuum hypothesis. *Equilibrium* thermodynamics is concerned with the properties of systems in thermodynamic equilibrium. Thermodynamic equilibrium is the state that an isolated system, i.e., a system that does not exchange mass, momentum, and energy with its surroundings, reaches after a sufficiently long time. The main properties of a system in thermodynamic equilibrium are:

1. The entropy of the system is a maximum.
2. Temperature, pressure, and chemical potentials are constant throughout the system, if the system is not exposed to an external potential.
3. The system may consist of one or more phases. In the latter case the system is not homogeneous but heterogeneous.
4. The state satisfies the thermodynamic inequalities. The temperature, specific heat at constant volume, and isothermal compressibility must be non-negative. The chemical potential difference must not decrease with salinity.
5. A system consisting of N components and M phases is completely described by the specification of $N - M + 2$ thermodynamic variables (Gibbs' phase rule). All other thermodynamic variables can be obtained from an appropriate thermodynamic potential that is a function of these $N - M + 2$ variables.
6. The thermodynamic variables satisfy thermodynamic relations, which are mathematical identities.

All these basic concepts are discussed in this appendix and are exemplified for an ideal gas, which may also serve as a model for air.

A.1 Thermodynamic variables

Consider a one-component, one-phase system that is confined to a volume, at rest, isolated from its surroundings and not exposed to external forces. In thermodynamic equilibrium this system is homogeneous and described by *thermodynamic variables* such as its:

- mass M [kg];
- volume V [m^3];
- pressure p [Pa $=$ N m^{-2}];
- temperature T [K];
- internal energy E [kg m^2 s^{-2}]; and
- entropy S [kg m^2 s^{-2} K^{-1}].

Specific values of the thermodynamic variables define the *thermodynamic state* of the system. The thermodynamic state is a point in the space spanned by the thermodynamic variables. The thermodynamic variables M, V, E, and S are *extensive* quantities. They are additive in thermodynamic equilibrium. If the system is divided into two subsystems having the internal energies E_1 and E_2 then the total system has the internal energy $E = E_1 + E_2$. The thermodynamic variables p and T are *intensive* quantities. In thermodynamic equilibrium their values are the same in any subsystem.

For extensive quantities one introduces *specific* variables that represent the amount per unit mass:

- specific volume

$$v := \frac{V}{M} \quad [\mathrm{m^3\,kg^{-1}}] \tag{A.1}$$

- specific internal energy

$$e := \frac{E}{M} \quad [\mathrm{m^2\,s^{-2}}] \tag{A.2}$$

- specific entropy

$$\eta := \frac{S}{M} \quad [\mathrm{m^2\,s^{-2}\,K^{-1}}] \tag{A.3}$$

and densities:

- mass density

$$\rho := \frac{M}{V} \quad [\mathrm{kg\,m^{-3}}] \tag{A.4}$$

- internal energy density

$$\tilde{e} := \frac{E}{V} = \rho e \quad [\mathrm{kg\,m^{-1}\,s^{-2}}] \tag{A.5}$$

- entropy density

$$\tilde{\eta} := \frac{S}{V} = \rho\eta \quad [\mathrm{kg\,m^{-1}\,s^{-2}\,K^{-1}}] \tag{A.6}$$

The specific quantities and densities are intensive quantities.

A.2 Thermodynamic processes

When the system under consideration is not isolated but interacts with surrounding systems, then changes in the surrounding systems lead to changes in the system under consideration. These changes are called *thermodynamic processes*. If these changes are so slow that the system under consideration always stays in thermodynamic equilibrium then these changes are called *equilibrium* or *quasi-static processes*. They can be represented by a line in the phase space spanned by the thermodynamic variables. Processes are called *reversible* if the system returns to its original state when the surrounding systems reverse to their original state. All other processes are called *irreversible*.

First law of thermodynamics Consider a *closed system*, i.e., a system that does not exchange mass but only energy with its surroundings. If δW is the amount of *mechanical* *work* done on the system and δQ is the amount of heat absorbed by the system then the first law of thermodynamics states that the internal energy changes by an amount

$$\mathrm{d}E = \delta Q + \delta W \tag{A.7}$$

The first law thus states that heat, mechanical energy, and internal energy are different forms of energy and that energy is conserved. The symbol "δ" indicates that Q and W are not thermodynamic variables. Their values depend on the way a certain thermodynamic state is reached. Especially, $\oint \delta Q$ and $\oint \delta W$ do not vanish for *cyclic processes*, i.e., for processes that return to their starting state. The symbol "d" indicates that E is a thermodynamic variable, i.e., its value depends only on the thermodynamic state, and not on the way the system reaches this state. The first law of thermodynamics therefore describes the existence of a thermodynamic variable, namely the internal energy E, and a way to determine it, up to a constant.

Though $\oint \delta Q$ does not necessarily vanish for a cyclic process it is found that $\oint \frac{\delta Q}{T} = 0$ for all cyclic reversible processes. This finding implies the existence of a thermodynamic quantity S, called *entropy*, which is obtained by integrating

$$\mathrm{d}S = \frac{\delta Q}{T} \tag{A.8}$$

along reversible paths. The entropy is also only defined up to an additive constant.

For reversible processes the mechanical work δW has the general form of force times displacement. Thus for a gas or fluid one has

$$\delta W = -p\,\mathrm{d}V \tag{A.9}$$

if surface effects (surface tension) are neglected.

Second law of thermodynamics Not all processes that conserve energy occur in nature. The second law of thermodynamics states that only processes occur for which

$$\oint \frac{\delta Q}{T} \leq 0 \tag{A.10}$$

The equality sign holds if, and only if, the process is reversible. Equivalent formulations are

$$\mathrm{d}S \geq \frac{\delta Q}{T} \tag{A.11}$$

or

$$\mathrm{d}S = \mathrm{d}S_\mathrm{e} + \mathrm{d}S_\mathrm{i} \tag{A.12}$$

where

$$\mathrm{d}S_\mathrm{e} = \frac{\delta Q}{T} \tag{A.13}$$

and

$$\mathrm{d}S_\mathrm{i} \geq 0 \tag{A.14}$$

The entropy changes through heat exchange with its surroundings, the term $\mathrm{d}S_\mathrm{e}$, and through internal processes, the term $\mathrm{d}S_\mathrm{i}$. The internal processes never decrease the entropy. For reversible processes $\mathrm{d}S_\mathrm{i} = 0$ and for irreversible processes $\mathrm{d}S_\mathrm{i} > 0$.

Processes are called:

- isochoric when $dV = 0$;
- adiabatic when $\delta Q = 0$;
- isothermal when $dT = 0$;
- isentropic when $dS = 0$; and
- isobaric when $dp = 0$.

Isothermal processes occur in a system that is immersed in a "heat bath." Adiabatic processes occur in a system that is thermally isolated. Reversible adiabatic processes are isentropic. The second law implies that the entropy of a thermally isolated system never decreases.

A.3 Thermodynamic potentials

If every thermodynamic state can be reached by a reversible process then the first law of thermodynamics implies that the internal energy E can be expressed as a function of V and S

$$E = E(V, S) \tag{A.15}$$

and that

$$p = -\left(\frac{\partial E}{\partial V}\right)_S \tag{A.16}$$

$$T = \left(\frac{\partial E}{\partial S}\right)_V \tag{A.17}$$

The internal energy function $E = E(V, S)$ thus *completely* describes the thermodynamic properties of the system. Such a function is called a *thermodynamic potential*.

The above results can be generalized to systems that are open, i.e., systems that exchange mass with their surroundings, and that consist of N different components. Let M_j ($j = 1, \ldots, N$) be the masses of the N components and $M = \sum_{j=1}^{N} M_j$ the total mass. For reversible processes the first law then takes the form

$$dE = T\,dS - p\,dV + \sum_{j=1}^{N} \mu_j\,dM_j \tag{A.18}$$

where μ_j ($j = 1, \ldots, N$) is the *chemical potential* of the jth component. The chemical potential μ_j describes the changes in the internal energy when mass dM_j is added to the system holding the entropy, the volume, and the masses of the other components constant. Now the thermodynamic state is determined by the values of the $N + 2$ variables $V, S, M_1, \ldots, M_N$. The internal energy function

$$E = E(V, S, M_1, \ldots, M_N) \tag{A.19}$$

again serves as a thermodynamic potential with

$$p = -\left(\frac{\partial E}{\partial V}\right)_{S,M_1,\ldots,M_N} \tag{A.20}$$

$$T = \left(\frac{\partial E}{\partial S}\right)_{V,M_1,\ldots,M_N} \tag{A.21}$$

$$\mu_j = \left(\frac{\partial E}{\partial M_j}\right)_{V,S,M_1,\ldots,M_{j-1},M_{j+1},\ldots,M_N} \tag{A.22}$$

The internal energy function must be a homogeneous first order function

$$E(aS, aV, aM_1, \dots, aM_N) = aE(S, V, M_1, \dots, M_N) \tag{A.23}$$

Euler's theorem for homogeneous functions[1] then implies

$$E = S\frac{\partial E}{\partial S} + V\frac{\partial E}{\partial V} + \sum_{j=1}^{N} M_j \frac{\partial E}{\partial M_j} \tag{A.24}$$

or

$$E = TS - pV + \sum_{j=1}^{N} \mu_j M_j \tag{A.25}$$

which is an identity between the thermodynamic variables, called *Euler's identity*. The total differential of Euler's identity and the first law of thermodynamics imply the Gibbs–Durham relation

$$S\,\mathrm{d}T - V\,\mathrm{d}p + \sum_{j=1}^{N} M_j\,\mathrm{d}\mu_j = 0 \tag{A.26}$$

These relations can also be expressed in terms of densities or specific quantities. To convert to a description in terms of densities one introduces the mass densities

$$\rho_j := \frac{M_j}{V} \qquad j = 1, \dots, N \tag{A.27}$$

and the (total) mass density $\rho = \sum_{i=1}^{N} \rho_i$. The internal energy density can then be expressed as

$$\tilde{e} = \tilde{e}(\tilde{\eta}, \rho_1, \dots, \rho_N) \tag{A.28}$$

with

$$T = \left(\frac{\partial \tilde{e}}{\partial \tilde{\eta}}\right)_{\rho_1,\dots,\rho_N} \tag{A.29}$$

$$\mu_j = \left(\frac{\partial \tilde{e}}{\partial \rho_j}\right)_{\tilde{\eta},\rho_1,\dots,\rho_{j-1},\rho_{j+1},\dots,\rho_N} \qquad j = 1, \dots, N \tag{A.30}$$

The intensive state of an N-component system is thus determined by $N + 1$ variables. The pressure p follows from Euler's identity and is given by

$$p = T\tilde{\eta} - \tilde{e} + \sum_{j=1}^{N} \mu_j \rho_j \tag{A.31}$$

To convert to a description in terms of specific quantities one introduces the concentrations

$$c_j := \frac{M_j}{M} \qquad j = 1, \dots, N \tag{A.32}$$

[1] If $f(x, y)$ satisfies the identity $f(tx, ty) = t^n f(x, y)$ for a fixed n then f is called a homogeneous function of degree n. Homogeneous functions satisfy $x\partial f/\partial x + y\partial f/\partial y = nf(x, y)$.

The specific internal energy function is then

$$e = e(v, \eta, c_2, \ldots, c_N) \tag{A.33}$$

since $\sum_{j=1}^{N} c_j = 1$. The equations for p, T, and $\mu_1, \ldots, \mu_N$ are listed in Table A.1. Note that the differentials of the thermodynamic potentials only determine the chemical potential differences. The individual potentials are inferred from Euler's identity.

Instead of the independent variables $(v, \eta, c_2, \ldots, c_N)$ one can choose other sets of $N + 1$ independent variables. This leads to alternative *thermodynamic representations*. The most common alternative representations and associated thermodynamic potentials are

$$\begin{array}{lll} (p, \eta, c_2, \ldots, c_N) & h := e + pV & \text{specific enthalpy} \\ (T, v, c_2, \ldots, c_N) & f := e - T\eta & \text{specific free energy} \\ (p, T, c_2, \ldots, c_N) & g := e + pV - T\eta & \text{specific free enthalpy} \end{array} \tag{A.34}$$

The basic formulae for these alternative representations are also given in Table A.1.

Instead of densities and specific quantities one can also introduce *partial contributions*. For every extensive quantity $Z = Z(p, T, M_1, \ldots, M_N)$ they are defined by

$$z_j := \left(\frac{\partial Z}{\partial M_j}\right)_{p,T,M_1,\ldots,M_{j-1},M_{j+1},\ldots,M_N} \qquad j = 1, \ldots, N \tag{A.35}$$

and are intensive quantities. Euler's theorem implies

$$z = \frac{Z}{M} = \sum_{j=1}^{N} z_j c_j \tag{A.36}$$

Thus the quantity z_j is indeed the partial contribution of the jth component to z. Specifically, the chemical potential μ_j is the partial contribution of the jth component to the specific free enthalpy

$$g = \sum_{j=1}^{N} \mu_j c_j \tag{A.37}$$

For a two-component system ($N = 2$) the difference between the partial contributions is given by

$$\Delta z = z_2 - z_1 = \left(\frac{\partial z}{\partial c_2}\right)_{p,T} \tag{A.38}$$

where $z = z(p, T, c_2)$. Specifically,

$$\begin{array}{ll} \Delta\mu = \left(\frac{\partial g}{\partial c_2}\right)_{p,T} & \text{chemical potential difference} \\ \Delta h = \left(\frac{\partial h}{\partial c_2}\right)_{p,T} & \text{partial enthalpy difference} \\ \Delta\eta = \left(\frac{\partial \eta}{\partial c_2}\right)_{p,T} & \text{partial entropy difference} \\ \Delta v = \left(\frac{\partial v}{\partial c_2}\right)_{p,T} & \text{partial volume difference} \end{array} \tag{A.39}$$

Table A.1. *The four most common thermodynamic potentials for an N-component system and some of their properties*

The partial derivatives are always taken at constant values of the other variables of the respective representation. (κ = isothermal compressibility, α = thermal expansion coefficient, c_p = specific heat at constant pressure)

Independent variables	$v, \eta, c_2, \ldots, c_N$	$p, \eta, c_2, \ldots, c_N$	$v, T, c_2, \ldots, c_N$	$p, T, c_2, \ldots, c_N$
Thermodynamic potential	specific internal energy e	specific enthalpy $h = e + pv$	specific free energy $f = e - T\eta$	specific free enthalpy $g = e + pv - T\eta$
Dependent variables	$p = -\frac{\partial e}{\partial v}$ $T = \frac{\partial e}{\partial \eta}$ $\mu_j - \mu_1 = \frac{\partial e}{\partial c_j}$	$v = \frac{\partial h}{\partial p}$ $T = \frac{\partial h}{\partial \eta}$ $\mu_j - \mu_1 = \frac{\partial h}{\partial c_j}$	$p = -\frac{\partial f}{\partial v}$ $\eta = -\frac{\partial f}{\partial T}$ $\mu_j - \mu_1 = \frac{\partial f}{\partial c_j}$	$v = \frac{\partial g}{\partial p}$ $\eta = -\frac{\partial g}{\partial T}$ $\mu_j - \mu_1 = \frac{\partial g}{\partial c_j}$
Euler's identity	$\mu_1 = e - T\eta + pv$ $- \sum_{j=2}^{N}(\mu_j - \mu_1)c_j$	$\mu_1 = h - T\eta$ $- \sum_{j=2}^{N}(\mu_j - \mu_1)c_j$	$\mu_1 = f + pv$ $- \sum_{j=2}^{N}(\mu_j - \mu_1)c_j$	$\mu_1 = g$ $- \sum_{j=2}^{N}(\mu_j - \mu_1)c_j$
Additional variables		$\tilde{\kappa} := -\frac{1}{v}\frac{\partial v}{\partial p}$ adiabatic compressibility $\Gamma := \frac{\partial T}{\partial p}$ adiabatic temperature gradient	$c_v := T\frac{\partial \eta}{\partial T}$ specific heat at constant volume	$\kappa := -\frac{1}{v}\frac{\partial v}{\partial p}$ $\alpha := \frac{1}{v}\frac{\partial v}{\partial T}$ $c_p := T\frac{\partial \eta}{\partial T}$
Thermodynamic relations		$\tilde{\kappa} = \kappa - \alpha\Gamma$ $\Gamma = \frac{\alpha v T}{c_p}$	$c_v = c_p - \frac{\alpha^2 T v}{\kappa}$ $c_v = \frac{\partial e}{\partial T}$	$c_p = \frac{\partial h}{\partial T}$

A.4 Thermodynamic relations

Given the thermodynamic potentials for the various thermodynamic representations one can define additional quantities. In Table A.1 we listed the definitions of adiabatic compressibility $\tilde{\kappa}$, the adiabatic temperature gradient Γ, the specific heat at constant volume c_{v}, the isothermal compressibility κ, the thermal expansion coefficient α, and the specific heat at constant pressure c_{p}. Not all of these quantities are independent. There exist relations among them, called *thermodynamic relations*, that are valid for all thermodynamic systems.

One set of thermodynamic relations follows from the fact that the order of differentiation can be exchanged, $\partial^2 f(x,y)/\partial x \partial y = \partial^2 f(x,y)/\partial y \partial x$. Thus for example

$$\alpha = \frac{1}{V}\left(\frac{\partial V}{\partial T}\right)_p = \frac{1}{V}\frac{\partial^2 g}{\partial T \partial p} = \frac{1}{V}\left(\frac{\partial \eta}{\partial p}\right)_T \tag{A.40}$$

Another set of thermodynamic relations is obtained when derivatives defined in one representation are expressed in terms of derivatives in a different representation. Partial derivatives can be expressed as

$$\left(\frac{\partial f}{\partial x}\right)_y = \frac{\partial(f,y)}{\partial(x,y)} \tag{A.41}$$

where

$$\frac{\partial(f,g)}{\partial(x,y)} = \frac{\partial f}{\partial x}\frac{\partial g}{\partial y} - \frac{\partial f}{\partial y}\frac{\partial g}{\partial x} \tag{A.42}$$

is the Jacobian. The transformation of a derivative from the (x, y)-representation to the (t, s)-representation is thus governed by the transformation rules for Jacobians

$$\frac{\partial(f,g)}{\partial(x,y)} = \frac{\partial(f,g)}{\partial(t,s)}\frac{\partial(t,s)}{\partial(x,y)} \tag{A.43}$$

When $\tilde{\kappa}$ and Γ (defined in the (p, η)-representation) and c_{v} (defined in the (v, T)-representation) are expressed in the (p, T)-representation one obtains the thermodynamic relations

$$\tilde{\kappa} = \kappa - \frac{T\alpha^2 v}{c_{\mathrm{p}}} \tag{A.44}$$

$$\Gamma = \frac{\alpha T v}{c_{\mathrm{p}}} \tag{A.45}$$

$$c_V = c_p - \frac{T\alpha^2 v}{\kappa} \tag{A.46}$$

These relations imply

$$\frac{c_{\mathrm{p}}}{c_V} = \frac{\kappa}{\tilde{\kappa}} \tag{A.47}$$

Additional thermodynamic relations are listed in Table A.1.

A.5 Extremal principles

The second law of thermodynamics states that $\mathrm{d}S_\mathrm{i} \geq 0$ in $\mathrm{d}S = \mathrm{d}S_\mathrm{e} + \mathrm{d}S_\mathrm{i}$. For a closed one-component system $\mathrm{d}S_\mathrm{e} = \delta Q/T$ and $\delta Q = \mathrm{d}E + p\mathrm{d}V$. The generalizations to an open multi-component system are

$$\mathrm{d}S_\mathrm{e} = \frac{\delta Q}{T} + \sum_{j=1}^{N} \eta_j \mathrm{d}M_j \tag{A.48}$$

where η_j are the partial entropies and

$$\begin{aligned}\delta Q &= \mathrm{d}E + p\mathrm{d}V - \textstyle\sum_{j=1}^{N} h_j \mathrm{d}M_j \\ &= \left(\tfrac{\partial H}{\partial p}\right)_{T,M_1,\ldots,M_N} \mathrm{d}p - V\mathrm{d}p + \left(\tfrac{\partial H}{\partial T}\right)_{p,M_1,\ldots,M_N} \mathrm{d}T\end{aligned} \tag{A.49}$$

where h_j are the partial enthalpies. Note that for open systems the amount of heat transferred to the system is not defined uniquely. Here it is defined so that it vanishes when mass is added to the system at constant p and T. Independent of this ambiguity the condition $\mathrm{d}S_\mathrm{i} \geq 0$ becomes

$$T\mathrm{d}S \geq \mathrm{d}E + p\mathrm{d}V - \sum_{j=1}^{N} \mu_j \mathrm{d}M_j \tag{A.50}$$

where $\mu_j = h_j - T\eta_j$ are the chemical potentials (partial free enthalpies). The entropy thus increases for processes at constant energy, volume, and masses and will be a maximum in thermodynamic equilibrium. This extremal property can be used to infer the conditions for thermodynamic equilibrium.

The maximum of S implies $\delta S = 0$ for admissible variations. Admissible variations are implemented most easily when the system is divided into two subsystems of equal energy, volume, and mass. Then

$$\begin{aligned}\delta S &= S(E+\delta E, V+\delta V, M+\delta M) + S(E+\delta E', V+\delta V', M+\delta M') - S(2E, 2M, 2V) \\ &= \frac{\partial S}{\partial E}\delta E + \ldots + \frac{\partial S}{\partial M'}\delta M'\end{aligned} \tag{A.51}$$

Energy, mass, and volume conservation imply

$$\delta E' = -\delta E \quad \delta M' = -\delta M \quad \delta V' = -\delta V \tag{A.52}$$

Thus

$$\delta S = \left(\frac{1}{T} - \frac{1}{T'}\right)\delta E + \left(\frac{p}{T} - \frac{p'}{T'}\right)\delta V + \left(\frac{\mu}{T} - \frac{\mu'}{T'}\right)\delta M \tag{A.53}$$

for arbitrary variations δE, δV, and δM. Necessary conditions for thermodynamic equilibrium are thus

$$T = \text{constant} \quad p = \text{constant} \quad \mu = \text{constant} \tag{A.54}$$

representing the conditions for thermal, mechanical, and chemical equilibrium. When a multi-component system is considered, all chemical potentials μ_j are constant.

Other equivalent extremal principles follow from (A.50) by rearrangement and by changing to other representations. For example, V is a minimum at constant S, E, and M and the free enthalpy G is a minimum at constant p, T, and M.

A.6 Thermodynamic inequalities

In thermodynamic equilibrium the entropy S of the system is a maximum. This does not only imply $\delta S = 0$ but also $\delta^2 S \leq 0$ for admissible variations. These variations are again implemented most easily when the system is divided into two subsystems of equal energy, volume, and mass. Then

$$\begin{aligned}
&\delta S + \delta^2 S + \ldots \\
&\quad = S(E+\delta E, V+\delta V, M+\delta M) + S(E+\delta E', V+\delta V', M+\delta M') - S(2E, 2M, 2V) \\
&\quad = \frac{\partial S}{\partial E}\delta E + \ldots + \frac{\partial S}{\partial M'}\delta M' + \frac{1}{2}\frac{\partial^2 S}{\partial E^2}(\delta E)^2 + \ldots + \frac{\partial^2 S}{\partial M'^2}(\delta M')^2 \qquad \text{(A.55)}
\end{aligned}$$

Mass and volume conservation imply

$$\begin{aligned}
\delta M' &= -\delta M \\
\delta V' &= -\delta V
\end{aligned} \qquad \text{(A.56)}$$

The admissible variations for $\delta E'$ follow directly from the conservation of momentum and total energy. Conservation of momentum

$$(M+\delta M)\,\delta \mathbf{u} + (M+\delta M')\,\delta \mathbf{u}' = 0 \qquad \text{(A.57)}$$

implies

$$\delta \mathbf{u} = -\delta \mathbf{u}' \qquad \text{(A.58)}$$

Here we have assumed that the complete system does not have any macroscopic velocity though such velocities must be allowed for in the subsystems as admissible variations. Conservation of energy

$$2E = E + \delta E + E + \delta E' + \frac{1}{2}(M+\delta M)\,\delta \mathbf{u}\cdot\delta \mathbf{u} + \frac{1}{2}(M+\delta M')\,\delta \mathbf{u}'\cdot\delta \mathbf{u}' \qquad \text{(A.59)}$$

then implies

$$\delta E' = -\delta E - M\,\delta \mathbf{u}\cdot\delta \mathbf{u} \qquad \text{(A.60)}$$

Hence

$$\begin{aligned}
\delta^2 S = &-\frac{\partial S}{\partial E}M\delta \mathbf{u}\cdot\delta \mathbf{u} + \frac{\partial^2 S}{\partial E^2}(\delta E)^2 + 2\frac{\partial^2 S}{\partial E\partial V}\delta E\delta V + 2\frac{\partial^2 S}{\partial E\partial M}\delta E\delta M \\
&+ \frac{\partial^2 S}{\partial V^2}(\delta V)^2 + 2\frac{\partial^2 S}{\partial V\partial M}\delta V\delta M + \frac{\partial^2 S}{\partial M^2}(\delta M)^2
\end{aligned} \qquad \text{(A.61)}$$

This is a quadratic form in the arbitrary variations $\delta\mathbf{u}$, δE, δV, and δM. For this form to be non-positive the principal diagonal minors must satisfy

$$\Delta_1 \leq 0, \quad \Delta_2 \geq 0, \quad \Delta_3 \leq 0, \quad \Delta_4 \geq 0 \qquad \text{(A.62)}$$

The first three conditions become

$$\Delta_1 = -M\frac{\partial S}{\partial E} = -M\frac{1}{T} \leq 0 \qquad \text{(A.63)}$$

$$\frac{\Delta_2}{\Delta_1} = \frac{\partial(\partial S/\partial E)}{\partial E} = -\frac{1}{T^2}\frac{\partial T}{\partial E} = -\frac{1}{T^2}\frac{1}{c_\mathrm{v}} \leq 0 \qquad \text{(A.64)}$$

$$\frac{\Delta_3}{\Delta_1} = \frac{\partial(\partial S/\partial E, \partial S/\partial V)}{\partial(E,V)} = -\frac{1}{T^3}\frac{\partial(T,p)}{\partial(E,V)} = -\frac{1}{T^3}\frac{1}{c_\mathrm{v}}\left(\frac{\partial p}{\partial V}\right)_{T,M} \geq 0 \qquad \text{(A.65)}$$

and imply

$$
\begin{aligned}
T &\geq 0 \\
c_v &\geq 0 \\
\kappa &\geq 0
\end{aligned}
\tag{A.66}
$$

The temperature, the specific heat at constant volume, and the isothermal compressibility must all be non-negative. The thermodynamic relations of Section A.5 then imply

$$
\begin{aligned}
0 \leq c_v \leq c_p \\
0 \leq \tilde{\kappa} \leq \kappa
\end{aligned}
\tag{A.67}
$$

The fourth condition becomes

$$
\begin{aligned}
\frac{\Delta_4}{\Delta_1} &= \frac{\partial(\partial S/\partial E, \partial S/\partial V, \partial S/\partial M)}{\partial(E, V)} = -\frac{1}{T^4}\frac{\partial(T, p, \mu)}{\partial(E, V, M)} \\
&= -\frac{1}{T}\left(\frac{\partial \mu}{\partial M}\right)_{p,T}\frac{\Delta_3}{\Delta_1} \leq 0
\end{aligned}
\tag{A.68}
$$

For a one-component system the Gibbs–Durham relation implies $(\partial\mu/\partial M)_{p,T} = 0$ identically. For a two-component system the relation implies

$$
M\left(\frac{\partial \mu_2}{\partial M_2}\right)_{p,T} = \frac{\partial(\Delta\mu)}{\partial c_2} c_1 \tag{A.69}
$$

and therefore

$$
\frac{\partial(\Delta\mu)}{\partial c_2} \geq 0 \tag{A.70}
$$

The thermodynamic inequalities prohibit processes that violate the second law. If $T < 0$ then the entropy of a system at rest would increase by converting internal into kinetic energy. The second law requires that heat flows from the higher to the lower temperature reservoir. If $c_v < 0$ then this process would increase the temperature difference. Similarly, a pressure difference exerts a force that expands the high pressure reservoir and compresses the low pressure reservoir. If $\kappa < 0$ then this process would increase the pressure difference. Finally, if $\frac{\partial(\Delta\mu)}{\partial c_2} < 0$ then the entropy would increase by increasing concentration gradients. Overall, the thermodynamic inequalities insure that mixing increases the entropy, as seen in the next section.

A.7 Mixing

Consider the mixing of two subsystems "A" and "B." Consider first the case that the systems are mixed adiabatically ($\delta Q = 0$) and without change of volume ($\delta V = 0$ and hence $\delta A = 0$). The mixed system "C" then has:

- mass $M_C = M_A + M_B$;
- volume $V_C = V_A + V_B$; and
- internal energy $E_C = E_A + E_B$.

The internal energies simply add because no work is done and no heat is transferred to the system. When specific quantities are introduced one finds that specific volume and specific energy mix in mass proportion

$$\begin{aligned} v_{\mathrm{C}} &= c_{\mathrm{A}} v_{\mathrm{A}} + c_{\mathrm{B}} v_{\mathrm{B}} \\ e_{\mathrm{C}} &= c_{\mathrm{A}} e_{\mathrm{A}} + c_{\mathrm{B}} e_{\mathrm{B}} \end{aligned} \tag{A.71}$$

The mixing occurs along a straight line in (e, v)-space.

If mixing occurs at constant pressure then the first law of thermodynamics implies that the specific enthalpy mixes in mass proportion (see Figure 2.6)

$$h_{\mathrm{C}} = c_{\mathrm{A}} h_{\mathrm{A}} + c_{\mathrm{B}} h_{\mathrm{B}} \tag{A.72}$$

For any thermodynamic variable a the derivation from such conservative mixing in mass proportions is given by (see Figure 2.6)

$$\Delta a = a_{\mathrm{C}} - (c_{\mathrm{A}} a_{\mathrm{A}} + c_{\mathrm{B}} a_{\mathrm{B}}) \tag{A.73}$$

If the two initial states differ only slightly from each other then this deviation can be obtained by a Taylor expansion of the initial states about the final state. Correct to the second order, one finds

$$\begin{aligned} \Delta a &= -\tfrac{1}{2} c_{\mathrm{A}} c_{\mathrm{B}} \left(\frac{\partial^2 a}{\partial h^2}\right)_p (h_{\mathrm{B}} - h_{\mathrm{A}})^2 \\ &= -\tfrac{1}{2} c_{\mathrm{A}} c_{\mathrm{B}} \left\{ \left(\frac{\partial^2 a}{\partial T^2}\right)_p - \frac{\partial a}{\partial T} \frac{1}{c_{\mathrm{p}}} \left(\frac{\partial c_{\mathrm{p}}}{\partial T}\right)_p \right\} (T_{\mathrm{B}} - T_{\mathrm{A}})^2 \\ &= \tfrac{1}{2} c_{\mathrm{A}} c_{\mathrm{B}} \frac{1}{\left(\frac{\partial h}{\partial a}\right)_p} \left(\frac{\partial^2 h}{\partial a^2}\right)_p (a_{\mathrm{B}} - a_{\mathrm{A}})^2 \end{aligned} \tag{A.74}$$

depending on the choice of independent variables. Using the last formula for $a = T, v,$ and η one finds

$$\begin{aligned} \Delta T &= \tfrac{1}{2} c_{\mathrm{A}} c_{\mathrm{B}} \frac{1}{c_{\mathrm{p}}} \left(\frac{\partial c_{\mathrm{p}}}{\partial T}\right)_p (T_{\mathrm{B}} - T_{\mathrm{A}})^2 \\ \Delta v &= \tfrac{1}{2} c_{\mathrm{A}} c_{\mathrm{B}} \frac{1}{\left(\frac{\partial v}{\partial \eta}\right)_p^2} \left[\frac{1}{T} \left(\frac{\partial T}{\partial \eta}\right)_p \left(\frac{\partial v}{\partial \eta}\right)_p - \left(\frac{\partial^2 v}{\partial \eta^2}\right)_p \right] (v_{\mathrm{B}} - v_{\mathrm{A}})^2 \\ \Delta \eta &= \tfrac{1}{2} c_{\mathrm{A}} c_{\mathrm{B}} \frac{1}{c_{\mathrm{p}}} (\eta_{\mathrm{B}} - \eta_{\mathrm{A}})^2 \end{aligned} \tag{A.75}$$

The second law of thermodynamics then requires $c_{\mathrm{p}} \geq 0$.

A.8 Phase transitions

A thermodynamic system can exist in different phases. Two phases "I" and "II" of a one-component system are in thermal equilibrium when pressure, temperature, and chemical potential in the two phases are equal

$$\begin{aligned} T^{\mathrm{I}} &= T^{\mathrm{II}} \\ p^{\mathrm{I}} &= p^{\mathrm{II}} \\ \mu^{\mathrm{I}} &= \mu^{\mathrm{II}} \end{aligned} \tag{A.76}$$

If one chooses p and T as the independent variables then

$$\mu^{\mathrm{I}}(p, T) = \mu^{\mathrm{II}}(p, T) \tag{A.77}$$

which constitutes a relation between T and p

$$T = T(p) \tag{A.78}$$

The equilibrium temperature depends on pressure, or vice versa, the equilibrium pressure depends on temperature. If two phases are in equilibrium the state of the system is thus determined by the specification of one variable. This is a special case of the Gibbs phase rule. The temperature at which the liquid and vapor phase are in equilibrium is called the *boiling* temperature. The temperature at which the solid and liquid phase are in equilibrium is called the *freezing* temperature. Boiling and freezing temperatures depend on pressure.

The amount of enthalpy to liquify or vaporize a unit amount of mass is called the *specific heat* of liquification or vaporization and given by

$$L := h^{\mathrm{I}} - h^{\mathrm{II}} \tag{A.79}$$

where $h^{\mathrm{I}} = e^{\mathrm{I}} + pv^{\mathrm{I}}$ is the specific enthalpy of the liquid (vapor) phase and $h^{\mathrm{II}} = e^{\mathrm{II}} + pv^{\mathrm{II}}$ is the specific enthalpy of the solid (liquid) phase. The Gibbs–Durham relation and the condition $\mu^{\mathrm{I}} = \mu^{\mathrm{II}}$ imply

$$L = T(\eta^{\mathrm{I}} - \eta^{\mathrm{II}}) \tag{A.80}$$

The dependence of the boiling or freezing temperature on pressure can be obtained by differentiating the implicit relation $\mu^{\mathrm{I}}(p, T) = \mu^{\mathrm{II}}(p, T)$ with respect to p. One obtains

$$\frac{\partial \mu^{\mathrm{I}}}{\partial p} + \frac{\partial \mu^{\mathrm{I}}}{\partial T}\frac{\partial T}{\partial p} = \frac{\partial \mu^{\mathrm{II}}}{\partial p} + \frac{\partial \mu^{\mathrm{II}}}{\partial T}\frac{\partial T}{\partial p} \tag{A.81}$$

which can be rewritten as the Clausius–Clapeyron relation

$$\frac{\partial T}{\partial p} = \frac{v^{\mathrm{I}} - v^{\mathrm{II}}}{\eta^{\mathrm{I}} - \eta^{\mathrm{II}}} = \frac{T(v^{\mathrm{I}} - v^{\mathrm{II}})}{L} \tag{A.82}$$

upon substitution of $\partial\mu/\partial p = v$ and $\partial\mu/\partial T = -\eta$. Since $v^{\mathrm{I}} > v^{\mathrm{II}}$ for a transition from a liquid to a gas the boiling temperature increases with pressure.

A.9 Ideal gas

A particular simple thermodynamic system is the *ideal gas*. It is a system that arises in statistical mechanics. It consists of particles whose interaction potential can be neglected when calculating the energy, but must of course be present for the system to approach thermodynamic equilibrium. It is a good approximation for a monatomic gas with no internal degrees of freedom.

Basic formulae The specific free energy of an ideal gas is derived in statistical mechanics and is given by

$$f(n, T) = \frac{k_{\mathrm{B}}T}{m}\left(\log\frac{n}{n_\rho} - 1\right) \tag{A.83}$$

where

k_{B} = Boltzmann constant
m = mass of particle
$n = \frac{N}{V}$ = number density
$n_\rho = \left(\frac{mk_{\mathrm{B}}T}{2\pi\hbar^2}\right)^{3/2}$ = quantum concentration
$\hbar$ = Planck's constant

The dependence on the quantum concentration arises by applying the concepts of statistical mechanics to the quantum states of the system.

In a purely phenomenological approach the specific free energy of an ideal gas is given by

$$f(v,T) = -RT\left(1 + \log v + \frac{3}{2}\log T + \frac{3}{2}\log A\right) \tag{A.84}$$

where $R := k_B/m$ is the specific gas constant and $A := k_B m^{5/2}/2\pi\hbar^2$ a (irrelevant) constant. From the free energy one obtains

$$\begin{aligned}
p &= -\left(\tfrac{\partial f}{\partial v}\right)_T = \tfrac{RT}{v}\\
\eta &= -\left(\tfrac{\partial f}{\partial T}\right)_v = R\left(\tfrac{5}{2} + \log v + \tfrac{3}{2}\log T + \tfrac{3}{2}\log A\right)\\
\mu &= f + pv = -RT\left(\log v + \tfrac{3}{2}\log T + \tfrac{3}{2}\log A\right)\\
e &= f + T\eta = \tfrac{3}{2}RT\\
c_V &= T\left(\tfrac{\partial \eta}{\partial T}\right)_V = \tfrac{3}{2}R
\end{aligned} \tag{A.85}$$

The specific free enthalpy of an ideal gas is given by

$$g(p,T) = RT\left(\log p - \frac{5}{2}\log T - \frac{3}{2}\log A - \log R\right) \tag{A.86}$$

and hence

$$c_p = T\left(\frac{\partial \eta}{\partial T}\right)_p = \frac{5}{2}R \tag{A.87}$$

$$\alpha = \frac{1}{v}\left(\frac{\partial v}{\partial p}\right)_T = \frac{1}{T} \tag{A.88}$$

$$\kappa = -\frac{1}{v}\left(\frac{\partial v}{\partial p}\right)_T = \frac{1}{p} \tag{A.89}$$

Thermodynamic relations imply

$$\Gamma = \frac{\alpha T v}{c_p} = \frac{2}{5}\frac{T}{p} \tag{A.90}$$

$$\tilde{\kappa} = \kappa - \alpha\Gamma = \frac{3}{5}\frac{1}{p} \tag{A.91}$$

The characteristic features of an ideal gas are thus:

- the internal energy depends only on the temperature;
- the specific heats are constant; and
- the equation of state has the form $pv = RT$.

The above results are for a monatomic gas. They can be generalized to polyatomic gases. The equipartition theorem of statistical mechanics states that each degree of freedom contributes $\frac{1}{2}k_B T$ to the internal energy, hence $e = f\frac{1}{2}RT$ where f is the number of degrees of freedom. A monatomic gas has three degrees of freedom, hence $e = \frac{3}{2}RT$. A diatomic molecule has five degrees of freedom, three translational and two rotational ones. The rotation about the axis connecting the two atoms does not contribute because it has no moment of inertia. Hence $e = \frac{5}{2}RT$ for an ideal gas of diatomic molecules. For polyatomic molecules one has $e = 3RT$ since now all rotational degrees of freedom contribute. Diatomic and polyatomic molecules also have vibrational degrees of freedom. To what extent these and other degrees of freedom are excited at a given temperature can only be assessed by quantum mechanical considerations. In general, the specific free enthalpy is given by

$$g(p, T) = RT\,(\log p - (f/2 + 1)\log T) \tag{A.92}$$

except for a (irrelevant) linear function of T.

Mixture of ideal gases Consider two ideal gases in a volume V at temperature T. Each of the gases has mass M_i, free energy F_i, pressure p_i, internal energy E_i, and entropy S_i $(i = 1, 2)$. Since the molecules do not interact the free energy of the mixture of two gases is given by their sum

$$F(V, T, M_1, M_2) = F_1(V, T, M_1) + F_2(V, T, M_2) \tag{A.93}$$

Hence

$$p = -\frac{\partial F}{\partial V} = -\frac{\partial F_1}{\partial V} - \frac{\partial F_2}{\partial V} = p_1 + p_2 \tag{A.94}$$

$$S = -\frac{\partial F}{\partial T} = -\frac{\partial F_1}{\partial T} - \frac{\partial F_2}{\partial T} = S_1 + S_2 \tag{A.95}$$

$$E = F + TS = F_1 + F_2 + T(S_1 + S_2) = E_1 + E_2 \tag{A.96}$$

$$G = F + pV = F_1 + F_2 + (p_1 + p_2)V = G_1 + G_2 \tag{A.97}$$

The free energy, the pressure, the entropy, the internal energy, and the free enthalpy are all additive.

When expressed in terms of (p, T, c_1, c_2) the specific free enthalpy becomes the thermodynamic potential

$$g(p, T, c_i) = \sum_{i=1}^{2} c_i R_i T \left[\log p - \left(\frac{f_i}{2} + 1 \right) \log T - \log \frac{c_1 R_1 + c_2 R_2}{c_i R_i} \right] \tag{A.98}$$

from which all quantities can be calculated. An example is given next where air is treated as a mixture of two ideal gases.

Air For many purposes air can be considered as a mixture of two ideal gases: dry air and water vapor. Dry air (index "d") is regarded as a diatomic gas with $f_{\mathrm{d}} =$ five degrees of freedom. Water vapor (index "v") is regarded as a triatomic gas with $f_{\mathrm{v}} =$ six degrees of freedom. The individual equations of state are

$$p_{\mathrm{d,v}} = \rho_{\mathrm{d,v}} R_{\mathrm{d,v}} T \tag{A.99}$$

with specific gas constants $R_{\mathrm{d,v}} = k_{\mathrm{B}}/m_{\mathrm{d,v}}$ and masses $m_{\mathrm{d,v}}$. The individual specific heats are

$$c_{\mathrm{v}}^{\mathrm{d,v}} = \frac{f_{\mathrm{d,v}}}{2} R_{\mathrm{d,v}} \tag{A.100}$$

For a mixture of ideal gases the pressures and densities are additive

$$p = p_{\mathrm{d}} + p_{\mathrm{v}} \tag{A.101}$$

$$\rho = \rho_{\mathrm{d}} + \rho_{\mathrm{v}} \tag{A.102}$$

If one introduces the specific humidity

$$q := \frac{\rho_{\mathrm{v}}}{\rho} \tag{A.103}$$

then these relations imply the equation of state

$$\rho = \frac{p}{T[(1-q)R_{\mathrm{d}} + qR_{\mathrm{v}}]} \tag{A.104}$$

from which one finds $\alpha = 1/T$ and $\kappa = 1/p$. Instead of q one may alternatively introduce the mixing ratio

$$r := \frac{\rho_{\mathrm{v}}}{\rho_{\mathrm{d}}} = \frac{q}{1-q} \tag{A.105}$$

or the relative humidity

$$\frac{r}{r_{\mathrm{s}}} = \frac{q(1-q_{\mathrm{s}})}{q_{\mathrm{s}}(1-q)} \tag{A.106}$$

where $q_{\mathrm{s}}(p, T)$ is the saturation humidity.

These and other relations can be inferred from the specific free enthalpy

$$\begin{aligned} g(p,T,q) = (1-q)R_{\mathrm{d}}T &\left[\log p - \left(\tfrac{f_{\mathrm{d}}}{2}+1\right)\log T - \log \tfrac{(1-q)R_{\mathrm{d}}+qR_{\mathrm{v}}}{1-q}\right] \\ qR_{\mathrm{v}}T &\left[\log p - \left(\tfrac{f_{\mathrm{v}}}{2}+1\right)\log T - \log \tfrac{(1-q)R_{\mathrm{d}}+qR_{\mathrm{v}}}{q}\right] \end{aligned} \tag{A.107}$$

From it we recover the equation of state

$$v = \frac{\partial g}{\partial p} = [(1-q)R_{\mathrm{d}} + qR_{\mathrm{v}}]\frac{T}{p} \tag{A.108}$$

and obtain

$$c_{\mathrm{p}} = T\left(\frac{\partial \eta}{\partial T}\right)_p = (1-q)\left(\frac{f_{\mathrm{d}}}{2}+1\right)R_{\mathrm{d}} + q\left(\frac{f_{\mathrm{v}}}{2}+1\right)R_{\mathrm{v}} \tag{A.109}$$

$$c_{\mathrm{v}} = c_{\mathrm{p}} - \frac{T\alpha^2 v}{\kappa} = (1-q)\frac{f_{\mathrm{d}}}{2}R_{\mathrm{d}} + q\frac{f_{\mathrm{v}}}{2}R_{\mathrm{v}} \tag{A.110}$$

$$e = g - pv + T\eta = c_{\mathrm{v}}T \tag{A.111}$$

$$h = g + t\eta = c_{\mathrm{p}}T \tag{A.112}$$

$$\mu_{\mathrm{d}} = R_{\mathrm{d}}T\left[\log p - \left(\frac{f_{\mathrm{d}}}{2}+1\right)\log T - \log\frac{(1-q)R_{\mathrm{d}}+qR_{\mathrm{v}}}{1-q}\right] \tag{A.113}$$

$$\mu_{\mathrm{v}} = R_{\mathrm{v}}T\left[\log p - \left(\frac{f_{\mathrm{v}}}{2}+1\right)\log T - \log\frac{(1-q)R_{\mathrm{d}}+qR_{\mathrm{v}}}{q}\right] \tag{A.114}$$

Mixing of ideal gases Consider two ideal gases in two separate volumes V_1 and V_2 at pressure p and temperature T. Each gas has mass M_i, internal energy E_i, and entropy S_i

($i = 1, 2$). The equation of state is given for each gas by $pV_i = M_i R_i T$. When these two gases are allowed to mix adiabatically with no work being done then the mixture fills the volume $V = V_1 + V_2$. The first law of thermodynamics implies $\mathrm{d}E = 0$ and hence $\mathrm{d}T = 0$ since the gases are ideal. Furthermore, $\mathrm{d}p = 0$ since

$$p = \frac{(M_1 R_1 + M_2 R_2)T}{V_1 + V_2} \tag{A.115}$$

The internal energies are additive. The temperature and pressure do not change. The entropy increases, however. The mixture has entropy

$$S(V, T, M_1, M_2) = \sum_{i=1}^{2} M_i R_i \left(\frac{5}{2} + \log \frac{V}{M_i} + \frac{3}{2} \log T + \frac{3}{2} \log A_i \right) \tag{A.116}$$

When $V = (M_1 R_1 + M_2 R_2)T/p$ is substituted then

$$S(V, T, M_1, M_2) = \sum_{i=1}^{2} M_i R_i \left(\frac{5}{2} + \log \frac{R_i T}{p} + \frac{3}{2} \log T + \frac{3}{2} \log A_i \right) + \delta S \tag{A.117}$$

where

$$\delta S = \sum_{i=1}^{2} M_i R_i \log \frac{V}{V_i} \geq 0 \tag{A.118}$$

When two different ideal gases mix at constant pressure and temperature the entropy increases by an amount δS. This excess entropy is called the *entropy of mixing*. It is positive as required by the second law of thermodynamics.

If the molecules of the two gases were identical the entropy of the mixture is

$$\begin{aligned} S(V, T, M_1, M_2) &= (M_1 + M_2)R \left(\frac{5}{2} + \log \frac{RT}{p} + \frac{3}{2} \log T + \frac{3}{2} \log A \right) \\ &= S_1(p, T, M_1) + S_2(p, T, M_2) \end{aligned} \tag{A.119}$$

and is just the sum of the entropies of the two mixed gases.

Ideal gas in gravitational field Thermal equilibrium in a gravitational field ϕ requires $\mu + \phi =$ constant and $T =$ constant. For $\phi = gz$ and μ from (A.85) one finds

$$v(z) = v(0) \exp \left\{ \frac{gz}{RT} \right\} \tag{A.120}$$

The specific volume increases exponentially with height, with e-folding scale RT/g. Density and pressure decrease exponentially with height. The e-folding scale contains the specific gas constant and is thus different for different components.

Appendix B

Vector and tensor analysis

B.1 Scalars, vectors, and tensors

The quantities that describe a fluid are either scalars, vectors, or tensors. Vectors and tensors are determined by their components in a specified coordinate system. A rectangular coordinate system is called *Cartesian*.

The following conventions are adopted:

1. Vectors and tensors are denoted by **boldface** symbols.
2. Italic indices $i, j, \ldots$ run from 1 to 3 and denote the Cartesian components of a three-dimensional vector or tensor. Thus a_i $(i = 1, 2, 3)$ denote the three Cartesian components of a three-dimensional vector **a** and T_{ij} $(i, j = 1, 2, 3)$ denote the nine Cartesian components of a three-dimensional second-order tensor **T**.
3. An index that occurs once in an expression is called a *free index* and can take any value in its range. An index that occurs twice is called a *dummy index* and is summed over its values (summation convention, thus $a_i b_i = \sum_{i=1}^{3} a_i b_i$). An index cannot occur three times in an expression.

Note that the index notation is only applied to the Cartesian components of a vector or tensor. The components in non-Cartesian curvilinear coordinate systems are denoted by special symbols.

Special tensors

Kronecker tensor:

$$\delta_{ij} = \begin{cases} 1 & \text{if } i = j \\ 0 & \text{if } i \neq j \end{cases} \tag{B.1}$$

Permutation tensor:

$$\epsilon_{ijk} = \begin{cases} 1 & \text{if } (i, j, k) \text{ is an even permutation of } (1, 2, 3) \\ -1 & \text{if } (i, j, k) \text{ is an odd permutation of } (1, 2, 3) \\ 0 & \text{otherwise} \end{cases} \tag{B.2}$$

Rule:

$$\epsilon_{ijk}\epsilon_{imn} = \delta_{jm}\delta_{kn} - \delta_{jn}\delta_{km} \tag{B.3}$$

Products of vectors and tensors

Scalar product: The scalar or dot product $\mathbf{a} \cdot \mathbf{b}$ between two vectors $\mathbf{a}$ and $\mathbf{b}$ is a scalar and defined by

$$\mathbf{a} \cdot \mathbf{b} = a_i b_i \tag{B.4}$$

Two vectors $\mathbf{a}$ and $\mathbf{b}$ are *orthogonal* if $\mathbf{a} \cdot \mathbf{b} = 0$.

Vector product: The vector or cross product $\mathbf{a} \times \mathbf{b}$ between two vectors $\mathbf{a}$ and $\mathbf{b}$ is a vector with components

$$(\mathbf{a} \times \mathbf{b})_i = \epsilon_{ijk} a_j b_k \tag{B.5}$$

and is perpendicular to the plane spanned by the vectors $\mathbf{a}$ and $\mathbf{b}$. Two vectors $\mathbf{a}$ and $\mathbf{b}$ are *parallel* if $\mathbf{a} \times \mathbf{b} = 0$.

Every vector can be written as the sum of a vector that is parallel and a vector that is perpendicular to a given unit vector $\mathbf{e}$

$$\mathbf{a} = (\mathbf{a} \cdot \mathbf{e})\,\mathbf{e} + \mathbf{e} \times (\mathbf{a} \times \mathbf{e}) \tag{B.6}$$

Rule:

$$\mathbf{a} \times (\mathbf{b} \times \mathbf{c}) = (\mathbf{a} \cdot \mathbf{c})\,\mathbf{b} - (\mathbf{a} \cdot \mathbf{b})\,\mathbf{c} \tag{B.7}$$

Tensor product: The tensor product $\mathbf{ab}$ of two vectors is a tensor with components

$$(\mathbf{ab})_{ij} = a_i b_j \tag{B.8}$$

Similarly, one defines for tensors $\mathbf{A}$ and $\mathbf{B}$ and vectors $\mathbf{a}$

$$\begin{aligned}
(\mathbf{A} \cdot \mathbf{B})_{ik} &= A_{ij} B_{jk} \\
\mathbf{A} : \mathbf{B} &= A_{ij} B_{ji} \\
(\mathbf{AB})_{ijkl} &= A_{ij} B_{kl} \\
(\mathbf{a} \cdot \mathbf{A})_j &= a_i A_{ij} \\
(\mathbf{A} \cdot \mathbf{a})_i &= A_{ij} a_j
\end{aligned} \tag{B.9}$$

B.2 Calculus

Scalars $\phi(t)$, vectors $\mathbf{a}(t)$, and tensors $\mathbf{T}(t)$ that depend on a continuous variable t, usually the time, are called scalar, vector, and tensor functions. The derivative of a vector function is again a vector function with components

$$\left(\frac{\mathrm{d}\mathbf{a}}{\mathrm{d}t}\right)_i := \frac{\mathrm{d}}{\mathrm{d}t} a_i \tag{B.10}$$

Similarly, the integral of a vector function is a vector with components

$$\left(\int_{t_1}^{t_2} \mathrm{d}t\, \mathbf{a}(t)\right)_i := \int_{t_1}^{t_2} \mathrm{d}t\, a_i(t) \tag{B.11}$$

Similar rules apply to tensor functions.

Scalars $\phi(\mathbf{x})$, vectors $\mathbf{a}(\mathbf{x})$, and tensors $\mathbf{T}(\mathbf{x})$ that depend on a continuous vector variable $\mathbf{x}$, usually the position vector, are called scalar, vector, and tensor *fields*. The gradient operator

$$\boldsymbol{\nabla} := \left(\frac{\partial}{\partial x_1}, \frac{\partial}{\partial x_2}, \frac{\partial}{\partial x_3}\right) \tag{B.12}$$

can be applied to these fields. If there is no ambiguity about the independent variables one also writes $\partial_i = \partial/\partial x_i$. Specific operations are

1. The gradient $\boldsymbol{\nabla}\phi$ of a scalar field is a vector field with components

$$(\boldsymbol{\nabla}\phi)_i = \partial_i \phi \tag{B.13}$$

2. The divergence $\boldsymbol{\nabla} \cdot \mathbf{a}$ of a vector field $\mathbf{a}$ is a scalar field given by

$$\boldsymbol{\nabla} \cdot \mathbf{a} = \partial_i a_i \tag{B.14}$$

3. The curl $\boldsymbol{\nabla} \times \mathbf{a}$ of a vector field $\mathbf{a}$ is a vector field with components

$$(\boldsymbol{\nabla} \times \mathbf{a})_i = \epsilon_{ijk}\partial_j a_k = \epsilon_{ijk} a_{k,j} \tag{B.15}$$

4. The Laplacian $\Delta\phi = \boldsymbol{\nabla} \cdot \boldsymbol{\nabla}\phi$ of a scalar field is a scalar field given by

$$\Delta\phi = \partial_i \partial_i \phi \tag{B.16}$$

5. The Laplacian $\Delta\mathbf{a}$ of a vector field is defined by

$$\Delta\mathbf{a} = \boldsymbol{\nabla}(\boldsymbol{\nabla} \cdot \mathbf{a}) - \boldsymbol{\nabla} \times (\boldsymbol{\nabla} \times \mathbf{a}) \tag{B.17}$$

and is a vector field with components

$$(\Delta\mathbf{a})_i = \partial_j \partial_j a_i \tag{B.18}$$

6. The gradient $\boldsymbol{\nabla}\mathbf{a}$ of a vector field is a tensor field with components

$$(\boldsymbol{\nabla}\mathbf{a})_{ij} = \partial_j a_i = a_{i,j} \tag{B.19}$$

7. The divergence $\boldsymbol{\nabla} \cdot \mathbf{A}$ of a tensor field $\mathbf{A}$ is a vector field with components

$$(\boldsymbol{\nabla} \cdot \mathbf{A})_i = \partial_j A_{ij} = A_{ij,j} \tag{B.20}$$

Expressions for these differential operators in spherical coordinates are given in Appendix F.

Rules:

$$\boldsymbol{\nabla} \cdot (\phi\mathbf{a}) = \boldsymbol{\nabla}\phi \cdot \mathbf{a} + \phi\boldsymbol{\nabla} \cdot \mathbf{a} \tag{B.21}$$

$$\boldsymbol{\nabla} \times (\phi\mathbf{a}) = \boldsymbol{\nabla}\phi \times \mathbf{a} + \phi\boldsymbol{\nabla} \times \mathbf{a} \tag{B.22}$$

$$\boldsymbol{\nabla} \cdot (\mathbf{a} \times \mathbf{b}) = \mathbf{b} \cdot (\boldsymbol{\nabla} \times \mathbf{a}) - \mathbf{a} \cdot (\boldsymbol{\nabla} \times \mathbf{b}) \tag{B.23}$$

$$\boldsymbol{\nabla} \times (\mathbf{a} \times \mathbf{b}) = (\mathbf{b} \cdot \boldsymbol{\nabla})\mathbf{a} - \mathbf{b}(\boldsymbol{\nabla} \cdot \mathbf{a}) - (\mathbf{a} \cdot \boldsymbol{\nabla})\mathbf{b} + \mathbf{a}(\boldsymbol{\nabla} \cdot \mathbf{b}) \tag{B.24}$$

$$\boldsymbol{\nabla}(\mathbf{a} \cdot \mathbf{b}) = (\mathbf{b} \cdot \boldsymbol{\nabla})\mathbf{a} + (\mathbf{a} \cdot \boldsymbol{\nabla})\mathbf{b} + \mathbf{b} \times (\boldsymbol{\nabla} \times \mathbf{a}) + \mathbf{a} \times (\boldsymbol{\nabla} \times \mathbf{b}) \tag{B.25}$$

$$\boldsymbol{\nabla} \times (\boldsymbol{\nabla}\phi) = 0 \tag{B.26}$$

$$\boldsymbol{\nabla} \cdot (\boldsymbol{\nabla} \times \mathbf{a}) = 0 \tag{B.27}$$

Comma notation: Differentiation is also often indicated by a comma as in $\phi_{,i} := \partial_i \phi$, $a_{i,j} := \partial_j a_i$, and $A_{ij,j} := \partial_j A_{ij}$. This notation is not used in this book but the rationale behind some of the above definitions is indicated. The comma notation determines the order of tensor indices, as in (B.19). In comma notation, summation is always over adjacent indices, as in (B.15) and (B.20).

B.3 Two-dimensional fields

Two-dimensional vectors and tensors are also denoted by boldface symbols, often with a subscript h. Their Cartesian components are denoted by Greek indices $\alpha, \beta, \ldots$, which run from 1 to 2. The summation convention is also applied. Most of the vector and tensor algebra and calculus carries directly over from three to two dimensions. Specifically:

1. Gradient operator

$$\nabla_h = (\partial_1, \partial_2) \tag{B.28}$$

2. Gradient of scalar field

$$(\nabla_h \phi)_\alpha = \partial_\alpha \phi \tag{B.29}$$

3. Divergence of vector field

$$\nabla_h \cdot \mathbf{a}_h = \partial_\alpha a_\alpha \tag{B.30}$$

4. Laplace operator

$$\Delta_h = \nabla_h \cdot \nabla_h \tag{B.31}$$

5. Permutation tensor

$$\epsilon_{\alpha\beta} = \begin{cases} 1 & \text{if} \quad (\alpha, \beta) = (1, 2) \\ -1 & \text{if} \quad (\alpha, \beta) = (2, 1) \\ 0 & \text{otherwise} \end{cases} \tag{B.32}$$

For each vector $\mathbf{a}_h$, we introduce the vector

$$\hat{\mathbf{a}}_h := (-a_2, a_1) \tag{B.33}$$

which is the vector obtained by rotating the vector $\mathbf{a}_h$ counterclockwise by $\pi/2$. In vector notation $\hat{\mathbf{a}}_h = (\mathbf{e}_3 \times \mathbf{a})_h$. Its components can also be written as

$$\hat{a}_\alpha = -\epsilon_{\alpha\beta} a_\beta \tag{B.34}$$

Rules:

$$\hat{\mathbf{a}}_h \cdot \mathbf{a}_h = \mathbf{a}_h \cdot \hat{\mathbf{a}}_h = 0 \tag{B.35}$$

$$\hat{\nabla}_h \cdot \nabla_h \phi = \nabla_h \cdot \hat{\nabla}_h \phi = 0 \tag{B.36}$$

$$\hat{\hat{\mathbf{a}}}_h = -\mathbf{a}_h \tag{B.37}$$

$$\hat{\mathbf{a}}_h \cdot \mathbf{b}_h = -\mathbf{a}_h \cdot \hat{\mathbf{b}}_h \tag{B.38}$$

$$\hat{\nabla}_h \cdot \mathbf{a}_h = -\nabla_h \cdot \hat{\mathbf{a}}_h \tag{B.39}$$

Note that $\hat{\nabla}_h \cdot \mathbf{a}_h = (\nabla \times \mathbf{a})_3$ is the "vertical" component of the curl.

B.4 Differential geometry

Curves A vector function $\mathbf{x}(s)$ with $s \in [a, b]$ defines a curve in position space. The curve is called *simple* if it does not cross itself. The curve is called closed if $\mathbf{x}(a) = \mathbf{x}(b)$. Further definitions are:

1. Differential of oriented arc length

$$\mathrm{d}\mathbf{x} = \frac{\mathrm{d}\mathbf{x}}{\mathrm{d}s}\,\mathrm{d}s \tag{B.40}$$

2. Differential of arc length

$$\mathrm{d}x = \left|\frac{\mathrm{d}\mathbf{x}}{\mathrm{d}s}\right|\,\mathrm{d}s \tag{B.41}$$

3. Unit tangent vector

$$\mathbf{t} = \frac{\frac{\mathrm{d}\mathbf{x}}{\mathrm{d}s}}{\left|\frac{\mathrm{d}\mathbf{x}}{\mathrm{d}s}\right|} \tag{B.42}$$

   These definitions imply $\mathrm{d}\mathbf{x} = \mathbf{t}\mathrm{d}x$.

4. Curvature vector

$$\mathbf{c} = \frac{\frac{\mathrm{d}\mathbf{t}}{\mathrm{d}s}}{\left|\frac{\mathrm{d}\mathbf{x}}{\mathrm{d}s}\right|} \tag{B.43}$$

   where $|\mathbf{c}|$ is called the *curvature* and $|\mathbf{c}|^{-1}$ is the radius of curvature.

Often the arclength is introduced as the parameter s.

Surfaces A vector function $\mathbf{x} = \mathbf{x}(r, s)$ with $(r, s) \in A \subset \mathbf{R}^2$ defines a surface S in position space. Surfaces can have one or two sides. A surface is called *closed* if it lies within a bounded region of space and has an inside and an outside, as for example a sphere. An open surface that is bounded by a closed curve and has two sides is called a *cap* of the curve, as for example a hemisphere with the equator as the boundary.

Further definitions:

1. Differential of oriented surface area

$$\mathrm{d}^2\mathbf{x} = \frac{\partial\mathbf{x}}{\partial r} \times \frac{\partial\mathbf{x}}{\partial s}\,\mathrm{d}r\,\mathrm{d}s \tag{B.44}$$

2. Differential of surface area

$$\mathrm{d}^2x = \left|\frac{\partial\mathbf{x}}{\partial r} \times \frac{\partial\mathbf{x}}{\partial s}\right|\,\mathrm{d}r\,\mathrm{d}s \tag{B.45}$$

3. Unit normal vector

$$\mathbf{n} = \frac{\frac{\partial\mathbf{x}}{\partial r} \times \frac{\partial\mathbf{x}}{\partial s}}{\left|\frac{\partial\mathbf{x}}{\partial r} \times \frac{\partial\mathbf{x}}{\partial s}\right|} \tag{B.46}$$

   These definitions imply $\mathrm{d}^2\mathbf{x} = \mathbf{n}\mathrm{d}^2x$.

Instead of the parametric definition $\mathbf{x} = \mathbf{x}(r, s)$ a surface can also be defined by a scalar function $G(\mathbf{x}) = 0$. The normal vector is then given by

$$\mathbf{n} = \frac{\boldsymbol{\nabla} G}{|\boldsymbol{\nabla} G|} \tag{B.47}$$

Integrals *Line integrals* along a curve C are defined by

$$\int_C \mathrm{d}\mathbf{x}\, F(\mathbf{x}) = \int_a^b \mathrm{d}s\, \frac{\mathrm{d}\mathbf{x}}{\mathrm{d}s}\, F(\mathbf{x}(s)) \tag{B.48}$$

where F can be a scalar, vector, or tensor field. Often $\mathrm{d}\mathbf{x} = \mathrm{d}x\mathbf{t}$ is introduced on the left-hand side. In general, a line integral has a value that depends on the particular path joining the points $\mathbf{x}(a)$ and $\mathbf{x}(b)$. However, when $F = \boldsymbol{\nabla}\phi$ then the line integral is independent of the path

$$\int_C \mathrm{d}\mathbf{x} \cdot \boldsymbol{\nabla}\phi = \phi(\mathbf{x}(b)) - \phi(\mathbf{x}(a)) \tag{B.49}$$

In particular, the line integral along a closed curve vanishes

$$\oint_C \mathrm{d}\mathbf{x} \cdot \boldsymbol{\nabla}\phi = 0 \tag{B.50}$$

Surface integrals over a surface S are defined by

$$\int\!\!\int_S \mathrm{d}^2\mathbf{x}\, F(\mathbf{x}) = \int\!\!\int_A \mathrm{d}r\, \mathrm{d}s\, \frac{\mathrm{d}\mathbf{x}}{\mathrm{d}r} \times \frac{\mathrm{d}\mathbf{x}}{\mathrm{d}s}\, F(\mathbf{x}(r,s)) \tag{B.51}$$

Often $\mathrm{d}^2\mathbf{x} = \mathrm{d}^2x\mathbf{n}$ is introduced on the left-hand side.

Integral theorems

1. Let δV be a closed surface with outward normal vector $\mathbf{n}$ enclosing a volume V then

$$\int\!\!\int\!\!\int_V \mathrm{d}^3x\, \boldsymbol{\nabla}F = \int\!\!\int_{\delta V} \mathrm{d}^2x\;\; \mathbf{n}F \tag{B.52}$$

Special cases are Gauss' theorem

$$\int\!\!\int\!\!\int_V \mathrm{d}^3x\; \boldsymbol{\nabla}\cdot\mathbf{b} = \int\!\!\int_{\delta V} \mathrm{d}^2x\;\; \mathbf{n}\cdot\mathbf{b} \tag{B.53}$$

and

$$\int\!\!\int\!\!\int_V \mathrm{d}^3x\; \boldsymbol{\nabla}\times\mathbf{b} = \int\!\!\int_S \mathrm{d}^2x\;\; \mathbf{n}\times\mathbf{b} \tag{B.54}$$

2. Stokes' theorem. Let C be a simple closed curve and S a cap of C then

$$\oint_C \mathrm{d}\mathbf{x}\cdot\mathbf{b} = \int\!\!\int_S \mathrm{d}^2x\;\; \mathbf{n}\cdot(\boldsymbol{\nabla}\times\mathbf{b}) \tag{B.55}$$

where the curve C is traversed in a counterclockwise direction when looking down onto the surface with upward normal vector $\mathbf{n}$.

B.5 Reynolds transport theorem

The curves, surfaces, volumes, and fields F in the above formulae may also depend on time t. In this case an important generalization of Leibnitz' theorem in calculus is

Reynolds' transport theorem. It states

$$\frac{\mathrm{d}}{\mathrm{d}t}\iiint_{V(t)} \mathrm{d}^3x\, F(\mathbf{x},t) = \iiint_{V(t)} \mathrm{d}^3x\, \partial_t F(\mathbf{x},t) + \iint_{\delta V(t)} \mathrm{d}^2x\, \mathbf{n}\cdot\mathbf{v}\, F(\mathbf{x},t) \tag{B.56}$$

for any scalar, vector, or tensor field $F(\mathbf{x}, t)$. Here V is a volume enclosed by a surface δV with outward normal vector $\mathbf{n}$ and velocity $\mathbf{v}$. The velocity of a surface is defined by

$$\mathbf{v} := \mathbf{n}\cdot\frac{\partial\mathbf{x}}{\partial t}\,\mathbf{n} \tag{B.57}$$

if the surface is given by $\mathbf{x} = \mathbf{x}(r, s, t)$ or by

$$\mathbf{v} = -\frac{\partial G}{\partial t}\,\frac{\mathbf{n}}{|\nabla G|} \tag{B.58}$$

if the surface is given by $G(\mathbf{x}, t) = 0$. For a fixed volume $\mathbf{v} = 0$ and the second term vanishes.

B.6 Isotropic and axisymmetric tensors

A tensor is said to be *isotropic* if its components do not change under:

- rotations about arbitrary axes through a point $\mathbf{x}$; and
- mirror reflections at arbitrary planes through $\mathbf{x}$.

Isotropic tensors are of the form

$$A_i = 0 \tag{B.59}$$

$$A_{ij} = a\,\delta_{ij} \tag{B.60}$$

$$A_{ijk} = 0 \tag{B.61}$$

$$A_{ijkl} = a_1\,\delta_{ij}\delta_{kl} + a_2\,\delta_{ik}\delta_{jl} + a_3\,\delta_{il}\delta_{jk} \tag{B.62}$$

where a_1, a_2, a_3 are arbitrary scalars.

A tensor is said to be *axisymmetric* if its components do not change under:

- rotations about an axis $\mathbf{k}$;
- mirror reflections at a plane containing $\mathbf{k}$; and
- mirror reflections at planes perpendicular to $\mathbf{k}$.

Axisymmetric tensors are of the form

$$A_i = 0 \tag{B.63}$$

$$A_{ij} = a_1\,\delta_{ij} + a_2 k_i k_j \tag{B.64}$$

$$A_{ijk} = 0 \tag{B.65}$$

$$\begin{aligned} A_{ijkl} = {} & b_1\,\delta_{ij}\delta_{kl} + b_2\,\delta_{ik}\delta_{jl} + b_3\,\delta_{il}\delta_{jk} + b_4\,\delta_{ij}k_k k_l + b_5\,\delta_{ik}k_j k_l + b_6\,\delta_{il}k_j k_k \\ & + b_7\,\delta_{jk}k_i k_l + b_8\,\delta_{jl}k_i k_k + b_9\,\delta_{kl}k_i k_j + b_{10}\,k_i k_j k_k k_l \end{aligned} \tag{B.66}$$

where $a_1, a_2, b_1, \ldots, b_{10}$ are arbitrary scalars.

The above formulae can be established by considering the scalar function

$$A(\mathbf{a}, \mathbf{b}, \cdots, \mathbf{c}) = A_{ij\ldots n} a_i b_j \cdots c_n \tag{B.67}$$

for arbitrary vectors **a**, **b**, ..., **c**. Then

$$A_{ij...n} = A\left(\mathbf{e}^{(i)}, \mathbf{e}^{(j)}, \cdots, \mathbf{e}^{(n)}\right) \tag{B.68}$$

where $\mathbf{e}^{(m)}$ $(m = 1, 2, 3)$ is the mth unit vector. It is a rigorous result of group theory that if A is invariant under a certain transformation group then A can be expressed in terms of the fundamental invariants of that group. In the case of isotropy the fundamental invariants are the scalar products $\mathbf{a} \cdot \mathbf{a}$, $\mathbf{a} \cdot \mathbf{b}$, etc., which are the lengths of the vectors and the angles between them. The triple products $\mathbf{a} \cdot [\mathbf{b} \times \mathbf{c}]$ are also invariant under rotation but not under reflection where they change sign. Hence

$$A(\mathbf{a}, \mathbf{b}, \cdots, \mathbf{c}) = A(\mathbf{a} \cdot \mathbf{a}, \mathbf{a} \cdot \mathbf{b}, \cdots, \mathbf{c} \cdot \mathbf{c}) \tag{B.69}$$

Since A is linear in the components of each of the vectors **a**, **b**,...,**c** it must be the sum of terms where each of the components $a_i, b_j, \ldots, c_n$ only occurs once. These requirements lead to the forms (B.59) to (B.62). In the case of axisymmetric tensors the fundamental invariants are $\mathbf{a} \cdot \mathbf{a}$, $\mathbf{a} \cdot \mathbf{b}$, ... and $\mathbf{k} \cdot \mathbf{a}$, $\mathbf{k} \cdot \mathbf{b}$, etc., where $\mathbf{k}$ is the unit vector of the axis of symmetry.

Appendix C

Orthogonal curvilinear coordinate systems

This appendix lists the most common differential operators in *orthogonal* curvilinear coordinate systems (q_1, q_2, q_3) with scale factors (h_1, h_2, h_3). It then specifies these general formulae for:

- spherical coordinates (φ, θ, r) with $(h_\varphi = r\cos\theta, h_\theta = r, h_r = 1)$;
- pseudo-spherical coordinates (φ, θ, z) with $(h_\varphi = r_0\cos\theta, h_\theta = r_0, h_z = 1)$; and
- two-dimensional spherical coordinates (φ, θ) with $(h_\varphi = r_0\cos\theta, h_\theta = r_0)$.

C.1 Curvilinear coordinate systems

Any point $\boldsymbol{x}$ can be represented by its Cartesian coordinates (x_1, x_2, x_3) or by curvilinear coordinates (q_1, q_2, q_3). The transformation from one set to the other is given by

$$\begin{aligned} x_1 &= x_1(q_1, q_2, q_3) \\ x_2 &= x_2(q_1, q_2, q_3) \\ x_3 &= x_3(q_1, q_2, q_3) \end{aligned} \tag{C.1}$$

As q_1 varies, while q_2 and q_3 are held constant, the position vector $\mathbf{x} = x_1\mathbf{i} + x_2\mathbf{j} + x_3\mathbf{k}$ describes a curve called *the* q_1 *coordinate*. The q_2 and q_3 coordinates are defined similarly. The vectors $\partial\mathbf{x}/\partial q_1$, $\partial\mathbf{x}/\partial q_2$, and $\partial\mathbf{x}/\partial q_3$ represent tangent vectors to the (q_1, q_2, q_3) coordinates. If one introduces the unit tangent vectors $\mathbf{e}_1$, $\mathbf{e}_2$, and $\mathbf{e}_3$ as basis vectors then

$$\frac{\partial\mathbf{x}}{\partial q_1} = h_1\mathbf{e}_1, \quad \frac{\partial\mathbf{x}}{\partial q_2} = h_2\mathbf{e}_2, \quad \frac{\partial\mathbf{x}}{\partial q_3} = h_2\mathbf{e}_3 \tag{C.2}$$

where

$$h_1 = \left|\frac{\partial\mathbf{x}}{\partial q_1}\right|, \quad h_2 = \left|\frac{\partial\mathbf{x}}{\partial q_2}\right|, \quad h_3 = \left|\frac{\partial\mathbf{x}}{\partial q_3}\right| \tag{C.3}$$

are called the *scale factors* or *Lamé* coefficients. Any vector $\boldsymbol{a}$ can be written

$$\boldsymbol{a} = a_1\boldsymbol{e}_1 + a_2\boldsymbol{e}_2 + a_3\boldsymbol{e}_3 \tag{C.4}$$

where $a_i = \boldsymbol{a} \cdot \boldsymbol{e}$ are the components of the vector in the curvilinear coordinate system. If $\mathbf{e}_1$, $\mathbf{e}_2$, and $\mathbf{e}_3$ are mutually orthogonal then the curvilinear coordinate system is called *orthogonal*. Alternatively, one can introduce coordinate surfaces $q_i =$ constant and their normal vectors $\hat{\mathbf{e}}_i = \nabla q_i / |\nabla q_i|$. The basis vectors $\hat{\mathbf{e}}_i$ are identical to $\mathbf{e}_i$ if the (q_1, q_2, q_3)-coordinate system is orthogonal. In this appendix we consider only such orthogonal coordinate systems.

C.2 Common differential operators

Differentials:

$$\begin{aligned} \mathrm{d}\mathbf{x} &= h_1 \mathrm{d}q_1 \mathbf{e}_1 + h_2 \mathrm{d}q_2 \mathbf{e}_2 + h_3 \mathrm{d}q_3 \mathbf{e}_3 \\ \mathrm{d}s^2 &= h_1^2 \mathrm{d}q_1^2 + h_2^2 \mathrm{d}q_2^2 + h_3^2 \mathrm{d}q_3^2 \\ \mathrm{d}V &= h_1 h_2 h_3 \, \mathrm{d}q_1 \mathrm{d}q_2 \mathrm{d}q_3 \end{aligned} \tag{C.5}$$

Velocity:

$$\boldsymbol{u} := \frac{\mathrm{d}\boldsymbol{x}}{\mathrm{d}t} = \sum_{i=1}^{3} h_i \dot{q}_i \boldsymbol{e}_i \tag{C.6}$$

Acceleration:

$$\boldsymbol{a} := \frac{\mathrm{d}^2 \boldsymbol{x}}{\mathrm{d}t^2} = \sum_{i,j=1}^{3} \frac{\partial h_i}{\partial q_j} \dot{q}_j \dot{q}_i \boldsymbol{e}_i + \sum_{i=1}^{3} h_i \ddot{q}_i \boldsymbol{e}_i + \sum_{i,j=1}^{3} h_i \dot{q}_i \frac{\partial \boldsymbol{e}_i}{\partial q_j} \dot{q}_j \tag{C.7}$$

Gradient of a scalar field:

$$(\nabla \phi)_i = \frac{1}{h_i} \frac{\partial \phi}{\partial q_i} \tag{C.8}$$

Divergence of a vector field:

$$\nabla \cdot \mathbf{a} = \frac{1}{h_1 h_2 h_3} \left\{ \frac{\partial (h_2 h_3 a_1)}{\partial q_1} + \frac{\partial (h_1 h_3 a_2)}{\partial q_2} + \frac{\partial (h_1 h_2 a_3)}{\partial q_3} \right\} \tag{C.9}$$

Curl of a vector field:

$$\begin{aligned} (\nabla \times \mathbf{a})_1 &= \frac{1}{h_2 h_3} \left\{ \frac{\partial (h_3 a_3)}{\partial q_2} - \frac{\partial (h_2 a_2)}{\partial q_3} \right\} \\ (\nabla \times \mathbf{a})_2 &= \frac{1}{h_1 h_3} \left\{ \frac{\partial (h_1 a_1)}{\partial q_3} - \frac{\partial (h_3 a_3)}{\partial q_1} \right\} \\ (\nabla \times \mathbf{a})_3 &= \frac{1}{h_1 h_2} \left\{ \frac{\partial (h_2 a_2)}{\partial q_1} - \frac{\partial (h_1 a_1)}{\partial q_2} \right\} \end{aligned} \tag{C.10}$$

Gradient of a vector field:

$$(\mathbf{b} \cdot \nabla \mathbf{a})_i = \sum_{k=1}^{3} \left(\frac{b_k}{h_k} \frac{\partial a_i}{\partial q_k} - \frac{b_k}{h_i} \frac{a_k}{h_k} \frac{\partial h_k}{\partial q_i} + \frac{b_i}{h_i} \frac{a_k}{h_k} \frac{\partial h_i}{\partial q_k} \right) \tag{C.11}$$

Divergence of a tensor field:

$$(\nabla \cdot R)_i = \frac{1}{h_i} \sum_{k=1}^{3} \left\{ \frac{1}{h_1 h_2 h_3} \frac{\partial}{\partial q_k} \left(\frac{h_1 h_2 h_3 h_i}{h_k} R_{ik} \right) - R_{kk} \frac{1}{h_k} \frac{\partial h_k}{\partial q_i} \right\} \tag{C.12}$$

Laplacian of a scalar field:

$$\Delta \varphi = \frac{1}{h_1 h_2 h_3} \left\{ \frac{\partial}{\partial q_1} \left(\frac{h_2 h_3}{h_1} \frac{\partial \varphi}{\partial q_1} \right) + \frac{\partial}{\partial q_2} \left(\frac{h_1 h_3}{h_2} \frac{\partial \varphi}{\partial q_2} \right) + \frac{\partial}{\partial q_3} \left(\frac{h_1 h_2}{h_3} \frac{\partial \varphi}{\partial q_3} \right) \right\} \tag{C.13}$$

Rate of strain:

$$D_{ij} = \frac{1}{2} \left\{ \frac{h_i}{h_j} \frac{\partial}{\partial q_j} \left(\frac{u_i}{h_i} \right) + \frac{h_j}{h_i} \frac{\partial}{\partial q_i} \left(\frac{u_j}{h_j} \right) \right\} \tag{C.14}$$

for $i \neq j$ and

$$D_{ii} = \frac{\partial}{\partial q_i} \left(\frac{u_i}{h_i} \right) + \sum_{k=1}^{3} \frac{u_k}{h_k} \frac{1}{h_i} \frac{\partial h_i}{\partial q_k} \tag{C.15}$$

for $i = j$ (no summation convention).

C.3 Spherical coordinates

Spherical coordinates (φ, θ, r) with longitude φ, latitude θ, and radial distance r have scale factors $(h_\varphi = r \cos\theta, h_\theta = r, h_r = 1)$.

Velocity:

$$\begin{aligned} u_\varphi &=: u = r \cos\theta \dot{\varphi} \\ u_\theta &=: v = r \dot{\theta} \\ u_r &=: w = \dot{r} \end{aligned} \tag{C.16}$$

Acceleration:

$$\begin{aligned} a_\varphi &= r \cos\theta \ddot{\varphi} + 2\dot{\varphi}(\dot{r} \cos\theta - r \sin\theta \dot{\theta}) \\ a_\theta &= 2\dot{r}\dot{\theta} + r\ddot{\theta} + r \sin\theta \cos\theta \dot{\varphi}^2 \\ a_r &= \ddot{r} - r\dot{\theta}^2 - r \cos^2\theta \dot{\varphi}^2 \end{aligned} \tag{C.17}$$

C.4 Pseudo-spherical coordinates

Pseudo-spherical coordinates (φ, θ, z) with longitude φ, latitude θ, and height z have scale factors $(h_\varphi = r_0 \cos\theta, h_\theta = r_0, h_z = 1)$.

Gradient of a scalar field:

$$(\nabla \phi)_\varphi = \frac{1}{r_0 \cos\theta} \frac{\partial \phi}{\partial \varphi} \tag{C.18}$$

$$(\nabla \phi)_\theta = \frac{1}{r_0} \frac{\partial \phi}{\partial \theta} \tag{C.19}$$

$$(\nabla \phi)_z = \frac{\partial \phi}{\partial z} \tag{C.20}$$

Divergence of a vector field:

$$\nabla \cdot \mathbf{a} = \frac{1}{r_0 \cos\theta}\frac{\partial a_\varphi}{\partial \varphi} + \frac{1}{r_0 \cos\theta}\frac{\partial}{\partial \theta}(a_\theta \cos\theta) + \frac{\partial}{\partial z} a_z \tag{C.21}$$

Curl of a vector field:

$$(\nabla \times \mathbf{a})_\varphi = \frac{1}{r_0}\frac{\partial a_z}{\partial \theta} - \frac{\partial a_\theta}{\partial z} \tag{C.22}$$

$$(\nabla \times \mathbf{a})_\theta = \frac{\partial a_\varphi}{\partial z} - \frac{1}{r_0 \cos\theta}\frac{\partial a_z}{\partial \theta} \tag{C.23}$$

$$(\nabla \times \mathbf{a})_z = \frac{1}{r_0 \cos\theta}\frac{\partial a_\theta}{\partial \varphi} - \frac{1}{r_0 \cos\theta}\frac{\partial (a_\varphi \cos\theta)}{\partial \theta} \tag{C.24}$$

Gradient of a vector field:

$$(\mathbf{b} \cdot \nabla \mathbf{a})_\varphi = \frac{b_\varphi}{r_0 \cos\theta}\frac{\partial}{\partial \varphi} a_\varphi + \frac{b_\theta}{r_0}\frac{\partial}{\partial \theta} a_\theta + b_z \frac{\partial a_\theta}{\partial z} + \frac{b_\varphi a_\varphi}{r_0} \tan\theta \tag{C.25}$$

$$(\mathbf{b} \cdot \nabla \mathbf{a})_z = \frac{b_\varphi}{r_0 \cos\theta}\frac{\partial}{\partial \varphi} a_z + \frac{b_\theta}{r_0}\frac{\partial}{\partial \theta} a_z + b_z \frac{\partial a_z}{\partial z} \tag{C.26}$$

Divergence of a tensor field:

$$(\nabla \cdot \mathbf{R})_\varphi = \frac{1}{r_0 \cos\theta}\frac{\partial R_{\varphi\varphi}}{\partial \varphi} + \frac{1}{r_0 \cos^2\theta}\frac{\partial}{\partial \theta}\left(\cos^2\theta R_{\varphi\theta}\right) + \frac{\partial}{\partial z} R_{\varphi z} \tag{C.27}$$

$$(\nabla \cdot \mathbf{R})_\theta = \frac{1}{r_0 \cos\theta}\frac{\partial R_{\theta\varphi}}{\partial \varphi} + \frac{\tan\theta}{r_0} R_{\varphi\varphi} + \frac{1}{r_0 \cos\theta}\frac{\partial}{\partial \theta}(\cos\theta R_{\theta\theta}) + \frac{\partial}{\partial z} R_{\theta z} \tag{C.28}$$

$$(\nabla \cdot \mathbf{R})_z = \frac{1}{r_0 \cos\theta}\frac{\partial R_{z\varphi}}{\partial \varphi} + \frac{1}{r_0 \cos\theta}\frac{\partial}{\partial \theta}(\cos\theta R_{z\theta}) + \frac{\partial}{\partial z} R_{zz} \tag{C.29}$$

Rate of strain:

$$D_{\varphi\varphi} = \frac{1}{r_0 \cos\theta}\frac{\partial u_\varphi}{\partial \varphi} - \frac{u_\theta}{r_0} \tan\theta \tag{C.30}$$

$$D_{\varphi\theta} = D_{\theta\varphi} = \frac{1}{2}\left\{\frac{1}{r_0 \cos\theta}\frac{\partial u_\theta}{\partial \varphi} + \frac{\cos\theta}{r_0}\frac{\partial}{\partial \theta}\left(\frac{u_\varphi}{\cos\theta}\right)\right\} \tag{C.31}$$

$$D_{z\varphi} = D_{\varphi z} = \frac{1}{2}\left\{\frac{\partial u_\varphi}{\partial z} + \frac{1}{r_0 \cos\theta}\frac{\partial u_z}{\partial \varphi}\right\} \tag{C.32}$$

$$D_{\theta\theta} = \frac{1}{r_0}\frac{\partial u_\theta}{\partial \theta} \tag{C.33}$$

$$D_{z\theta} = D_{\theta z} = \frac{1}{2}\left\{\frac{\partial u_\theta}{\partial z} + \frac{1}{r_0}\frac{\partial u_z}{\partial \theta}\right\} \tag{C.34}$$

$$D_{zz} = \frac{\partial u_z}{\partial z} \tag{C.35}$$

Laplacian of a scalar field:

$$\Delta\phi = \frac{1}{r_0 \cos\theta}\left\{\frac{\partial}{\partial \varphi}\left(\frac{1}{r_0 \cos\theta}\frac{\partial}{\partial \varphi}\phi\right) + \frac{\partial}{\partial \theta}\left(\frac{\cos\theta}{r_0}\frac{\partial}{\partial \theta}\phi\right) + \frac{\partial}{\partial z}\frac{\partial}{\partial z}\phi\right\} \tag{C.36}$$

C.5 Two-dimensional spherical coordinates

The two-dimensional spherical coordinates (φ, θ) have scale factors $(h_\varphi = r_0 \cos\theta, h_\theta = r_0)$. Formulae for the differential operators are obtained simply by ignoring the z-component in the formulae for pseudo-spherical coordinates.

The operator $\hat{\boldsymbol{\nabla}}_{\mathrm{h}}$ is given by

$$\hat{\boldsymbol{\nabla}}_{\mathrm{h}}\phi = \left(-\frac{1}{r_0}\frac{\partial \phi}{\partial \theta}, \quad \frac{1}{r_0 \cos\theta}\frac{\partial \phi}{\partial \varphi} \right) \tag{C.37}$$

$$\hat{\boldsymbol{\nabla}}_{\mathrm{h}} \cdot \mathbf{a}_{\mathrm{h}} = \frac{1}{r_0 \cos\theta}\frac{\partial a_\theta}{\partial \varphi} - \frac{1}{r_0 \cos\theta}\frac{\partial}{\partial \theta}(a_\varphi \cos\theta) \tag{C.38}$$

The rules listed in Appendix B apply.

Appendix D
Kinematics of fluid motion

This appendix summarizes various ways in which fluid motion can be represented, described, and characterized. There are two basic frameworks: the Lagrangian and the Eulerian description. The Lagrangian description follows the paths of (marked) fluid particles. It is the continuous limit of discrete particle physics. The Eulerian description is a field description. Variables are considered at points in space, independent of which fluid particles occupy this point. The central quantity of the Lagrangian description is the deformation tensor; the central quantity of the Eulerian description is the velocity gradient tensor.

D.1 Lagrangian description

In the Lagrangian framework, the fluid motion is described by the position **x** of a fluid particle as a function of its label **s** and time τ.

$$\mathbf{x} = \mathbf{x}(\mathbf{s}, \tau) \tag{D.1}$$

In order to avoid ambiguities time is denoted by τ in the Lagrangian and by t in the Eulerian description. The fluid velocity is given by

$$\mathbf{u}(\mathbf{s}, \tau) = \frac{\partial \mathbf{x}(\mathbf{s}, \tau)}{\partial \tau} \tag{D.2}$$

and the fluid acceleration by

$$\mathbf{a}(\mathbf{s}, \tau) = \frac{\partial^2 \mathbf{x}(\mathbf{s}, \tau)}{\partial \tau^2} \tag{D.3}$$

The label coordinates $\mathbf{s} = (s_1, s_2, s_3)$ are assumed to form a continuous three-dimensional manifold. Often the initial position $\mathbf{r} := \mathbf{x}(\mathbf{s}, \tau = 0)$ is used as a label. The Lagrangian description of the fluid flow represents a time-dependent mapping from label (initial position) space to position space.

D.2 Displacement gradient tensor

The tensor **F** with components

$$F_{ij} := \frac{\partial x_i}{\partial r_j} \tag{D.4}$$

is called *the displacement gradient* tensor. It describes the deformation that a fluid element undergoes when it moves from its initial position $\mathbf{r}$ to its current position $\mathbf{x}$. Specifically, one finds for:

1. Line elements. A differential line element $\mathrm{d}\mathbf{r}$ is deformed into a differential line element $\mathrm{d}\mathbf{x}$ with components

$$\mathrm{d}x_i = F_{ij}\,\mathrm{d}r_j \tag{D.5}$$

The orientation of the line elements

$$\mathbf{M} := \frac{\mathrm{d}\mathbf{r}}{|\mathrm{d}\mathbf{r}|}, \qquad \mathbf{m} := \frac{\mathrm{d}\mathbf{x}}{|\mathrm{d}\mathbf{x}|} \tag{D.6}$$

and the length stretch

$$\lambda := \frac{|\mathrm{d}\mathbf{x}|}{|\mathrm{d}\mathbf{r}|} \tag{D.7}$$

are given by

$$\lambda = (F_{ij}F_{ik}M_jM_k)^{1/2} \tag{D.8}$$

$$m_i = \frac{1}{\lambda}F_{ij}M_j \tag{D.9}$$

2. Area elements. An oriented differential area element $\mathrm{d}\mathbf{A} = \mathrm{d}\mathbf{r}^{(1)} \times \mathrm{d}\mathbf{r}^{(2)}$ is deformed into an oriented differential area element $\mathrm{d}\mathbf{A} = \mathrm{d}\mathbf{x}^{(1)} \times \mathrm{d}\mathbf{x}^{(2)}$ with components

$$\mathrm{d}a_i = J F_{ji}^{-1}\,\mathrm{d}A_j \tag{D.10}$$

where $J = \partial(\mathbf{x})/\partial(\mathbf{r})$ is the Jacobian of the mapping $\mathbf{x} = \mathbf{x}(\mathbf{r}, \tau)$ and $\mathbf{F}^{-1}$ the inverse of the deformation gradient tensor (which is identical to the deformation gradient tensor of the inverse mapping $\mathbf{r} = \mathbf{r}(\mathbf{x}, \tau)$). The orientations

$$\mathbf{N} := \frac{\mathrm{d}\mathbf{A}}{|\mathrm{d}\mathbf{A}|}, \qquad \mathbf{n} := \frac{\mathrm{d}\mathbf{A}}{|\mathrm{d}\mathbf{A}|} \tag{D.11}$$

and the area stretch

$$\eta := \frac{|\mathrm{d}\mathbf{A}|}{|\mathrm{d}\mathbf{A}|} \tag{D.12}$$

are given by

$$\eta = J\left(F_{ji}^{-1}F_{ki}^{-1}N_jN_k\right)^{1/2} \tag{D.13}$$

$$n_i = \frac{J}{\eta}\,F_{ji}^{-1}N_j \tag{D.14}$$

3. Volume element. A volume element $\mathrm{d}V = (\mathrm{d}\mathbf{r}^{(1)} \times \mathrm{d}\mathbf{r}^{(2)}) \cdot \mathrm{d}\mathbf{r}^{(3)}$ is deformed into a volume element $\mathrm{d}v = (\mathrm{d}\mathbf{x}^{(1)} \times \mathrm{d}\mathbf{x}^{(2)}) \cdot \mathrm{d}\mathbf{x}^{(3)}$ with

$$\mathrm{d}v = J\mathrm{d}V \tag{D.15}$$

The Jacobian thus describes the volume "stretch" (or the dilation, or the expansion).

D.3 Eulerian description

In the Eulerian framework, the flow is specified by the velocity **u** as a function of position **x** and time t

$$\mathbf{u} = \mathbf{u}(\mathbf{x}, t) \tag{D.16}$$

Eulerian and Lagrangian descriptions differ in their choice of dependent and independent variables. The two descriptions are, however, equivalent. To transform from the Lagrangian description $\mathbf{x} = \mathbf{x}(\mathbf{s}, \tau)$ to the Eulerian description $\mathbf{u} = \mathbf{u}(\mathbf{x}, t)$ one has to:

1. differentiate $\mathbf{x} = \mathbf{x}(\mathbf{r}, \tau)$ with respect to τ to obtain $\mathbf{u} = \mathbf{u}(\mathbf{r}, \tau)$;
2. invert $\mathbf{x} = \mathbf{x}(\mathbf{r}, \tau)$ to obtain $\mathbf{r} = \mathbf{r}(\mathbf{x}, t)$; and
3. substitute $\mathbf{r} = \mathbf{r}(\mathbf{x}, t)$ into $\mathbf{u}(\mathbf{r}, \tau)$ to obtain $\mathbf{u}(\mathbf{x}, t)$.

To transform from the Eulerian to the Lagrangian description one has to solve the ordinary differential equations

$$\frac{\mathrm{d}\mathbf{x}}{\mathrm{d}t} = \mathbf{u}\,(\mathbf{x}(t), t) \tag{D.17}$$

subject to the initial condition $\mathbf{x}(t = 0) = \mathbf{r}$ for all $\mathbf{r}$.

Lagrangian and Eulerian derivatives are related by

$$\frac{\partial \phi(\mathbf{r}, \tau)}{\partial r_i} = \frac{\partial \phi(\mathbf{x}, t)}{\partial x_j} F_{ji} \tag{D.18}$$

$$\frac{\partial \phi(\mathbf{r}, \tau)}{\partial \tau} = \frac{\mathrm{D}}{\mathrm{D}t}\phi(\mathbf{x}, t) \tag{D.19}$$

where

$$\frac{\mathrm{D}}{\mathrm{D}t} := \frac{\partial}{\partial t} + u_i \frac{\partial}{\partial x_i} \tag{D.20}$$

is called the Lagrangian, advective, material, or particle derivative. The acceleration **a** is therefore given in the Eulerian description by

$$\mathbf{a}(\mathbf{x}, t) = \frac{\mathrm{D}}{\mathrm{D}t}\mathbf{u}\,(\mathbf{x}, t) \tag{D.21}$$

D.4 Velocity gradient tensor

The tensor **G** with components

$$G_{ij} := \frac{\partial u_i}{\partial x_j} \tag{D.22}$$

is called the velocity gradient tensor. Its symmetric part

$$D_{ij} := \frac{1}{2}(G_{ij} + G_{ji}) \tag{D.23}$$

is called the rate of deformation (or the rate of strain) tensor. Its antisymmetric part

$$V_{ij} := \frac{1}{2}(G_{ij} - G_{ji}) \tag{D.24}$$

is called the vorticity tensor.

Since the rate of deformation tensor is symmetric there exists an orthonormal coordinate system in which the tensor has non-zero components only in the diagonal. This coordinate system defines the principal axes of the rate of deformation tensor. The diagonal elements are called the principal rates of deformation. The principal axes $\mathbf{b}^{(s)}$ and the rates of deformation $\lambda^{(s)}$ are the solution of the eigenvalue problem

$$D_{ij}b_j^{(s)} = \lambda^{(s)}b_i^{(s)} \qquad s = 1, 2, 3 \tag{D.25}$$

The rate of deformation tensor can further be decomposed into

$$D_{ij} = N_{ij} + S_{ij} \tag{D.26}$$

where

$$N_{ij} = \frac{1}{3}\mathrm{d}\,\delta_{ij} \tag{D.27}$$

is the rate of normal deformation tensor,

$$S_{ij} = D_{ij} - \frac{1}{3}\mathrm{d}\,\delta_{ij} \tag{D.28}$$

the rate of shear deformation tensor. The divergence of the velocity field, $\partial_i u_i$, is the trace of the rate of normal deformation tensor. The rate of shear deformation tensor has zero trace.

The vorticity tensor $\mathbf{V}$ is related to the vorticity vector

$$\boldsymbol{\omega} := \boldsymbol{\nabla} \times \mathbf{u} \tag{D.29}$$

by the relations

$$V_{ij} = -\frac{1}{2}\epsilon_{ijk}\omega_k = \frac{1}{2}\begin{pmatrix} 0 & -\omega_3 & \omega_2 \\ \omega_3 & 0 & -\omega_1 \\ -\omega_2 & \omega_1 & 0 \end{pmatrix} \tag{D.30}$$

$$\omega_k = \epsilon_{kji}V_{ij} \tag{D.31}$$

The motion near a point is given by

$$u_i(\mathbf{x} + \mathrm{d}\mathbf{x}) = u_i(\mathbf{x}) + D_{ij}\mathrm{d}x_j + V_{ij}\mathrm{d}x_j \tag{D.32}$$

The first term describes a uniform translation. The second term describes the dilations along the three principle axes of the rate of deformation tensor. A sphere is thus transformed into an ellipsoid. The third term can be rewritten as

$$V_{ij}\mathrm{d}x_j = \frac{1}{2}\epsilon_{ijk}\omega_j\mathrm{d}x_k \tag{D.33}$$

and describes a solid body rotation with angular velocity $\boldsymbol{\omega}/2$. The motion near a point thus consists of a uniform translation, dilations along three mutually perpendicular axes, and solid body rotations about these axes.

The velocity gradient tensor also determines the time changes of:

- line elements

$$\frac{\mathrm{D}}{\mathrm{D}t}(\mathrm{d}x_i) = G_{ij}\,\mathrm{d}x_j \tag{D.34}$$

- area elements

$$\frac{\mathrm{D}}{\mathrm{D}t}(\mathrm{d}a_i) = D_{jj}\,\mathrm{d}a_i - G_{mi}\mathrm{d}a_m \tag{D.35}$$

- volume elements

$$\frac{\mathrm{D}}{\mathrm{D}t}(\mathrm{d}v) = D_{jj}\,\mathrm{d}v \tag{D.36}$$

- line stretch

$$\frac{\mathrm{D}}{\mathrm{D}t}(\ln\lambda) = D_{ij}\,m_i m_j \tag{D.37}$$

- area stretch

$$\frac{\mathrm{D}}{\mathrm{D}t}(\ln\eta) = D_{jj} - D_{ij}\,n_i n_j \tag{D.38}$$

- volume stretch

$$\frac{\mathrm{D}}{\mathrm{D}t}(\ln J) = D_{jj} \quad \text{(Euler's formula)} \tag{D.39}$$

and time changes of the integrals

$$\frac{\mathrm{d}}{\mathrm{d}t}\iiint_V \mathrm{d}^3\mathbf{x}\,F\ldots(\mathbf{x},t) = \iiint_V \mathrm{d}^3\mathbf{x}\left[\frac{\mathrm{D}}{\mathrm{D}t}F\ldots + F\ldots D_{jj}\right] \tag{D.40}$$

$$\frac{\mathrm{d}}{\mathrm{d}t}\iint_S \mathrm{d}a_i\,F\ldots(\mathbf{x},t) = \iint_S \mathrm{d}a_i\left[\frac{\mathrm{D}}{\mathrm{D}t}F\ldots + F\ldots D_{jj}\right] - \iint_S \mathrm{d}a_m\,G_{mi}F\ldots \tag{D.41}$$

$$\frac{\mathrm{d}}{\mathrm{d}t}\int_C \mathrm{d}x_i\,F\ldots(\mathbf{x},t) = \int_C \mathrm{d}x_i\,\frac{\mathrm{D}}{\mathrm{D}t}F\ldots + \int_C \mathrm{d}x_j\,F\ldots G_{ij} \tag{D.42}$$

where $F\ldots(\mathbf{x},t)$ is a scalar, vector, or tensor field and V, A, and C are material volumes, surfaces, and lines. Other quantities related to the velocity gradient tensor are the helicity $\boldsymbol{\omega}\cdot\mathbf{u}$ and the Lamb vector $\boldsymbol{\omega}\times\mathbf{u}$, which enters the identity

$$\mathbf{u}\cdot\boldsymbol{\nabla}\mathbf{u} = \boldsymbol{\omega}\times\mathbf{u} + \boldsymbol{\nabla}\left(\frac{1}{2}\mathbf{u}\cdot\mathbf{u}\right) \tag{D.43}$$

The circulation along a (simple) closed circuit C is defined by

$$\Gamma := \oint_C \mathrm{d}\mathbf{x}\cdot\mathbf{u} \tag{D.44}$$

If C can be capped by a surface A that lies fully in the fluid volume then the circulation is related to the vorticity vector by Stokes' theorem

$$\oint_C \mathrm{d}\mathbf{x}\cdot\mathbf{u} = \iint_A \mathrm{d}^2x\,\mathbf{n}\cdot(\boldsymbol{\nabla}\times\mathbf{u}) \tag{D.45}$$

D.5 Classification and representations of velocity fields

Classes of velocity fields

1. Irrotational fields: A velocity field $\mathbf{u}(\mathbf{x},t)$ is called irrotational if its vorticity is zero everywhere

$$\boldsymbol{\nabla}\times\mathbf{u} = 0 \tag{D.46}$$

The velocity field can then be represented by a velocity potential $\varphi(\mathbf{x}, t)$ such that

$$\mathbf{u} = \boldsymbol{\nabla}\varphi \tag{D.47}$$

The velocity potential is given by

$$\varphi(\mathbf{x}, t) - \varphi(\mathbf{x}_0, t) = \int_{\mathbf{x}_0}^{\mathbf{x}} \mathrm{d}\mathbf{x}' \cdot \mathbf{u}(\mathbf{x}', t) \tag{D.48}$$

For a simply connected domain φ is uniquely defined up to a constant since the line integral between $\mathbf{x}_0$ and $\mathbf{x}$ is independent of the path according to Stokes' theorem. For a doubly connected domain $\varphi(\mathbf{x})$ is a multi-valued function, $\varphi(\mathbf{x}) - \varphi(\mathbf{x}_0) = \varphi_0(\mathbf{x}) + n\Gamma_0$ where Γ_0 is the circulation around the "island" and n an integer.

2. Solenoidal fields: A velocity field $\mathbf{u}(\mathbf{x}, t)$ is called solenoidal if its divergence vanishes everywhere

$$\boldsymbol{\nabla} \cdot \mathbf{u} = 0 \tag{D.49}$$

In this case the velocity field can be represented by the scalar fields η and χ such that

$$\mathbf{u} = \boldsymbol{\nabla}\eta \times \boldsymbol{\nabla}\chi = \boldsymbol{\nabla} \times (\eta\boldsymbol{\nabla}\chi) \tag{D.50}$$

or by a vector potential $\mathbf{A}$ such that

$$\mathbf{u} = \boldsymbol{\nabla} \times \mathbf{A} \tag{D.51}$$

The vector potential is not unique. The gradient of an arbitrary scalar function can be added to it without affecting $\mathbf{u}$. This function can be chosen such that $\boldsymbol{\nabla} \cdot \mathbf{A} = 0$. The vector potential is given in this case by (D.57).

3. Irrotational and solenoidal fields: If the velocity field $\mathbf{u}(\mathbf{x}, t)$ is both irrotational and solenoidal then the velocity potential must satisfy the Laplace equation

$$\Delta\varphi = 0 \tag{D.52}$$

For a simply connected domain $\mathbf{u} = \boldsymbol{\nabla}\varphi$ is determined uniquely if the normal velocity $\mathbf{u} \cdot \mathbf{n}$ is given on the boundary. For multiply connected domains one must additionally specify the circulations around the "islands."

Additional special classes of velocity fields include Beltrami fields for which $\boldsymbol{\omega} \times \mathbf{u} = 0$ and complex lamellar fields for which $\boldsymbol{\omega} \cdot \mathbf{u} = 0$.

Representation of velocity fields

1. Velocity fields for specified divergence and vorticity: A velocity field that has specified divergence d can be constructed from Poisson's equation

$$\Delta\varphi = d \tag{D.53}$$

which has the solution

$$\varphi = -\iint_V \mathrm{d}^3\mathbf{x}' \, \frac{d(\mathbf{x}')}{4\pi|\mathbf{x} - \mathbf{x}'|} \tag{D.54}$$

The velocity field is then given by

$$\mathbf{u}^{(1)} = \nabla\varphi = -\iint_V d^3\mathbf{x}' \, \frac{d(\mathbf{x}')}{4\pi|\mathbf{x}-\mathbf{x}'|^3}\,(\mathbf{x}-\mathbf{x}') \tag{D.55}$$

Similarly, a velocity field that has specified vorticity ω can be constructed from

$$\nabla(\nabla\cdot\mathbf{A}) - \Delta\mathbf{A} = \omega \tag{D.56}$$

Assuming $\nabla\cdot\mathbf{A} = 0$ the solution is

$$\mathbf{A} = \iint_V d^3\mathbf{x}' \, \frac{\omega(\mathbf{x}')}{4\pi|\mathbf{x}-\mathbf{x}'|} \tag{D.57}$$

and gives the velocity field

$$\mathbf{u}^{(2)} = \nabla\times\mathbf{A} = -\iint_V d^3\mathbf{x}' \, \frac{(\mathbf{x}-\mathbf{x}')\times\omega(\mathbf{x}')}{4\pi|\mathbf{x}-\mathbf{x}'|^3} \tag{D.58}$$

The vector potential is solenoidal provided that $\omega\cdot\mathbf{u} = 0$ on the boundary. If this is not the case for the real boundary one needs to extend the volume V to a new artificial boundary where the condition $\omega\cdot\mathbf{u} = 0$ can be imposed.

From the above it follows that every velocity field can be written in the form

$$\mathbf{u} = \mathbf{u}^{(1)} + \mathbf{u}^{(2)} + \mathbf{u}^{(3)} \tag{D.59}$$

where $\mathbf{u}^{(1)}$ and $\mathbf{u}^{(2)}$ are given by the above expressions, and where $\mathbf{u}^{(3)}$ is both irrotational and solenoidal. The latter component is thus described by a velocity potential that satisfies Laplace's equation and is determined by the values of $\mathbf{u} - \mathbf{u}^{(1)} - \mathbf{u}^{(2)}$ on the boundary.

2. Helmholtz decomposition: The representations for irrotational and solenoidal fields imply that any continuously differentiable velocity field can be written in the form

$$\mathbf{u} = \nabla\varphi + \nabla\times\mathbf{A} \tag{D.60}$$

with scalar potential φ and vector potential $\mathbf{A}$. This representation is called the Helmholtz decomposition. The first term is irrotational. The second term is solenoidal. The vector potential can be chosen such that $\nabla\cdot\mathbf{A} = 0$.

3. Clebsch representation: At points where the vorticity does not vanish the velocity field can be represented in the form

$$\mathbf{u} = \nabla\varphi + \eta\nabla\chi \tag{D.61}$$

The first term is irrotational. The second term is complex lamellar.

Two-dimensional fields For two-dimensional velocity fields $\mathbf{u}_h(\mathbf{x}_h, t) = (u_1(x_1, x_2, t), u_2(x_1, x_2, t))$ one specifically finds:

Velocity gradient tensor:

$$G_{\alpha\beta} = \partial u_\alpha/\partial x_\beta = \begin{pmatrix} \partial u_1/\partial x_1 & \partial u_1/\partial x_2 \\ \partial u_2/\partial x_1 & \partial u_2/\partial x_2 \end{pmatrix} \tag{D.62}$$

Rate of deformation tensor:

$$D_{\alpha\beta} = \frac{1}{2}(G_{\alpha\beta} + G_{\beta\alpha})$$
$$= \begin{pmatrix} \partial u_1/\partial x_1 & \frac{1}{2}(\partial u_1/\partial x_2 + \partial u_2/\partial x_1) \\ \frac{1}{2}(\partial u_1/\partial x_2 + \partial u_2/\partial x_1) & \partial u_2/\partial x_2 \end{pmatrix} \tag{D.63}$$

Rate of normal deformation tensor:

$$N_{\alpha\beta} = \frac{1}{2} D_{\gamma\gamma}\, \delta_{\alpha\beta} = \frac{1}{2}\begin{pmatrix} \partial_\alpha u_\alpha & 0 \\ 0 & \partial_\alpha u_\alpha \end{pmatrix} = \frac{1}{2}\begin{pmatrix} d & 0 \\ 0 & d \end{pmatrix} \tag{D.64}$$

with eigenvalues $\lambda_{1,2} = \frac{1}{2}d$.

Rate of shear deformation tensor:

$$S_{\alpha\beta} = D_{\alpha\beta} - N_{\alpha\beta}$$
$$= \frac{1}{2}\begin{pmatrix} \frac{\partial u_1}{\partial x_1} - \frac{\partial u_2}{\partial x_2} & \frac{\partial u_1}{\partial x_2} + \frac{\partial u_2}{\partial x_1} \\ \frac{\partial u_1}{\partial x_2} + \frac{\partial u_2}{\partial x_1} & -\left(\frac{\partial u_1}{\partial x_1} - \frac{\partial u_2}{\partial x_2}\right) \end{pmatrix} = \frac{1}{2}\begin{pmatrix} \sigma_n & \sigma_s \\ \sigma_s & -\sigma_n \end{pmatrix} \tag{D.65}$$

with eigenvalues $\lambda_{1,2} = \pm\frac{1}{2}(\sigma_n^2 + \sigma_s^2)^{1/2}$.

Vorticity tensor:

$$V_{\alpha\beta} = \frac{1}{2}(G_{\alpha\beta} - G_{\beta\alpha})$$
$$= \begin{pmatrix} 0 & -\left(\frac{\partial u_2}{\partial x_1} - \frac{\partial u_1}{\partial x_2}\right) \\ \frac{\partial u_2}{\partial x_1} - \frac{\partial u_1}{\partial x_2} & 0 \end{pmatrix} = \frac{1}{2}\begin{pmatrix} 0 & -\xi \\ \xi & 0 \end{pmatrix} \tag{D.66}$$

with eigenvalues $\lambda_{1,2} = \pm i\frac{1}{2}\xi$.

If the flow has zero vorticity, $\xi = 0$, then there exists a velocity potential φ such that

$$\mathbf{u}_\mathrm{h} = \nabla_\mathrm{h}\varphi \tag{D.67}$$

where

$$\varphi(\mathbf{x}_\mathrm{h}) - \varphi(\mathbf{x}_{\mathrm{h},0}) = \int_{\mathbf{x}_{\mathrm{h},0}}^{\mathbf{x}_\mathrm{h}} \mathrm{d}\mathbf{x}'_\mathrm{h} \cdot \mathbf{u}_\mathrm{h}(\mathbf{x}'_\mathrm{h}, t) \tag{D.68}$$

If the flow has zero divergence then there exists a streamfunction ψ such that

$$\mathbf{u}_\mathrm{h} = \hat{\nabla}_\mathrm{h}\psi \tag{D.69}$$

The flow is everywhere tangent to the isolines of the streamfunction and in a direction such that higher values of the streamfunction are to the right. The streamfunction is the negative of the third component of the vector potential **A**. An arbitrary two-dimensional flow field can be represented in the Helmholtz form

$$\mathbf{u}_\mathrm{h} = \nabla_\mathrm{h}\varphi + \hat{\nabla}_\mathrm{h}\psi \tag{D.70}$$

Warning The mathematical proofs of the existence of the various representations may require continuously differentiable velocity fields, may exclude points where the vorticity vanishes, and other subtle assumptions in order that the potentials of the various representations are single-valued and non-singular. For details see, e.g., Aris (1962).

D.6 Global description

Streamlines At every time instant one can define streamlines $\mathbf{a}(s;t)$ of the velocity field. Streamlines are everywhere tangent to the velocity field

$$\frac{\mathrm{d}\mathbf{a}(s;t)}{\mathrm{d}s} \times \mathbf{u} = 0 \tag{D.71}$$

The streamline through a point $\mathbf{x}_0$ is obtained by solving the ordinary differential equations

$$\frac{\mathrm{d}\mathbf{a}(s,t)}{\mathrm{d}s} = \mathbf{u}(\mathbf{x} = \mathbf{a}(s), t) \tag{D.72}$$

subject to the condition $\mathbf{a}(s = 0) = \mathbf{x}_0$. Note that s is the parameter describing the streamline. The time t enters as an external parameter. For two-dimensional fields, the isolines of the streamfunction are streamlines.

There are various theorems regarding the topological structure of streamlines, such as:

- In steady 3-D problems streamlines need not be closed.
- In steady 2-D problems streamlines are closed and not chaotic.

Fluid trajectory The trajectory $\mathbf{b}(t)$ of a fluid particle that was at position $\mathbf{x}_0$ at time $t = 0$ is obtained by solving the ordinary differential equations

$$\frac{\mathrm{d}\mathbf{b}(t)}{\mathrm{d}t} = \mathbf{u}(\mathbf{x} = \mathbf{b}(t), t) \tag{D.73}$$

subject to the initial condition $\mathbf{b}(t = 0) = \mathbf{x}_0$. If the flow is steady then streamlines and trajectories are identical.

Streaklines A streakline is the curve traced out by particles continuously injected at a point $\mathbf{x}_0$. At time t the streakline $\mathbf{c}(s)$ is thus the line formed by all fluid particles that were at position $\mathbf{x}_0$ at some previous time t' with $0 < t' < t$. Given the fluid trajectories $\mathbf{b}(t;\mathbf{r})$ that passed through $\mathbf{r}$ at $t = 0$ one can construct the streakline originating at $\mathbf{x}_0$ by the following operations:

1. Invert $\mathbf{b}(t';\mathbf{r}) = \mathbf{x}_0$ to obtain $\mathbf{r} = \mathbf{r}(\mathbf{x}_0, t')$.
2. Substitute $\mathbf{r} = \mathbf{r}(\mathbf{x}_0, t')$ into $\mathbf{b}(t, \mathbf{r})$.

The equation for the streakline is then given by

$$\mathbf{c}(t') = \mathbf{b}(t, \mathbf{r}(\mathbf{x}_0, t')) \tag{D.74}$$

where the parameter t' lies in the interval $[0, t]$.

Material lines, surfaces, and volumes Material lines, surfaces, and volumes are lines, surfaces, and volumes that always consist of the same fluid particles.

For a line $\mathbf{x} = \mathbf{x}(s, t)$ that depends on time the velocity is defined by

$$\mathbf{v} := \mathbf{t} \times \left[\frac{\partial \mathbf{x}}{\partial t} \times \mathbf{t}\right] \tag{D.75}$$

where $\mathbf{t} = \frac{\partial \mathbf{x}}{\partial s}/|\frac{\partial \mathbf{x}}{\partial s}|$ is the unit tangent vector of the line. For the line always to consist of the same fluid particles, the velocity component perpendicular to the line has to be equal to the fluid velocity

$$\mathbf{t} \times [\mathbf{v} \times \mathbf{t}] = \mathbf{t} \times [\mathbf{u} \times \mathbf{t}] \tag{D.76}$$

For a surface $\mathbf{x} = \mathbf{x}(r, s; t)$ that depends on time the velocity is defined by

$$\mathbf{v} := \mathbf{n} \cdot \frac{\partial \mathbf{x}}{\partial t} \mathbf{n} \tag{D.77}$$

where $\mathbf{n}$ is the normal vector of the surface. For the surface always to consist of the same fluid particles, the velocity component normal to the surface has to be equal to the fluid velocity

$$\mathbf{v} \cdot \mathbf{n} = \mathbf{u} \cdot \mathbf{n} \tag{D.78}$$

If the surface is defined by $G(\mathbf{x}, t) = 0$ then the velocity of the surface is given by

$$\mathbf{v} = -\frac{\partial G}{\partial t} \frac{\mathbf{n}}{|\boldsymbol{\nabla} G|} \tag{D.79}$$

and the above condition becomes

$$\partial_t G + \mathbf{u} \cdot \boldsymbol{\nabla} G = 0 \tag{D.80}$$

Vortex line A vortex line is a line $\mathbf{d}(s)$ that is everywhere tangential to the vorticity field

$$\frac{\mathrm{d}\mathbf{d}(s)}{\mathrm{d}s} \times \boldsymbol{\omega} = 0 \tag{D.81}$$

A vortex line through $\mathbf{x}_0$ can be constructed from solving

$$\frac{\mathrm{d}\mathbf{d}(s)}{\mathrm{d}s} = \boldsymbol{\omega}(\mathbf{x} = \mathbf{d}(s), t) \tag{D.82}$$

subject to the condition $\mathbf{d}(s = 0) = \mathbf{x}_0$. The vortex lines are the "streamlines" of the vorticity field.

Vortex tube A vortex tube is formed by all the vortex lines that pass through a closed curve C_1 (see Figure D.1). The circulation $\Gamma = \oint_C \mathrm{d}\mathbf{x} \cdot \mathbf{u} = \iint_S \mathrm{d}^2x\ \mathbf{n} \cdot \boldsymbol{\omega}$ is called the strength of the vortex tube. Since $\boldsymbol{\nabla} \cdot \boldsymbol{\omega} = 0$ one finds

$$\begin{aligned} 0 = \iiint_V \mathrm{d}^3\mathbf{x}\ \boldsymbol{\nabla} \cdot \boldsymbol{\omega} &= \iint_S \mathrm{d}^2\mathbf{x}\ \mathbf{n} \cdot \boldsymbol{\omega} \\ &= \iint_{S_1} \mathrm{d}^2\mathbf{x}\ \mathbf{n}_1 \cdot \boldsymbol{\omega} + \iint_{S_2} \mathrm{d}^2\mathbf{x}\ \mathbf{n}_2 \cdot \boldsymbol{\omega} \end{aligned} \tag{D.83}$$

because $\mathbf{n} \cdot \boldsymbol{\omega} = 0$ on the other surfaces (see Figure D.1). The strength of the vortex tube is the same for every cross-section.

D.7 Conservative tracers

The concentration c of a conservative tracer satisfies

$$\frac{\mathrm{D}}{\mathrm{D}t} c = 0 \tag{D.84}$$

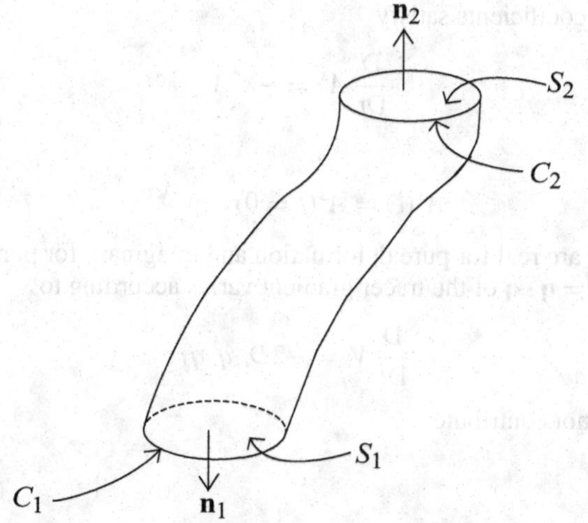

Figure D.1. Two cross-sections of vortex tube with areas $S_{1,2}$, encircling curves $C_{1,2}$ and normal vectors $\mathbf{n}_{1,2}$.

If the initial distribution is given by $c(\mathbf{x}, t = 0) = c_0(\mathbf{x})$, then the solution is

$$c(\mathbf{x}, t) = c_0(\mathbf{r}(\mathbf{x}, t), t) \tag{D.85}$$

where $\mathbf{r}$ is the initial position.

The tracer gradient $\mathbf{q} = \boldsymbol{\nabla} c$ changes according to

$$\frac{\mathrm{D}}{\mathrm{D}t} q_i = -G_{ji} q_j \tag{D.86}$$

where $\mathbf{G}$ is the velocity gradient tensor. The evolution of the tracer gradient at each point is determined by the eigenvalues of $\mathbf{G}$, as can be seen by expressing the tracer gradients in terms of the eigenvectors of $\mathbf{G}$. Explicitly, let λ^s, $\mathbf{a}_h^s$, and $\hat{\mathbf{a}}_h^s$ ($s = 1, 2, 3$) be the eigenvalues and right and left eigenvectors of $\mathbf{G}$

$$G_{ij}\, a_j^s = \lambda^s a_i^s \qquad s = 1, 2, 3 \tag{D.87}$$

$$\hat{a}_j^s G_{ij} = \lambda^s \hat{a}_i^s \qquad s = 1, 2, 3 \tag{D.88}$$

Then

$$\hat{a}_i^s a_i^{s'} = \delta_{ss'} \qquad \text{orthogonality} \tag{D.89}$$

$$\sum_s a_i^s \hat{a}_j^s = \delta_{ij} \qquad \text{completeness} \tag{D.90}$$

If the tracer gradient is expanded

$$\mathbf{q} = \sum_s A^s \hat{\mathbf{a}}^s \tag{D.91}$$

then the expansion coefficients satisfy

$$\frac{\mathrm{D}}{\mathrm{D}t} A^s = -\lambda^s A^s \tag{D.92}$$

which is solved by

$$A^s(t) = A^s(t=0) \exp^{-\lambda^s t} \tag{D.93}$$

The eigenvalues λ^s are real for pure deformation and imaginary for pure rotation.

The variance $V = \mathbf{q} \cdot \mathbf{q}$ of the tracer gradient varies according to

$$\frac{\mathrm{D}}{\mathrm{D}t} V = -2D_{ij} q_i q_j \tag{D.94}$$

The vorticity does not contribute.

Appendix E

Kinematics of waves

This appendix summarizes the basic *kinematic* properties of propagating *free linear* waves. The dynamic processes that generate and dissipate waves and the nonlinear interaction among waves are not considered. Linear waves are formally of infinitesimal amplitude. Free linear waves come about as solutions of homogeneous linear equations of motion. Free linear wave solutions can be superimposed.

E.1 Elementary propagating waves

An unbounded, homogeneous, and stationary medium supports elementary propagating waves

$$\psi(\mathbf{x}, t) = \Re\left\{a \exp^{\mathrm{i}(\mathbf{k}\cdot\mathbf{x}-\omega t)}\right\} \tag{E.1}$$

with

$$a = \text{amplitude} \tag{E.2}$$

$$\theta = \mathbf{k}\cdot\mathbf{x} - \omega t = \text{phase} \tag{E.3}$$

$$\mathbf{k} = (k_1, k_2, k_3) = \text{wavenumber vector} \tag{E.4}$$

$$\omega = \text{frequency} \tag{E.5}$$

$$\lambda := \frac{2\pi}{|\mathbf{k}|} = \text{wavelength} \tag{E.6}$$

$$T := \frac{2\pi}{\omega} = \text{period} \tag{E.7}$$

Important properties of such elementary propagating waves are:

Dispersion relation The homogeneity of the governing equations of motion imply the existence of a dispersion relation that relates the frequency to the wavenumber vector

$$\omega = \Omega(\mathbf{k}) \tag{E.8}$$

The dispersion relation is the most important property of a wave.

Group velocity The group velocity is defined by

$$\mathbf{c}_\mathrm{g} := \frac{\partial \Omega}{\partial \mathbf{k}} = \left(\frac{\partial \Omega}{\partial k_1}, \frac{\partial \Omega}{\partial k_2}, \frac{\partial \Omega}{\partial k_3}\right) \tag{E.9}$$

and is a vector. It is the velocity with which "influence" propagates.

Phase speed The phase speed is the speed at which a surface of constant phase moves in a specific direction. The phase speed in the direction of the unit vector **e** is given by

$$c_\mathrm{p} = \frac{\omega}{\mathbf{k}\cdot\mathbf{e}} \tag{E.10}$$

The phase speed is a scalar. The value of c_p is not the component of a vector.

Sometimes one defines the vector $\mathbf{c}_\mathrm{p} := \frac{\omega}{|k^2|}\mathbf{k}$. This definition follows the convention that the velocity of a surface, in this case the surface of constant phase, is always in its normal direction.

The slowness vector is defined by $\mathbf{s} := \frac{\mathbf{k}}{\omega}$. The inverse of the components of the slowness vector are the phase speeds in the direction of that component.

Initial value problem If

$$\psi(\mathbf{x}, 0) = \int\!\!\int\!\!\int \mathrm{d}^3k\, \hat{\psi}(\mathbf{k}, 0)\, \exp^{\mathrm{i}\mathbf{k}\cdot\mathbf{x}} \quad \text{at} \quad t = 0 \tag{E.11}$$

then

$$\psi(\mathbf{x}, t) = \int\!\!\int\!\!\int \mathrm{d}^3k\, \hat{\psi}(\mathbf{k}, 0)\, \mathrm{e}^{\mathrm{i}(\mathbf{k}\cdot\mathbf{x}-\omega t)} \quad \text{at} \quad t \geq 0 \tag{E.12}$$

Example 1. Wave packets:

A wave packet is defined by

$$\psi(\mathbf{x}, 0) = a(\epsilon\mathbf{x})\, \exp^{\mathrm{i}\mathbf{k}_0\cdot\mathbf{x}} \tag{E.13}$$

where the envelope $a(\epsilon\mathbf{x}) \approx 0$ outside a sphere of radius ϵ^{-1} and $\epsilon \ll 1$. Then

$$\hat{\psi}(\mathbf{k}, 0) = \frac{1}{\epsilon^3}\hat{a}\left(\frac{\mathbf{k}-\mathbf{k}_0}{\epsilon}\right) \tag{E.14}$$

is approximately zero outside a sphere centered at $\mathbf{k}_0$ with radius ϵk_0. Within this sphere

$$\Omega(\mathbf{k}) = \Omega(\mathbf{k}_0) + \frac{\partial \Omega}{\partial \mathbf{k}}(\mathbf{k}-\mathbf{k}_0) + O(\epsilon^2) \tag{E.15}$$

Thus

$$\begin{aligned}\psi(\mathbf{x}, t) &\approx \exp^{\mathrm{i}(\mathbf{k}_0\cdot\mathbf{x}-\omega_0 t)} \int\!\!\int\!\!\int \mathrm{d}^3k \frac{1}{\epsilon^3}\hat{a}\left(\frac{\mathbf{k}-\mathbf{k}_0}{\epsilon}\right) \exp^{\mathrm{i}(\mathbf{k}-\mathbf{k}_0)\cdot(\mathbf{x}-\mathbf{c}_g^0 t)} \\ &= a\left(\epsilon\left(\mathbf{x}-\mathbf{c}_\mathrm{g}^0 t\right)\right) \exp^{\mathrm{i}(\mathbf{k}_0\cdot\mathbf{x}-\omega_0 t)}\end{aligned} \tag{E.16}$$

where $\omega_0 = \Omega(\mathbf{k}_0)$ and $\mathbf{c}_\mathrm{g}^0 = \frac{\partial \Omega}{\partial \mathbf{k}}(\mathbf{k}_0)$. The envelope or wave packet moves with the group velocity.

Example 2. Non-dispersive waves in one dimension:

If $\omega = ck$ then

$$\psi(x,t) = \int_{-\infty}^{+\infty} \mathrm{d}k\ \hat{\psi}(x,0)\ \exp^{ik(x-ct)} = \psi(x-ct,0) \tag{E.17}$$

Any initial perturbation propagates at phase speed c without change of shape.

Example 3. Dispersion in one dimension:

Consider an initial perturbation $\psi(x,0) = \delta(0)$. Then $\hat{\psi}(k,0) = 1/2\pi$ and

$$\psi(x,t) = \frac{1}{2\pi}\int_{-\infty}^{+\infty} \mathrm{d}k\ \exp^{\mathrm{i}[k(x/t)-\Omega(k)]t} \tag{E.18}$$

In the limit $t \to \infty$ with $\gamma_0 = x/t =$ constant, contributions to the integral come primarily from the vicinity of the point k_0 where the phase is stationary

$$\frac{\mathrm{d}}{\mathrm{d}k}\left[k\gamma_0 - \Omega(k)\right] = 0 \tag{E.19}$$

Anywhere else the phase is varying so rapidly that contributions cancel out. Expansion about k_0 yields

$$\Omega(k) = \Omega(k_0) + \Omega'(k-k_0) + \frac{1}{2}\Omega''(k-k_0)^2 + \ldots \tag{E.20}$$

where the primes denote differentiation with respect to k. Thus

$$\begin{aligned} \psi(x,t) &\approx \exp^{\mathrm{i}(k_0x-\omega_0t)} \int_{-\infty}^{+\infty} \mathrm{d}k\ \exp^{-\mathrm{i}\frac{1}{2}\Omega''(\mathbf{k}-\mathbf{k}_0)^2 t} \\ &\sim \exp^{\mathrm{i}(k_0x-\omega_0t)} \frac{1}{\sqrt{t}} \end{aligned} \tag{E.21}$$

The disturbance thus breaks up into a slowly varying wave train. For a fixed position the amplitude decays as $1/\sqrt{t}$. For a fixed time one observes at position x_1 a wavenumber k_1 and at position $x_2 = x_1 + t\Omega''\Delta k$ a wavenumber $k_2 = k_1 + \Delta k$.

E.2 Standing waves

In a bounded domain wave solutions have to satisfy boundary conditions. The solutions take the form of standing waves where the nodes and maxima are fixed in space. These solutions can be thought of as superpositions of "incident" and "reflected" propagating waves, as demonstrated in the following example.

Assume a homogeneous and stationary medium and propagating waves with dispersion relation $\omega^2 = c^2k^2$. Insert a horizontal plane at $x_3 = 0$ where the waves have to satisfy the boundary condition $\psi = 0$. Since the system remains horizontally homogeneous and stationary the horizontal wavenumber and frequency do not change upon reflection. The dispersion relation then allows for two vertical wavenumbers

$$k_3^{\pm} = \pm\left[\frac{\omega^2}{c^2} - \left(k_1^2 + k_2^2\right)\right]^{1/2} \tag{E.22}$$

Superposition of these two waves gives

$$\psi = a_+ \exp^{ik_3^+ x_3} \exp^{i(k_1x_1+k_2x_2-\omega t)} + a_- \exp^{ik_3^- x_3} \exp^{i(k_1x_1+k_2x_2-\omega t)} \tag{E.23}$$

The boundary condition $\psi = 0$ at $x_3 = 0$ requires $a_- = -a_+$ and hence

$$\psi = 2ia_+ \sin k_3^+ x_3 \exp^{i(k_1x_1+k_2x_2-\omega t)} \tag{E.24}$$

The wave is standing in x_3-direction, i.e., it has fixed nodes and maxima.

If another infinite plane at $z = -H_0$ is introduced with the same boundary condition then the vertical wavenumber can only assume discrete values $k_3^+ = n\pi/H$ with $n = 0, 1, 2, \dots$

If the medium between the planes is not homogeneous then one obtains non-sinusoidal standing wave functions.

E.3 Geometric optics

Wave train solutions In a slowly varying background one may have wave train solutions of the form

$$\psi(\mathbf{x}, t) = a(\mathbf{x}, t) \exp^{i\theta(\mathbf{x},t)} \tag{E.25}$$

with phase function $\theta(\mathbf{x}, t)$. The phase function is constant along wave crests. A local wavenumber and frequency can then be defined by

$$\mathbf{k}(\mathbf{x}, t) := \frac{\partial \theta}{\partial \mathbf{x}} \tag{E.26}$$

$$\omega(\mathbf{x}, t) := -\frac{\partial \theta}{\partial t} \tag{E.27}$$

These definitions imply

$$\nabla \times \mathbf{k} = 0 \tag{E.28}$$

$$\frac{\partial \mathbf{k}}{\partial t} = -\frac{\partial \omega}{\partial \mathbf{x}} \tag{E.29}$$

The wavenumber vector field is irrotational. Wave crests have no ends and are not created or destroyed.

The underlying assumption is that the amplitude a, wavenumber $\mathbf{k}$, and frequency ω do not change significantly over a wavelength and period.

Local dispersion relation Assume that the dynamics imply a local dispersion relation

$$\omega = \Omega(\mathbf{k}; \mathbf{x}, t) \tag{E.30}$$

Again, it is implied that the dependence on $\mathbf{x}$ and t is on scales larger than the wavelength and period. The group velocity is then given by

$$\mathbf{c}_g(\mathbf{k}; \mathbf{x}, t) = \frac{\partial \Omega(\mathbf{k}; \mathbf{x}, t)}{\partial \mathbf{k}} \tag{E.31}$$

Ray equations A point that always moves with the local group velocity of the wave train moves along a path $\mathbf{x}(t)$, called the ray path and defined by

$$\frac{d\mathbf{x}}{dt} = \frac{\partial \Omega}{\partial \mathbf{k}} \tag{E.32}$$

The wavenumber $\mathbf{k}(t)$ changes along the ray path by

$$\frac{\mathrm{d}\mathbf{k}}{\mathrm{d}t} = -\frac{\partial \Omega}{\partial \mathbf{x}} \tag{E.33}$$

since

$$\begin{aligned}\frac{\mathrm{d}}{\mathrm{d}t}k_i &= \left(\partial_t + \frac{\partial \Omega}{\partial k_j}\frac{\partial}{\partial x_j}\right)\frac{\partial \theta}{\partial x_i} \\ &= -\frac{\partial \omega}{\partial x_i} + \frac{\partial \Omega}{\partial k_j}\frac{\partial}{\partial x_i}\frac{\partial \theta}{\partial x_j} \\ &= -\frac{\partial \Omega}{\partial k_j}\frac{\partial k_j}{\partial x_i} - \frac{\partial \Omega}{\partial x_i} + \frac{\partial \Omega}{\partial k_j}\frac{\partial k_j}{\partial x_i} \\ &= -\frac{\partial \Omega}{\partial x_i}\end{aligned} \tag{E.34}$$

The ray equations (E.32) and (E.33) constitute a set of six ordinary differential equations for $\mathbf{x}(t)$ and $\mathbf{k}(t)$ that can be solved with appropriate initial conditions. Changes of the frequency $\omega(t)$ along the ray path are given by

$$\frac{\mathrm{d}\omega}{\mathrm{d}t} = \frac{\partial \Omega}{\partial \mathbf{k}}\frac{\mathrm{d}\mathbf{k}}{\mathrm{d}t} + \frac{\partial \Omega}{\partial \mathbf{x}}\frac{\mathrm{d}\mathbf{x}}{\mathrm{d}t} + \frac{\partial \Omega}{\partial t} = \frac{\partial \Omega}{\partial t} \tag{E.35}$$

The above equations are equivalent to the motion of a particle with position $\mathbf{x}$, momentum $\mathbf{k}$, energy ω, and Hamiltonian $\Omega(\mathbf{x}, \mathbf{k}, t)$.

If the amplitude vanishes outside a volume V over which $\mathbf{k}$ and ω do not vary much then the wave train becomes a *wave packet*. Viewed locally, the waves are sinusoidal with wavenumber $\mathbf{k}$ and frequency ω. Viewed globally the wave packet appears as a point that has a position $\mathbf{x}$ and wavenumber $\mathbf{k}$, which change according to the ray equations.

Appendix F

Conventions and notation

F.1 Conventions

The following conventions are adopted:

1. Vectors and tensors are denoted by **boldface** symbols.
2. Italic indices $i, j, \ldots$ run from 1 to 3 and denote the Cartesian components of a three-dimensional vector or tensor. Greek indices $\alpha, \beta, \ldots$ run from 1 to 2 and denote the Cartesian components of a two-dimensional vector or tensor. Summation convention applies to indices that occur twice.
3. A subscript 0 denotes a constant value.
4. In $a := b$ the colon means that a is defined by b.

F.2 Notation

Latin symbols

a = amplitude
$\mathbf{a}$ = acceleration
$\mathbf{A}$ = vector potential
= eddy viscosity tensor

b = thermobaric coefficient
= buoyancy
B = Bernoulli function
= surface buoyancy flux

c = concentration
= sound speed
c_d = drag coefficient
c_p = specific heat at constant pressure
= phase speed
c_v = specific heat at constant volume
C = curve
$\mathbf{c}_\mathrm{g}$ = group velocity

d = cabbeling coefficient
= Ekman depth
= half distance between foci of ellipse
D = vertical length (depth) scale
= salt diffusion coefficient
= Lagrangian rate of change
D' = thermal diffusion coefficient
$\mathrm{D}/\mathrm{D}t$ = Lagrangian derivative
$\mathbf{D}$ = rate of deformation tensor

e = specific energy
E = rate of evaporation
Ek = Ekman number
$\mathbf{e}$ = unit vector

f = specific free energy
= number of degrees of freedom
= Coriolis frequency
= function

F = volume force

g = specific free enthalpy
= gravitational acceleration
G = gravitational constant
= scalar function defining a surface
G = surface force
= velocity gradient tensor

h = specific enthalpy
= scale factor, Lamé coefficient
= thickness
H = depth
= Hermite polynomial

I = flux vector

J = Jacobian

k = eastward component of wavenumber vector
k = wavenumber vector
K = eddy diffusion tensor

l = northward component of wavenumber vector
L = latent heat
= horizontal length scale
= Laguerre polynomial

m = zonal wavenumber
m = momentum density
M = mass
= acceleration potential
M = (volume) transport

n = normal vector
N = buoyancy (Brunt-Väisälä) frequency
N = rate of normal deformation tensor

p = pressure
p = modified spherical coordinates
P = horizontal eigenfunction of pressure
= rate of precipitation
= associated Legendre function of the first kind

q = (Ertel's) potential vorticity
q = oblate spheroidal coordinates
= heat flux
Q = surface heat flux
= heat

r = radial component of position vector
= bottom friction coefficient
r = initial position vector
= bottom friction tensor
R = gas constant
= Rossby radius of deformation
Ro = Rossby number

s = generalized vertical coordinate
s = label vector
S = salinity
= surface
= source/sink term
= entropy
S = rate of shear deformation tensor

t = time
T = time scale
= temperature
= period
T = depth-integrated force

u = eastward velocity component
u = velocity vector
U = horizontal velocity scale
= horizontal eigenfunction of zonal velocity

v = specific volume
= northward velocity component
v = velocity of surface
V = volume
= horizontal eigenfunction of meridional velocity
V = translation velocity
= vorticity tensor

w = vertical velocity component
W = scale of vertical velocity

x = eastward component of position vector
x = position vector

y = northward component of position vector

z = upward component of position vector
Z = enstrophy
= scale of vertical displacement

Greek symbols

α = thermal expansion coefficient

β = haline contraction coefficient
= beta parameter

γ = isopycnal slope in (T, S)-plane
= thermodynamic equilibrium gradient of salinity
= ratio of horizontal length scale to Earth's radius

Γ = adiabatic temperature gradient
= circulation

δ = aspect ratio
δ_{ij} = Kronecker tensor
δV = surface enclosing volume V
δS = line encircling surface S
Δh = partial enthalpy difference
$\Delta\mu$ = chemical potential difference
ΔP = scale of deviation pressure
ΔR = scale of deviation density

ϵ = dimensionless expansion parameter
= separation constant
= Rayleigh friction coefficient
ϵ_{ijk} = permutation tensor

ζ = enstrophy density

η = specific entropy
= vertical displacement
= stochastic process

θ = potential temperature
= phase
= latitude

κ = isothermal compressiblity coefficient

$\tilde{\kappa}$ = adiabatic compressibility coefficient

λ = thermal conduction coefficient
= wavelength
= damping coefficient

μ = chemical potential
= dynamic shear viscosity coefficient
= sine of latitude
μ' = dynamic expansion viscosity coefficient

ν = kinematic shear viscosity coefficient
ν' = kinematic expansion viscosity coefficient

$\mathbf{\Pi}$ = stress tensor

ρ = density

σ = relative height
$\boldsymbol{\sigma}$ = molecular viscous stress tensor

$\boldsymbol{\tau}$ = eddy stress tensor

φ = longitude
= velocity potential
ϕ = potential
= vertical eigenfunction

χ = spiciness

ψ = velocity streamfunction
= vertical eigenfunction
= function
Ψ = transport streamfunction

ω = frequency
$\boldsymbol{\omega}$ = relative vorticity
$\mathbf{\Omega}$ = angular velocity vector

Sub- and superscripts

a = absolute
= adiabatic
= antisymmetric
= atmospheric
= acoustic
adv = advective

b = boiling
= bottom

bt = barotropic
bc = baroclinic

c = centrifugal
= center
= critical
chem = chemical

d = deviation
= dry air

= diapycnal
= diabatic

e = equatorial
= Earth
= external
= energy
= epipycnal

equ = equilibrium

f = freezing
= frictional
= fast

g = gravitational

h = horizontal
= heat

i = inertial
= internal

int = internal

kin = kinetic

l = liquid
= liquification
= lunar
= load

m = Moon
= mass

m = momentum

mech = mechanical

mol = molecular

n = normal

o = ocean

p = polar
= planetary

pot = potential

r = reference
= relative

rad = radiative

s = salt
= Sun
= slow
= solid
= saturation
= symmetric

T = tidal

tot = total

v = vapor
= vaporization
= vertical

w = water

References

Apel, J. R. (1987). *Principles of Ocean Physics*. Academic Press.

Aris, R. (1962). *Vectors, Tensors and the Basic Equations of Fluid Dynamics*. Available as Dover Publication, Prentice Hall.

Del Grosso, V. A. (1974). New equation for the speed of sound in natural waters (with comparisons to other equations). *Journal of the Acoustical Society of America*, **56**, 1084–1091.

Dushaw, B. D., Worcester P. F., Cornuelle B. D., and Howe, B. M. (1993). On equations for the speed of sound in seawater. *Journal of the Acoustical Society of America*, **93**, 255–275.

Flament, P. (2002). A state variable for characterizing water masses and their diffusive stability: "spiciness". *Progress in Oceanography*, **54**, 493–501.

Fofonoff, N. P. (1962). Physical properties of sea-water. In: *The Sea*, Vol. 1. ed. M. N. Hill. Interscience Publishers, pp. 3–30.

(1985). Physical properties of seawater: a new salinity scale and equation of state for seawater. *Journal of Geophysical Research*, **90**, 3332.

Frankignoul, C. and Müller, P. (1979). Quasi-geostrophic response of an infinite beta-plane ocean to stochastic forcing by the atmosphere. *Journal of Physical Oceanography*, **9**, 105–127.

Kamenkovich, V. M. (1977). *Fundamentals in Ocean Dynamics*. Elsevier Scientific Publishing Company.

LeBlond, P. H. and Mysak L. A. (1978). *Waves in the Ocean*. New York, Elsevier Scientific Publishing Company.

Lighthill, M. J. (1952). On sound generated aerodynamically, I, General theory. *Proceedings of the Royal Society A*, **211**, 564–587.

Longuet-Higgins, M. S. (1968). The eigenfunctions of Laplace's tidal equations over a sphere. *Proceedings of the Royal Society*, **262**, A1132, 511–607.

MacKenzie, K. V. (1981). Nine-term equation for sound speed in the oceans. *Journal of the Acoustical Society of America*, **70**, 807–812.

Matsuno, T. (1966). Quasi-geostrophic motions in the equatorial area. *Journal of the Meteorological Society of Japan*, **44**, 25–43.

McWilliams, J. C. (1977). A note on a consistent quasigeostrophic model in a multiply connected domain. *Dynamics of Atmospheres and Oceans*, **1**, 427–441.

Montgomery, R. B. (1957). Oceanographic Data. In *American Institute of Physics Handbook*. Sect. 2, *Mechanics*. McGraw-Hill.

Müller, P. and Willebrand, J. (1989). Equations for oceanic motions. In: *Landolt/Börnstein, 5.3b, Oceanography*. Springer Verlag, pp. 1–14.

Siedler, G. and Peters, H. (1986). Properties of sea water. In: *Landolt/Börnstein, 5.3a, Oceanography*. Springer Verlag, pp. 233–264.

UNESCO (1981). *The Practical Salinity Scale 1978 and the International Equation of State of Seawater 1980. Tenth Report of the Joint Panel on Oceanographic Tables and Standards*. UNESCO Technical Papers in Marine Science, **36**.

Veronis, G. (1972). On properties of seawater defined by temperature, salinity and pressure. *Journal of Marine Research*, **30**, 227–255.

Index

Printed in the United States
By Bookmasters